Springer Series in Electrophysics
Volume 5

Edited by Walter Engl

Springer Series in Electrophysics

Editors: Günter Ecker Walter Engl Leopold B. Felsen

Volume 1 **Structural Pattern Recognition**
2nd Printing
By T. Pavlidis

Volume 2 **Noise in Physical Systems**
Editor: D. Wolf

Volume 3 **The Boundary-Layer Method in Diffraction Problems**
By V. M. Babič, N. Y. Kirpičnikova

Volume 4 **Cavitation and Inhomogeneities**
in Underwater Acoustics
Editor: W. Lauterborn

Volume 5 **Very Large Scale Integration (VLSI)**
Fundamentals and Applications
2nd Edition
Editor: D. F. Barbe

Volume 6 **Parametric Electronics**
An Introduction
By K.-H. Löcherer, C.-D. Brandt

Volume 7 **Insulating Films on Semiconductors**
Editors: M. Schulz, G. Pensl

Volume 8 **Theoretical Fundamentals of the Ocean Acoustics**
By L. Brekhovskikh, Y. P. Lysanov

Volume 9 **Principles of Plasma Electrodynamics**
By A. F. Alexandrov, L. S. Bogdankevich,
A. A. Rukhadze

Very Large Scale Integration (VLSI)

Fundamentals and Applications

Editor: D. F. Barbe

With Contributions by
D. F. Barbe R. C. Eden P. Losleben J. L. Prince
R. I. Scace E. E. Swartzlander, Jr. E. C. Urban
R. K. Watts B. M. Welch B. H. Whalen

Second Corrected and Updated Edition

With 147 Figures

Springer-Verlag Berlin Heidelberg New York 1982

David F. Barbe, Ph.D.
Office of ASN (R,E&S), The Pentagon
Washington, DC 20350, USA

Series Editors:

Professor Dr. Günter Ecker
Ruhr-Universität Bochum, Theoretische Physik, Lehrstuhl I, Universitätsstrasse 150, D-4630 Bochum-Querenburg, Fed. Rep. of Germany

Professor Dr. Walter Engl
Institut für Theoretische Elektrotechnik, Rhein.-Westf. Technische Hochschule, Templergraben 55, D-5100 Aachen, Fed. Rep. of Germany

Professor Leopold B. Felsen Ph.D.
Polytechnic Institute of New York, 333 Jay Street, Brooklyn, NY 11201, USA

ISBN-13: 978-3-642-88642-3 e-ISBN-13: 978-3-642-88640-9
DOI: 10.1007/978-3-642-88640-9

Softcover reprint of the hardcover 2nd edition 1982

Offset printing: Beltz Offsetdruck, 6944 Hemsbach/Bergstr.

2153/3130-543210

Preface to the Second Edition

Even elementary school students of today know that electronics can do fantastic things. Electronic calculators make arithmetic easy. An electronic box connected to your TV set provides a wonderful array of games. Electronic boxes can translate languages! Electronics has even changed watches from a pair of hands to a set of digits. Integrated circuit (IC) chips, which use transistors to store information in binary form and perform binary arithmetic, make all of this possible. In just a short twenty years, the field of integrated circuits has progressed from chips containing several transistors performing simple functions such as OR and AND functions to chips presently available which contain thousands of transistors performing a wide range of memory, control and arithmetic functions.

In the late 1970's Very Large Scale Integration (VLSI) caught the imagination of the industrialized world. The United States, Japan and other countries now have substantial efforts to push the frontier of microelectronics across the one-micrometer barrier and into sub-micrometer features. The achievement of this goal will have tremendous implications, both technological and economic for the countries involved.

The substantial demand for the first edition of this book is evidence of the importance of the field. The principal difference between the first edition and the second, aside from the updating of some of the information and the addition of new references, is the addition of a chapter on Very High Speed Integrated Circuits (VHSIC). VHSIC is both a technology and a program. As a technology, it is a subset of VLSI which emphasizes speed. It is also a program funded by the U.S. Department of Defense. Chapter 8 describes both the technology and the program.

The number of special journal issues, conferences, workshops, seminars, etc., related to the field of IC's is large. While no single volume could adequately summarize the field, this volume attempts to provide a summary of some of the important issues and factors for Very Large Scale Integration from the perspective of several authors deeply involved in the field.

In the field of VLSI, composed of many facets and disciplines, the demand for engineers, physicists and chemists trained in IC skills, exceeds supply. It is hoped that this volume will be of value to students entering the IC field and to scientists and engineers, trained in other fields, who wish to make the transition into the IC field.

Washington, D.C.
December 1981

D.F. Barbe

Contents

Index of Contributors

Barbe, David F.

Navy Department, Washington, DC 20350, USA

Eden, Richard C.

Vedette Energy Research, Inc., Newbury Park, CA 91320, USA

Losleben, Paul

Defense Advanced Research Projects Agency, Arlington, VA 22209, USA

Prince, John L.

Intermedics, Inc., Freeport, TX, 77541 and
Zymos Corporation, Sunnyvale, CA, 94088, USA

Scace, Robert I.

National Bureau of Standards, Washington, DC 20234, USA

Swartzlander, Earl E., Jr.

TRW Systems, Redondo Beach, CA 90278, USA

Urban, Ellison C.

Navy Department, Washington, DC 20350, USA

Watts, R.K.

Bell Laboratories, Murray Hill, NJ 07974, USA

Welch, Bryant M.

Rockwell International/Electronics Research Center,
Thousand Oaks, CA 91360, USA

Whalen, Barry H.

TRW Systems, Redondo Beach, CA 90278, USA

1. Introduction

D. F. Barbe

The era of integrated circuits (ICs) began in the late 1950s with the development of the planar silicon transistor technology. The development of ICs has progressed through small scale integration (SSI), medium scale integration (MSI) and large scale integration (LSI). The delineation between these levels of integration is not based on fundamental definitions; however, the generally accepted conventions are given in Table 1.

Table 1 Evolution of Integrated Circuits

Level of integration	Abbreviation	Number of logic gates/chip	Time period of development
Small Scale Integration	SSI	3-30	Early 1960s
Medium Scale Integration	MSI	30-300	Mid-late 1960s
Large Scale Integration	LSI	300-3000	Early-mid 1970s
Very Large Scale Integration	VLSI	> 3000	Beyond late 1970s

SSI, introduced in the early 1960s, was characterized by 3-30 bipolar transistors on a chip. The MSI phase began in the mid 1960s when the metal-oxide-semiconductor (MOS) transistor technology matured. A typical example of a mature MSI chip was the 256-bit memory. The LSI phase began about 1970 and extended through the mid 1970s. Examples of early and late LSI chips are the 1 K-bit and 16 K-bit random access memory (RAM) chips, respectively. The first commercial chip generally accepted as being in the VLSI class is the 64 K-bit MOS RAM introduced late in the 1970s [1].

1.1 Outlook

Since the first planar transistor was fabricated in 1959, the number of transistors on memory chips has followed the trend of doubling every year. This phenomenal growth is due to steadily improving lithographic and mask generation techniques, innovative fabrication techniques and innovative circuit designs. Architectural innovation has provided highly functional chips for numerous commercial applications such as watches, calculators, speech processors and microprocessors. While the processing capability of LSI chips is great, the potential capability and impact of VLSI chips is staggering.

In the past decade, semiconductor technology has been the driving force in the removal of three of the four barriers to pervasive application of electronics in our society. As the barriers of cost, reliability and complexity were removed, the semiconductor market grew to over $10 billion. The principal remaining barrier to pervasive application of electronics in our society is the specialized knowledge that a user is required to have in order to apply and use electronics [2]. VLSI will contribute greatly to overcoming this barrier by providing enough signal processing capability for speech and image recognition, thus allowing inputs to systems to be made in familiar forms.

While the terms SSI, MSI and LSI in the past have been used in connection to silicon as the semiconductor substrate, gallium arsenide (GaAs) technology for ICs has matured over the past several years. Owing to the semi-insulating substrate of GaAs and the higher mobility of electrons in GaAs, high speed digital circuits have evolved to the point where LSI and even VLSI in GaAs substrates are of considerable potential importance.

1.2 Scope of This Volume

As the lateral dimensions of a device structure are reduced, other parameters such as oxide thickness, doping levels and applied voltages must also be scaled appropriately to keep fundamental quantities such as electric fields and current densities from increasing. In Chapter 2, the effects of scaling device features down to one micrometer (micron) and below are discussed. Theory indicates that conventional MOS and bipolar devices can function with dimensions much smaller than those used in LSI and smaller than those which can be achieved with optical lithography. Three lithographic techniques, electron beam lithography, X-ray lithography and ion-beam lithography, are currently undergoing intense development as replacements for optical lithography for submicron ICs. Chapter 3 is an excellent review of optical, electron-beam, X-ray and ion-beam lithographic techniques.

With ICs becoming more complex, design cost and time are becoming prohibitively large [3]. The widespread use of VLSI will depend critically on future reduction of design cost and time. This reduction requires increasingly sophisticated computer aided design (CAD) techniques and widespread use of them. In Chapter 4, the use of computers as design tools for ICs is discussed.

The progress made in the density of silicon integrated circuits over the past twenty years is truly remarkable, generating great confidence in the continued increase in density into the VLSI area. As the fundamental limits of device size are approached it will be increasingly difficult to increase the density of silicon integrated circuits. For the past ten years, the maximum switching speed of silicon integrated circuits has not increased significantly. As the functional throughput limits are approached in silicon, integrated circuits in gallium arsenide substrates are likely to become increasingly important. The two reasons for this are (1) the higher mobility of electrons in GaAs and (2) the semi-insulating of GaAs which reduces parasitic capacitance and thus allows higher switching speeds and improved dynamic switching energies in logic circuits. For these reasons, Chapter 5 on GaAs ICs is included in this volume which otherwise concentrates on silicon technology.

Effective chip architecture development requires the coordination of the technological capabilities with external interfaces to provide the needed circuit functions. Chapter 6 treats internal chip architecture, systems architecture and their interrelation. A case study of VLSI architecture selection for an advanced radar digital beamformer is included to demonstrate the interrelation of systems architecture and chip development.

The potential for new applications provided impetus to the development of LSI circuits; similarly, there are a large number of circuit functions awaiting the development of VLSI circuits. Chapter 7 discusses these functions and applications. Testing of VLSI chips is also discussed in Chapter 7.

While most of this volume is written from the perspective of authors working in the United States and accordingly it describes VLSI work in this country, other countries have significant programs in VLSI. A discussion of these efforts is contained in Chapter 8.

1.3 Summary

The field of ICs requires several considerations and techniques at micron and submicron dimensions. Among the most important considerations for VLSI are the theory of device scaling, lithography, fabrication and computer aided design. The first part of this volume, composed of four chapters, treats these topics. The fifth chapter deals with circuits in GaAs substrates. The sixth chapter deals with chip architecture. The seventh chapter deals with VLSI applications and the testing of VLSI chips. Finally, the eighth chapter summarizes VLSI programs in other parts of the world.

References

1. Hans Friedrich, Walter F. Kosnocky and Takuo Sugano, "Special Issue on VLSI - Foreword," IEEE Trans. on Electron Devices, Vol. ED-26, p. 257, Apr. 1979.

2. J. Fred Bucy, "The Semiconductor Industry Challenges in the Decade Ahead," Proc. IEEE Solid-State Circuits Conf., p. 53, San Fran., CA 1980.

3. G. E. Moore, "VLSI: Some Fundamental Challenges," IEEE Spectrum, pp. 30-37, Apr. 1979.

2. VLSI Device Fundamentals

J. L. Prince

With 26 Figures

That electronics technology will soon enter a new era, that of Very Large Scale Integration, is clear. In this era the "VLSIC's" will be integrated systems rather than integrated circuits. Although this differentiation may seem trivial, examination of the economics and the capabilities, and the limitations, of VLSI chips results in the conclusion that profound changes in fabrication, design, and product definition must occur before VLSI chips can become pervasive. Fig. 1 shows the well-known integrated circuit complexity increase with time, known by some as "Moore's Law" [1,2]. The historical data in this figure, which gives the envelope of integrated circuit complexity growth, shows a doubling of complexity every year up to the present time, with a projected decrease of the rate of growth to a doubling every two years. This decrease in rate of growth is projected as being due to the practical elimination of wasted, or non-functional, space on the silicon chip at or about the 1980 time [1], with no further progress to be made

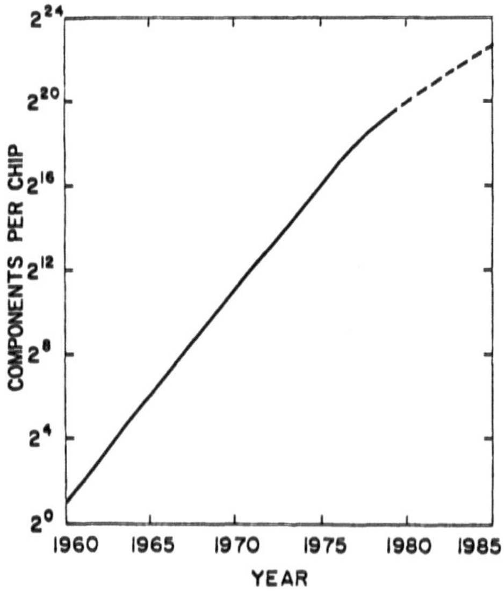

Fig. 1 Integrated circuit complexity growth with time, following MOORE [1]

in this area. The projection seems somewhat conservative. More important than the decrease in the complexity growth rate envelope due to this effect, are the possible limitations on the available product spectrum due to a crisis in product definition coupled with the exponentially growing cost of design. The question is just this: will the VLSI chip set consist only of ever-larger memories, and longer-word-length microprocessors, with other functional blocks at the LSI level or below? The answer to this question, and in fact the driving force for VLSI, is implicit in the future systems requirements which will drive the semiconductor industry.

One source of future systems requirements, a source more open to view than strictly commercial thrusts but one which does allow insight into IC improvements required in the future, is defense systems. Table 1 shows both estimated current status and projected future needs of a variety of computationally intensive military equipments. Examples of these include acoustic surveillance systems (beam formers, spectrum analyzers and large associative memories for target signature identification), high resolution synthetic aperture radar (SAR) processors for surveillance, message encoding and encryption devices for secure and reliable communication, the implementation of guidance and target identification algorithms for "smart" bombs and missiles, high precision automatic navigation equipment, and image processors for data rate reduction. These equipments address important if not vital military capabilities, and their effectiveness has evoked experimental and conceptual designs of a complexity (as indicated in the table) demanding substantially more advanced integrated circuits than those in current use. That the answer to this problem is not simply squeezing more LSI chips into a platform is implicit in the

<u>Table 1</u> Military system processing requirements, current and (projected) future, MIPS = Million Instructions Per Second for a digital processing system

		SIGNAL PROCESSING REQUIRED (MIPS)	
APPLICATION	PLATFORM	CURRENT	FUTURE
ARMY TACTIAL SIGNAL INTELLIGENCE	LAND BASED MOBILE	0.4	40
CRUISE MISSILE TERMINAL GUIDANCE	SMALL MISSILE	0.1	50
DATA CORRELATION FOR OTHT/FIRE CONTROL RADAR	SHIP/LAND BASED	1.0	50-100
AIRBORNE SAR RADAR	AIRCRAFT/SPACECRAFT	3.0	100-500
EW RADAR PULSE PROCESSOR (1-10 MPPS)	AIRCRAFT/SPACECRAFT	2.0	200-300
ANTI-JAM SPREAD-SPECTRUM COMM	SMALL MISSILE/RPV AIRCRAFT/SPACECRAFT	5.0	500
WIDEBAND DATA LINKS (1 GBPS)	SPACECRAFT	10.0	500
UNDERSEA GLOBAL SEARCH	SHIP/LAND BASED	0.5	2,000
ELINT/ESM PROCESSOR	AIRCRAFT/SPACECRAFT	10.0	10,000

platform/chip limit data of Table 2. In considering the information in this table it should be kept in mind that large systems often entail support or special components (SSIC's, MSIC's, discretes, I/O chips, etc.) in a number 5 to 20 times the number of LSIC's utilized. It is clear that, for military systems at the least, an improvement in IC speed, or in complexity, of two orders of magnitude or more will be necessary in the next few years in order to meet projected systems performance needs.

Table 2 Limitations on the number of LSIC's in a subsystem, by platform

TYPE OF PLATFORM	TYPICAL LIMIT, NUMBER OF LSI CHIPS PER SUBSYSTEM	ORIGIN OF LIMIT
MANPACK	30	SIZE, POWER
SMALL MISSILE	100	SIZE
SPACECRAFT	300	SIZE, POWER
AIRCRAFT, SUBMARINE	1,000	SIZE, COST
SHIP	3,000	SIZE
LAND BASED VEHICLE	10,000	COST

It is obvious that cost and reliability benefits, on a per-function basis, will also be important factors in VLSI, since the support components mentioned have first-order impact on system reliability and maintainability as well as on size, cost and power. The manufacturing cost of integrated circuits and their assembly into subsystems is generally small relative to the aggregate cost of qualification of the integrated circuit (for operating temperature range, tolerance to nuclear radiation, etc.), documentation, special test equipment, logistics and operational support and, above all, the life cycle costs of the host system which are attributable to its integrated circuit subsystems. Typically, the latter are referred to collectively as system support costs and include the incremental cost of prime power, deck space, air conditioning and so on, which in the case of submarines, missiles, high performance aircraft and satellites, considerably exceeds the total procurement cost of the integrated circuit subsystems. Projected reliability improvements alone (on a failure rate per function basis) to be gained from the transition to VLSIC's have astonishing and far-reaching implications, particularly with the ensuing ability to implement fault tolerant and built-in-test techniques.

Analysis of benefits obtainable from IC fabrication technology development (notably a capability to decrease the minimum dimension) discussed in later sections shows that the likelihood of having available VLSI chips which will allow the required systems improvement be realized is high, given that research and development funds are channeled in the proper directions. Fundamental chip capabilities can be measured by chip Functional Throughput Rate (FTR), defined here as the product of the equivalent gates per chip times the maximum clock rate of a gate F_{max}. F_{max} is taken to be equal to $1/4T_d$ where T_d is the gate delay. The processing rate of any complex system-on-a-chip is heavily influenced by architectural decisions and techniques (e.g., parallel processing). The use of FTR as a figure of merit thus comprehends the amenability of a device technology to improved performance through complexity-related architectural techniques as well as by raw gate speed. Note that in considering possible improvements in the FTR of an IC technology it is necessary to consider power per gate separately; as feasible complexity scales up (and FTR presumably scales up) it is necessary that power dissipation per chip scale down in such a way that total power dissipation per package remain within reasonable limits, e.g., 2W per chip. Currently available LSI chips have FTR per chip in the range 1×10^4 Gate-MHz to 2×10^5 Gate-MHz, as shown in Fig. 2. With scaling of the critical minimum dimensions to the range 0.5 - 0.8 micrometer, and with technology development to realize the potential high functional density of VLSI chips (e.g., reliable multilevel interconnection technology), projections to be discussed show that FTR per chip in the neighborhood of 10^7 Gate-MHz can be obtained, as shown in Fig. 2.

Although earlier discussion should have made clear the importance of architectural, design (as in layout), and product definition considerations to the eventual pervasiveness of VLSIC's, the orders of magnitude improvement in FTR required for future systems and other benefits in life cycle costs will not be achieved without fundamental progress in device and circuit technology. Subsequent sections examine the prospects, sources, and amounts of improvement in power, speed and complexity of VLSI chips.

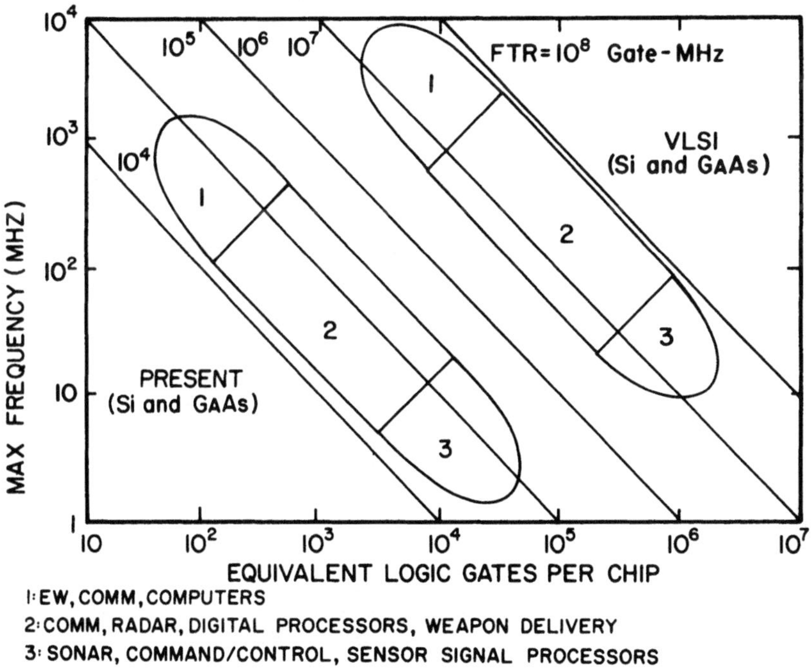

Fig. 2 Present and future (VLSI) Functional Throughput Rate capabilities, and domains of applicability to military systems

2.1 Fundamentals of VLSI Device Improvements

2.1.1 Relative Importance of Chip Area, Circuit Innovation, and Minimum Dimension

Methods of increasing IC FTR per chip by fabrication and circuit technology alone are limited to three. These methods are increase of chip area, use of circuit and device innovation ("cleverness"), and decrease of the minimum fabrication dimension. The relative contributions of these factors to the progress in IC complexity shown in Fig. 1 has been analyzed [1]. For the future, the first method is very limited in potential, perhaps giving an increase in gates per chip (and thus FTR) of a factor of two, due to the coupling between chip area and probe yield. Several yield models exist, and all predict yield-area behavior similar to the experimental data shown in Fig. 3. As the minimum dimension is scaled down by the factor 1/K (K>1), the minimum significant yield-reducing defect size also scales down by the same factor and the area density of significant defects scales up by the factor K, all other things being equal. Considerable process development work will have to be done for VLSI processes in order to achieve acceptable yield at no increase in chip size. The second method, "cleverness", has contributed substantially to FTR improvements in the past. However, it is difficult to invent to schedule, and future contributions to FTR per chip from cleverness are not predictable, although worth addressing. Certainly, innovations involving devices which require simpler processing, have higher packing density, and which are more tolerant of the type of defects which limit yield, will improve FTR independently and through the ability to utilize larger chips with acceptable yield. Step-function improvements, however, of order-of-magnitude size in FTR per chip are not in sight. The potential for increase in FTR per chip due to scaling down of the minimum dimension is large, and much previous progress has been based on this same factor. Fig. 4 shows the increase in MOS dynamic RAM

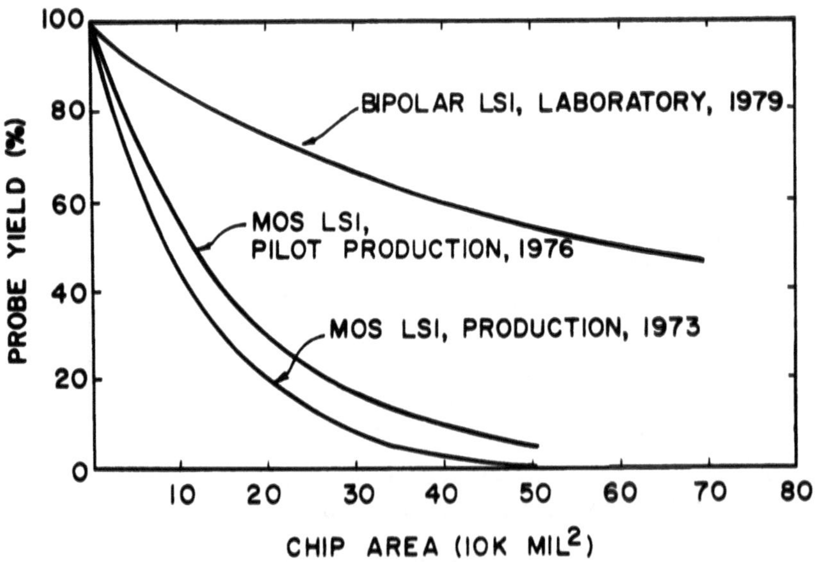

Fig. 3 Experimental behavior of IC probe yield versus chip area

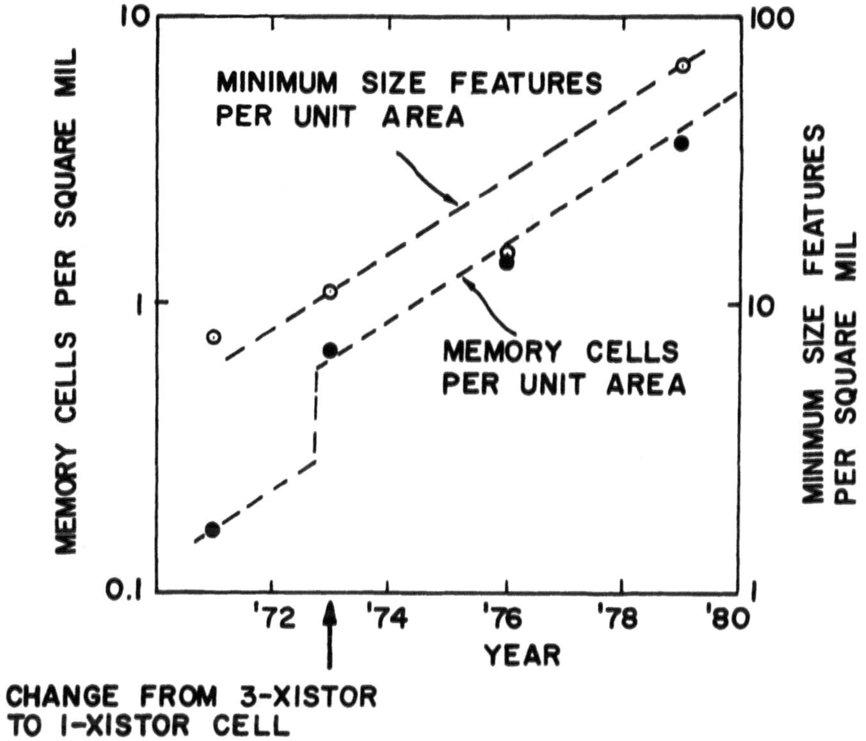

Fig. 4 Evolution of memory cell density and minimum-size feature density with time for MOS dynamic RAM chips [3]

cell density over the span of time corresponding to evolution from 1K bits/chip to 64K bits/chip [3]. The figure also shows the increase in minimum-size density over the period. Although the increased memory cell density certainly has contributions from circuit and device "cleverness" contained within it (note, for example, the step in the memory cell density caused by the transition from a 3-transistor memory cell to a 1-transistor memory cell), the correlation between minimum-size feature density increase and memory cell density increase is strong. An appreciation of the relative contribution of chip size to the total complexity increase can be obtained from Fig. 5, which shows the evolution in both over the 8-year period [3]. Thus in the evolution of the MOS dynamic RAM from 1K to 64K bits, increase in chip area has accounted for a 1.65 increase in complexity, circuit innovation for at least a factor of three increase in complexity, and reduction in minimum dimension somewhat less than a factor of ten increase in complexity.

The significance of scaling of the minimum dimension to VLSI progress is the greater due to the predictability (or at least the high potential) for practical progress in this area. Past progress in reduction of the minimum dimension, based strictly on optical or near-UV lithography, is shown in Fig. 6. For years the capability of patterning micron or submicron features using electron-beam direct slice writing has been known and many workers have re-

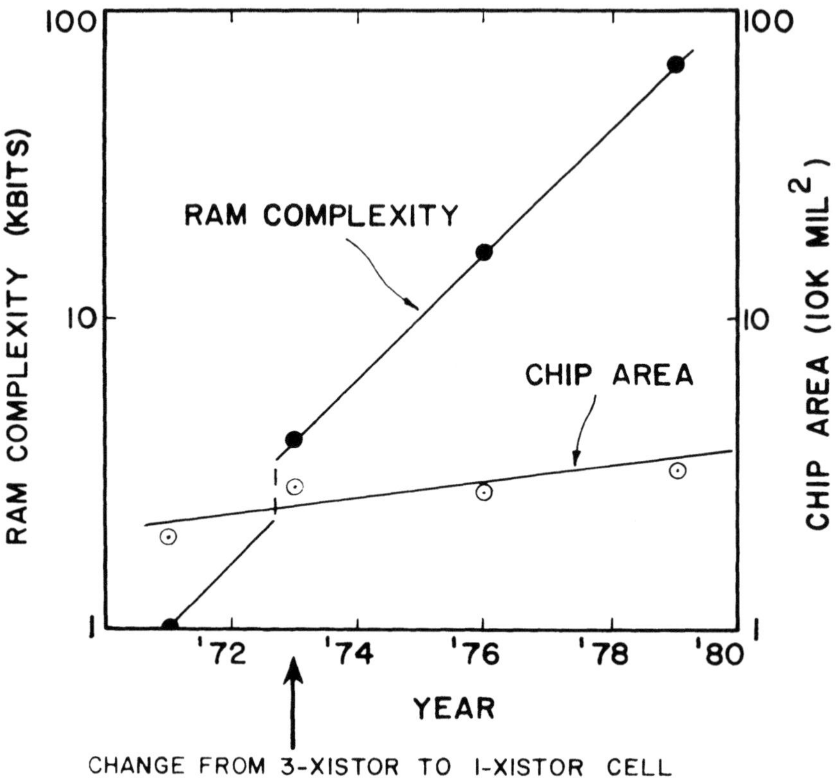

Fig. 5 Evolution of MOS dynamic RAM complexity and chip size with time [3]

ported laboratory results; recent reports from workers at IBM [4] have shown the existence of a well-researched VLSI design/process combination based on 1.25 micron minimum features obtained using electron-beam direct slice writing. Work is also in progress on deep-UV and x-ray lithographic techniques which have some promise for the micron/submicron patterning regime. A recent U. S. Department of Defense initiative in very high speed VLSIC's [5] has among its goals the achievement of (economical) minimum dimensions in the range 0.5 micron to 1 micron. Both the IBM work and the DoD VHSIC program goals are indicated on Fig. 6, and jointly appear to indicate at least a high probability of "staying on the curve" in terms of future progress in decrease in the minimum dimension. From these developments it is possible that the rate of progress will be temporarily accelerated.

Based on the earlier contributions of scaling the minimum dimension to IC complexity increase, and the predictability of continuing decreases in the future, an examination of the implications of scaling on device performance and complexity is warranted. The remainder of this section is devoted to such an examination, both of scaling principles and some details of scaled device characteristics.

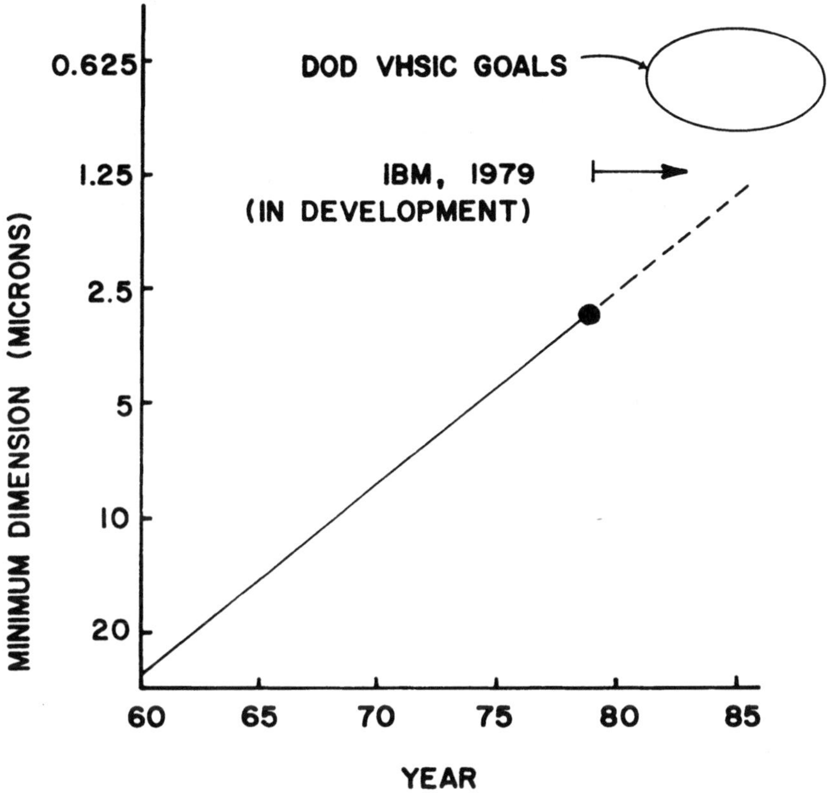

Fig. 6 Decrease of IC minimum dimension with time, with recent developments [4,5] indicated

2.1.2 Device Scaling Fundamentals

Detailed results of scaling device minimum dimension can be obtained through use of device and process imulation models implemented on digital computers. However, use of such models tends to obscure first-order trends and implications of scaling which become clear when examined in the light of first-order principles. This section explores first-order results and implications of device/circuit scaling.

MOS Scaling Principles

A simple, coherent set of scaling principles, so-called constant field, originally put forth by DENNARD, et al, [6] exists for MOS devices. They have proven to be effective in scaling from the range 5-10 micron minimum dimension to the range 1-3 micron minimum dimension, although they do not give an optimized device at small dimensions. The straightforward application of these principles is rarely done; instead various scaled device parameters are optimized to fit each application. However, they do give a very clear picture of the fundamental improvements caused by scaling, and also show potential problems. Table 3 shows the basic principles for scaling by factor K, K>1.

Table 3 Basic MOS device scaling principles, after DENNARD, et al [6]

PARAMETER	SCALING FACTOR
DEVICE DIMENSION-T_{OX}, L, W	1/K
DOPING CONCENTRATION N_A	K
VOLTAGE V	1/K
CURRENT I	1/K

In fact the scaling relationships shown in the table are not all independent. If the first three quantities (device dimensions including oxide thickness t_{ox}, channel width W, channel length L, and junction depths; doping concentration in the substrate N_a; and applied voltages) are scaled as shown, then device current I_d, depletion region width X_d, and the threshold voltage V_t all scale approximately as shown in the table. This can be seen by applying the scaling principles to the simple long-channel device equations in any standard textbook (e.g., SZE [7]). For the depletion layer thickness, with zero substrate-source bias voltage V_{sx},

$$X_d = \left[2\epsilon_s(V+V_{bi})/N_a\right]^{1/2} \quad \rightarrow \quad X_d' = X_d/k \tag{1}$$

where ϵ_s is the relative dielectric constant of silicon. In (1), the effect of the "improper" scaling of V_{bi} (i.e., $V_{bi}' = V_{bi}/K$) is taken to be small. In fact, V_{sx} can be scaled somewhat more than the supply voltage in order to keep the sum $V_{bi} + V_{sx}$ constant, if that is important [6]. For the device V_t,

$$V_t = \frac{t_{ox}}{\epsilon_{ox}}\left\{-Q_{eff} + \sqrt{2\epsilon_s q N_a (V_{sx}+\psi_{si})}\right\} + (\phi_{ms} + \psi_{si})$$
$$V' = \rightarrow \quad V_t' = V_t/k \tag{2}$$

where Q_{eff} is the net oxide charge, ϵ_{ox} is the relative dielectric constant of the oxide, ϕ_{ms} is the gate metal-semiconductor work junction difference, and ψ_{si} is the strong-inversion silicon band-bending at the surface. Here again some of the terms in (2) do not scale properly, and approximations have been made. The quantity ψ_{sl} is approximately equal to V_{bi} and scales as in K. Stronger scaling of V_{SX}, as mentioned above for scaling of depletion layer widths, can be used if necessary to adjust V_t slightly. The last term in (2) also scales improperly; however, as pointed out by DENNARD [6], the last term often has net value near zero and its scaling behavior is not usually important. Finally, device currents in the linear region scale as

$$I_d = \mu \frac{\epsilon_{ox}}{t_{ox}} \frac{W}{L} (V_{gs} - V_t - V_d/2) V_d \rightarrow I'_d = I_d / k \qquad (3)$$

where μ is the surface mobility, and device currents in the saturation region scale as

$$I_d = \mu \frac{\epsilon_{ox}}{2 t_{ox}} \frac{W}{L} (V_{gs} - V_t)^2 \rightarrow I'_d = I_d / k \qquad (4)$$

An example of the degree of agreement of experimental results with the predicted behavior of scaled V_t is shown in Fig. 7 [6]. For the scaled device

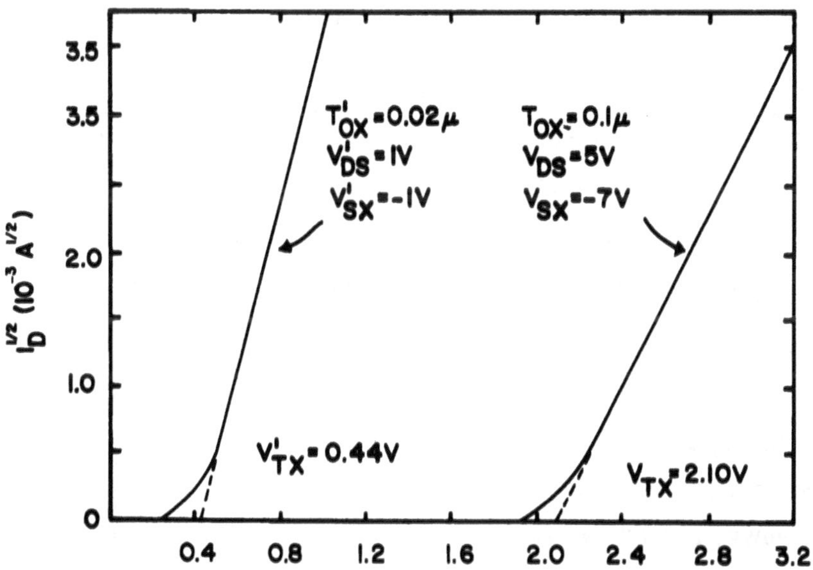

Fig. 7 Experimental determination of extrapolated threshold voltage V_{tx} for scaled and unscaled MOSFET's; scaling factor K=5 (after DENNARD, et al [6])

of Fig. 7, N'_a = 2.5 x 10^{16} cm^{-3} and the effective channel length is 1 micron. From an emperical relationship which predicts the onset of short-channel effects on subthreshold conduction characteristics [8], the device of Fig. 7 is on the margin of, or just outside of, these short channel effects.

From the above it is clear that nowhere in the scaling laws are fundamental or practical limits, such as have been pointed out [9,10], inserted. Thus, the laws would permit scaling proportionally to zero dimension, or to zero threshold voltage, when practical and theoretical considerations do not allow such things. For example, for silicon surface concentrations greater than 1×10^{19} cm^{-3} the gate oxide will theoretically and practically break down before the surface inverts to form the channel; the scaling laws do not comprehend this and other limits. The laws also clearly do not explicitly take into account parameters which do not scale (e.g., minority carrier diffusion length) or those which scale poorly (e.g., V_{bi}). They do, however, maintain all electric field strengths constant under scaling ($V \rightarrow V/K$, $X \rightarrow X/K$), and electric field distributions are maintained constant except for a scale factor in linear dimension. This is true, of course, only as long as limits due to the discreteness of charge, doping distributions, and the silicon lattice are not approached too closely. Current density in the scaled device is also maintained constant, although this is not true of current density in interconnections or at contacts. In the scaling all nodal voltages are reduced proportionally; thus scaled circuits should function in terms of scaled voltage levels just as the unscaled circuit functions. Based on these factors, some characteristics of scaled MOS circuits using constant chip area can be predicted as shown in Table 4. All lengths and thicknesses are scaled by 1/K; thus all areas scale by $1/K^2$ and circuit complexity scales up as K^2. Capacitance per unit area, both device (oxide and depletion region) and parasitic interconnection, scales up as K ($t_{ox} \rightarrow t'_{ox} = t_{ox}/K$); total capacitance, both device and parasitic, scales as 1/K. Since device currents in both the linear and saturation region of operation are scaled down by 1/K, as are all voltage levels and capacitances, the differential equations governing the transient response of simple MOS gates are invariant under scaling except for the use

Table 4 Circuit implications of MOS device scaling

PARAMETER	SCALING FACTOR
LOGIC GATE AREA	$1/K^2$
DEPLETION AND OXIDE CAPACITANCE ($\epsilon A/T$)	$1/K$
DELAY TIME ($R_{ON}C = T_D$)	$1/K$
DC POWER DISSIPATION / GATE ($V \cdot I$)	$1/K^2$
CHIP DC POWER DENSITY ($V \cdot I/A$)	1
POWER-DELAY PRODUCT	$1/K^3$
FREQUENCY-DEPENDENT POWER DISSIPATION (CV^2/T_D)	$1/K^2$
CHIP FREQUENCY-DEPENDENT POWER DISSIPATION ($CV^2/T_D)/A$	1

of a scaled time $t' = t/K$. Thus the delay time for these simple gates scales down as $1/K$. For static logic circuits the dc power dissipation per gate scales down as $1/K^2$; with the increased packing density of the scaled circuit the dc power density (per unit chip area) is constant. For dynamic logic circuits operated at the maximum clock frequency ($\sim 1/T_d$), or for the frequency dependent component of power dissipation for static logic circuits, the power dissipation per gate also scales down as $1/K^2$ and the power density remains constant.

The combination of the scaling behavior of gate power and propagation delay leads to the conclusion that the scaling behavior of the power-delay product is

$$PT_d \rightarrow P'T_d' = PT_d / k^3 \tag{5}$$

and that the Functional Throughput Rate scales as

$$FTR \rightarrow FTR' = FTR \cdot k^3 \tag{6}$$

The strong scaling behavior shown in (5) and (6), taken together with the power density scaling shown in Table 4, is of great importance, because it predicts a sort of scaling "bonus" which differentiates MOS circuit scaling behavior from bipolar circuit scaling behavior. For I^2L-type bipolar circuits, scaling methods discussed in subsequent sections result in power-delay product scaling as $1/K^2$, and FTR scaling as $\sim K^2$ if the total chip power is held constant, or $\sim K^4$ if the power per gate is held constant (and the power per chip scaled up by K^2). The advantage of MOS over bipolar technology in this respect is due to the ability to scale voltages for MOS circuits from present LSI levels ($V_{DD} \simeq 5V$) down to perhaps 0.5V - 1V (K = 5 - 10). Bipolar I^2L-type circuits already operate near a physical minimum voltage ($V_{BE} \simeq 0.75V$) for that technology, and the additional degree of freedom afforded MOS circuits is not available. However, present LSI I^2L-type bipolar circuits generally have lower power-delay products than MOS LSI circuits [11].

Note that MOS scaling options other than those shown in Table 3 are available, including not scaling the supply voltages. Although there are good reasons why certain scaling steps shown in Table 3 might need to be modified [6], not scaling supply voltages at least by a certain amount leads to scaling behavior of the dc power-delay product as $1/K^2$ or worse and generally a scaling up of total chip power.

Although application of the Dennard scaling principles gives many circuit improvements, there are circuit characteristics which exhibit perverse scaling behavior. Many of these are associated with interconnections and contacts. Table 5 shows some implications of scaling according to the prescriptions of Table 3. Scaling of interconnection layer thicknesses down by a factor $1/K$ has been assumed. This degree of scaling is perhaps stronger than will be necessary; however, characteristics of etching techniques and requirements on surface planarity for multi-level metalization systems will require some scaling of interconnection film thickness. Note also that is has been assumed that the average interconnection line length scales down as $1/K$. This assumption is reasonable, but it is also reasonable that for constant chip size the length of at least some of the longest interconnection lines will not scale down at all. This has very adverse effects on the scaling of some interconnection-related parameters. For example, for worst-case interconnection lines (length not scaled)

$$\text{worst-case}\left\{\begin{array}{ll} R_L \rightarrow R'_L = k^2 R_L & C_L \rightarrow C'_L = C_L \\ R_L C_L \rightarrow R'_L C'_L = k^2 R_L C_L & \end{array}\right. \quad (7)$$

and the worst-case line response time normalized to scaled gate delay behaves as

$$R_L C_L / T_d \rightarrow R'_L C'_L / T'_d = k^3 R_L C_L / T_d \quad (8)$$

Thus the poor scaling behavior shown in Table 5 for normalized interconnection line voltage drop and response time is much better than the reasonably worst case for these quantities. This makes it difficult to take advantage of the

Table 5 Implications of scaling for "average" interconnection lines and contacts

PARAMETER	SCALING FACTOR
INTERCONNECTION LINE RESISTANCE, ($R_L = \rho L/WT$)	K
NORMALIZED INTERCONNECTION LINE VOLTAGE DROP (IR/V)	K
INTERCONNECTION LINE RESPONSE TIME ($R_L \cdot C_L$)	1
INTERCONNECTION LINE CURRENT DENSITY (I/A)	K
CONTACT RESISTANCE (R_C)	K^2
CONTACT VOLTAGE DROP (V_C)	K
NORMALIZED CONTACT VOLTAGE DROP (V_C/V)	K^2
NORMALIZED LINE RESPONSE TIME ($R_L \cdot C_L/T_D$)	K

higher switching speeds inherent in the scaled devices when signal propagation over long lines is involved. This is particularly so for lines fabricated using doped polycrystalline silicon, which has a minimum resistivity of about 300 micro-ohm cm as opposed to aluminum or gold which have resistivity two orders of magnitude less.

Poor scaling behavior of normalized interconnection line voltage drop shown in Table 5 also suffers from the "longest-line" problem. The resistance of an average line increases directly with the scaling factor K. The IR drop in such a line is therefore constant (with the decreased current levels), but is K times greater relative to the scaled operating voltages. However, for the worst-case lines the normalized voltage drop also scales as K^2 instead of K, similar to the normalized line response time. The behavior of contact resistance and normalized contact voltage drop with scaling is strongly improper because the contact voltage drop, relative to scaled circuit voltage levels, scales up as K^2. Note that specific contact resistance is taken to be constant with scaling. This behavior at the contacts is particularly of concern in light of the fact that even for LSI geometries a substantial percentage of the total contact area on a chip is not "open", i.e. making good contact to metal (see, for example, [12]). Thus the K^2 behavior is a best case, for scaling according to the principles of Table 3. The increase in interconnection line current density shown in Table 5 is also a potential source of reliability problems due to electromigration. This problem will be discussed in more detail in section 2.2.

Another parameter which does not scale properly is the subthreshold turn-on characteristic of MOS devices. The scaled expression for log I_d in the subthreshold region is [6]

$$\log_{10} I'_d = V'_{gs}/\alpha' \tag{9}$$

where

$$\alpha \rightarrow \alpha' = 2.3\,\beta\left[1 + \frac{\epsilon_s t_{ox}/k}{\epsilon_{ox} X_d/k}\right] = \alpha \tag{10}$$

In the above, β is the thermal voltage k T/q. The scaled value of α is equal to the unscaled value, meaning that the (logarithmic) rate of decrease of I'_D with decreasing V'_{gs} is the same before and after scaling. An example of the consequences of the invariance of α with scaling is shown in Fig. 8. In this figure, scaling of V_t by K = 3 was assumed, and as an approximation the current level at which strong-inversion characteristics begin was also scaled down by K. From Fig. 8 the problem is clear. The absolute variation of gate voltage "available" for turn-off of the device is reduced by a factor 1/K in the scaled device but the rate of decrease of log I_d with decreasing V_{gs} is the same for both devices, albeit the turn-off begins at a somewhat lower current level for the scaled device. It would be preferable for to be larger, implying a more complete or rapid turn-off of the device with decreasing V_{gs} below threshold. The poor scaling characteristic has led to the conclusion by some [6,13] that MOS dynamic logic or memory circuits will not be feasible in the VLSI era, at least for room temperature operation.

Bipolar Scaling

There is no accepted set of fundamental scaling principles for bipolar devices analogous to the MOS scaling laws [14]. This is because bipolar scaling is considerably more complicated than MOS scaling, since the basic device itself is more complicated due to involvement of the third dimension and the basic dependence on minority carrier injection and collection. Also, it is in part because the bipolar circuit technology spectrum has been much wider than that

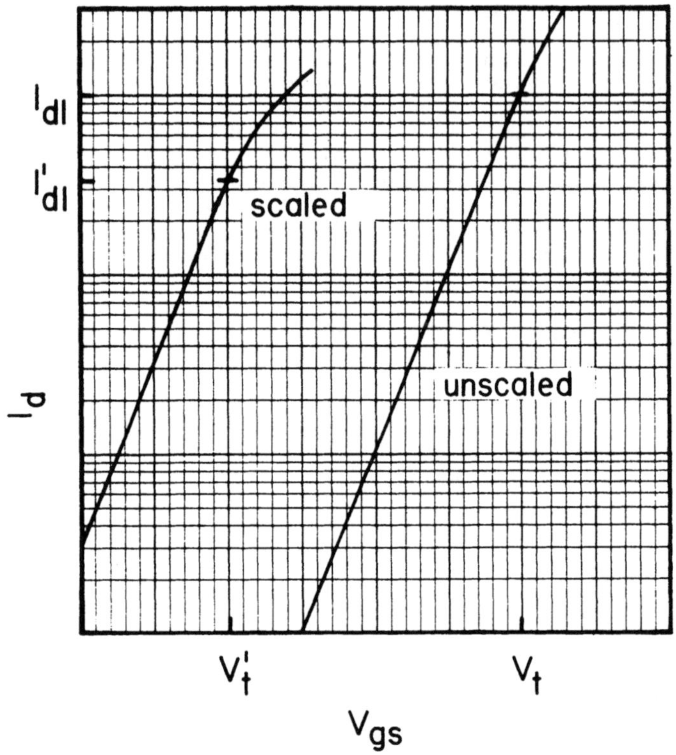

Fig. 8 Illustration of scaling of MOSFET subthreshold conduction characteristics

of MOS circuit technology. Some work has been done recently in an attempt to clarify the situation [14,15], and some examination of limits on bipolar scaling has been done [16]. The options open in terms of not exceeding physical limits varies considerably depending on the bipolar circuit technology which is involved in the scaling. In this section we will take the approach of putting forth a set of scaling "rules" appropriate to I^2L-type circuit technologies [17] and in a subsequent section will show some results of detailed scaling of I^2L circuits. This particular technology, I^2L, or a version of it will almost certainly be the primary VLSI bipolar technology due to its packing density and power-delay product advantages [11]. Aside from considering broadly applicable physical principles, discussion of detailed scaling of other bipolar technologies is not justified.

Scaling of voltages in bipolar technology is generally not applicable since they are for most high-complexity cases close to the physical limit of one V_{be} (0.6 - 0.7V). For the specific case of I^2L, the supply voltage is currently only slightly greater than 0.7V and can not be reduced much if at all. Thus <u>constant-voltage</u> scaling will be used in bipolar technology. Ironically, application of MOS scaling rules and use of constant voltage bipolar scaling results ultimately in a common VLSI MOS/bipolar power supply voltage of approximately 0.5 - 1.0V. This is because practical tolerances on MOS V_t and consideration of the relationship of V_{dd} and V_t for the proper functioning of basic MOS gates [15] lead to the limiting value of V_{dd} of 0.5 - 1.0V. This constant voltage scaling for bipolar circuits reduces an important degree of freedom in scaling as discussed in section 2.2, and results in increase in electric field strengths, and generally in device current density and circuit power density as dimensions are scaled down. Table 6 shows a set of I^2L scaling "rules" which give maximum scaling down of propagation delay (thus the unscaled current). This method of scaling has strong benefits in terms of the scaling behavior of the Functional Throughput Rate ($FTR' = K^4$ FTR) but as shown in the table results in rapid increase of total chip power ($\sim K^2$).

In this method of scaling device lateral and vertical dimensions are scaled by 1/K, with care where punchthrough might occur (e.g., injector base width, epitaxial layer thickness). This gives scaling up of the gate packing density by K^2. Doping concentrations are scaled as necessary; the situation is so

Table 6 A set of I^2L scaling "rules" for maximum scaling of propagation delay

DEVICE/CIRCUIT PARAMETER	SCALING FACTOR
DEVICE LATERAL AND VERTICAL DIMENSIONS L	1/K (SOME EXCEPTIONS)
VOLTAGE V, CURRENT I	1
LOGIC GATE AREA (A)	$1/K^2$
DOPING CONCENTRATIONS	? (VARIABLE)
STORED INTRINSIC CHARGE Q_{SAT} EXTRINSIC CHARGE Q_{EXT}	$1/K^2$
LOGIC GATE DELAY ($\sim Q_{SAT}/I$, Q_{EXT}/I)	$1/K^2$
LOGIC GATE POWER ($V \cdot I$)	1
CHIP POWER DENSITY ($V \cdot I/A$)	K^2
INTRINSIC POWER-DELAY PRODUCT ($P \times T_D$)	$1/K^2$

complicated that no general rule can be given except that most concentrations go up some. Since areas are scaled as $1/K^2$ and lengths as $1/K$, all charge storage elements are taken to be scaled down by $1/K^2$ as well. At constant injector current, propagation delays, which are controlled by depletion region capacitances in the low injector current case and by diffusion capacitances at various locations in the device in the high current case [18,19,20], are scaled down by $1/K^2$. The dc power dissipation per gate is constant, while the chip power density and the chip total power scales up as K^2. The result of all this is that the scaling behavior of the power-delay product is

$$P T_d \rightarrow P' T_d' = P T_d / k^2 \tag{11}$$

and that of the Functional Throughput Rate is

$$FTR \rightarrow FTR' = FTR \cdot k^4 \tag{12}$$

The poorer scaling behavior PT_d for bipolar technology compared to MOS was discussed earlier, and is due to the inability to scale voltage. The strong scaling behavior of the FTR shown in (12) is purchased at the price of increased chip power dissipation ($\sim K^2$). An alternative method of scaling, in which gate current is scaled by $1/K$, would result in a more reasonable chip power scaling behavior ($\sim K$), power-delay scaling as in (11), and FTR scaling as K^3 instead of K^4.

Just as in the case for MOS devices, it is found that certain parasitics do not scale properly in bipolar technology. The same types of interconnection related scaling problems discussed in section 2.2.1 exist for bipolar technology. For the set of scaling rules of Table 6, normalized interconnection and contact voltage drops scale just as for the MOS scaling rules, since voltage is not scaled in the prescription of Table 6. However, the increase in interconnection line current density is stronger ($\approx K^2$) than for the MOS case, since for this bipolar case gate current was not scaled.

2.2 Problem Areas for VLSIC's

In this section three factors which are common to MOS and bipolar VLSI devices and circuits are discussed. These factors may be speed or complexity-limiting, and in any case will be design considerations in the VLSI era.

2.2.1 Interconnections

Two items are of prime concern in the general interconnection area. These are electromigration characteristics, particularly in the very thin - very narrow stripe case, and contact characteristics, particularly in the very shallow junction regime. Other aspects, such as electrical resistivity and etchability, are also of interest for the VLSI case and are being pursued through investigation of alternative metalization systems [21], but appear to be of less immediate concern than these.

Electromigration has been established as a primary failure mode for pure Al-metalized IC's subjected to very high metalization current densities ($\simeq 10^6 A/cm^2$) [22]. Although present IC design rules guard against electromigration problems with no severe penalty in packing density, the situation with VLSI circuits may well be different because the metal line current density does not scale properly (see Table 5). The mechanism of atomic transport of metal atoms in a thin film carrying relatively high current is well understood, but the predictability of the location and the time of occurrence of open circuits due to electromigration is not a well understood factor. However, the gross MTF of pure Al conductors is sometimes fairly well predicted by the early work of BLACK [22]. In general, the MTF of large-grain (several micron) aluminum lines is given by

$$MTF = C A_x J^{-N} \exp(1.2/q\beta) \tag{13}$$

where C = constant (1×10^9 cm^4/A for aluminum), A_x = conductor crossectional area, J = current density, and $2 \leq N \leq 9$. The value N = 2 is usually accepted for aluminum. Fig. 9 shows the MTF for large grain aluminum stripes with crossectional area 10 micron^2, for which N = 2 [22]. The potentially

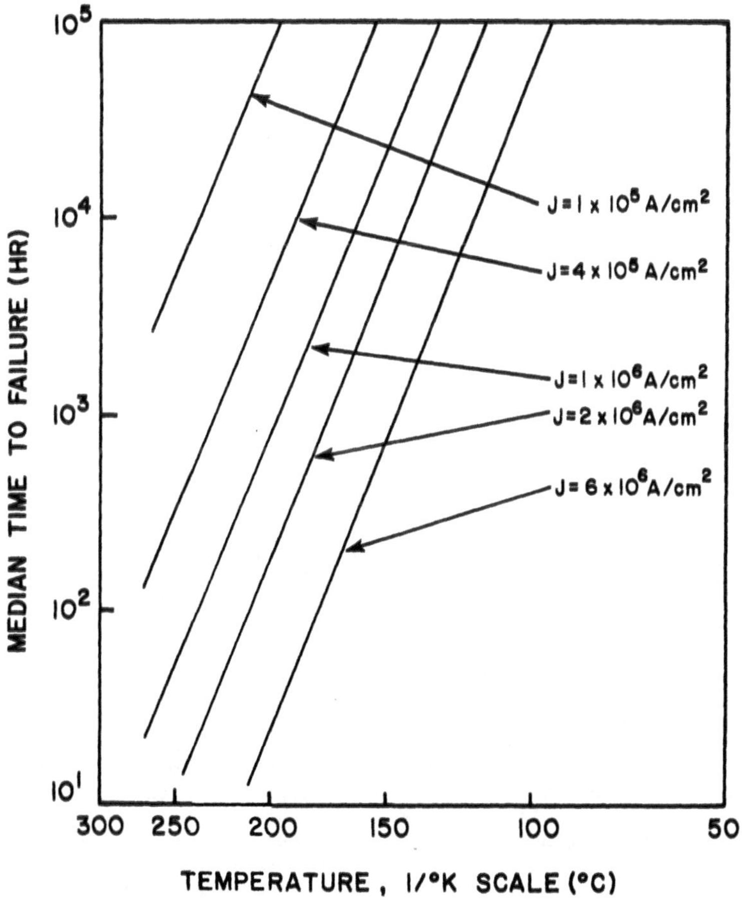

Fig. 9 Median time to failure of large-grain aluminum stripes, A_x = 10 micron^2, after BLACK [22]

serious penalty paid in MTF for allowing the interconnection current density to scale as K is clear. In fact, if the MOS scaling principles are used and if $N = 2$, then $MTF \longrightarrow MTF' = MTF/K^4$.

If Fig. 9 is to be used by scaling A_x for VLSI, there is question as to its absolute accuracy due to stripe width/grain size effects. In the region W_n = stripe width/median grain size > 4, MTF has been observed to depend linearly on A_x for constant stripe thickness. However, for $W_n < 4$ the MTF of Al-Cu and Al-Cu-Si metal stripes has been observed to vary sublinearly with A_x, and in fact to begin to increase for $W_n > 2$ [23]. Fig. 10 shows be-

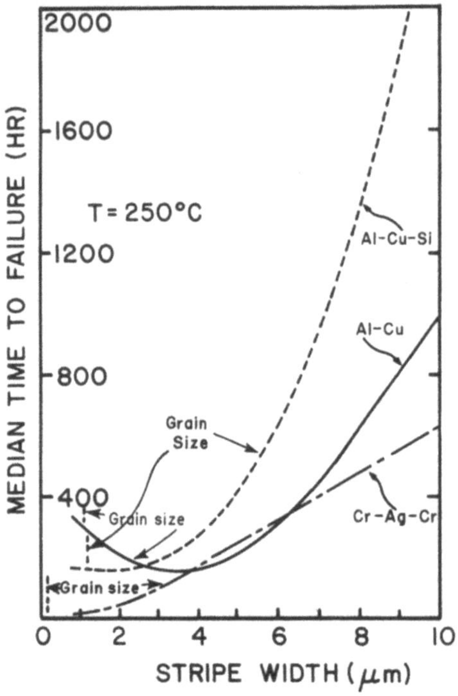

Fig. 10 Electromigration MTF versus interconnection stripe width, after SCOGGAN, et al [23]

havior of MTF of conductors of fixed grain size and varying stripe width, at 250°C, showing the deviation from linear dependence on A_x at small values of W_n. Since even small-grain aluminum normally has median grain size in the area of 0.5 - 1.5 micron, any dependence of MTF on W_n such as shown in Fig. 10 is relevant. However, very little work has been done in the area of electromigration in relatively narrow stripes. The interpretation of the W_n-dependence of electromigration MTF is that as W_n approaches 1, the metal stripes approach the limit of a string of single grains, and single grain boundaries extend across the width of the stripe. The electromigration properties of the stripe should thus approach the properties of a single crystal by this argument. However, other workers argue that grain boundaries which span the full stripe width serve as a diffusion barrier to metal atoms [24]. This would result in relatively early depletion of metal atoms on the electron downstream side of a grain boundary and thus result in low MTF. This unresolved point is a key one for VLSIC's. Grain size control and understanding of grain size effects will be key in preventing electromigration regardless of what metalization system is used.

One practical factor which is seldom accounted for in electromigration MTF calculations is the influence of the metallurgical transition at contacts. At the positive terminal of IC devices (transistors, resistors, etc.) electrons leave the Si and in so doing generally cause both Si and metal atoms to migrate in the direction of electron transport. Note that there is some disagreement even over the direction of material transport for inadequately studied metalization systems and in certain cases for Al. In any case, the lack of a source of Al atoms (or Al + Cu atoms, etc.) on the electron upstream side of any metal-Si contact means that the region of such a contact is weak in its electromigration resistance. Factors such as Si alloying and grain size uniformity would appear to be of limited use in preventing electromigration failure at contacts. The process should proceed relatively rapidly and result in a MTF determined primarily by median grain size and possibly by overcoating. Thus many of the steps taken to increase electromigration MTF (Cu alloying, etc.) will be effective up to a point; beyond that point MTF will be limited by electromigration-induced open circuits at contacts and this upper limit will be lower than that obtained experimentally using simple metal stripe resistors. One factor which should be mentioned is the effect of whisker or hillock growth in regions of metal accumulation. These whiskers or hillocks can grow at relatively negative device contacts at which electrons enter the Si device, or at other points on the stripe, and can extend several microns. Two problems can be caused by this, cracking of overcoats or interlevel dielectrics and shorting between adjacent metalization stripes.

Mechanisms of electromigration failure in Au-based metalization systems are apparently identical to the mechanisms for Al-based systems. The transport of metal atoms is in the direction of electron transport, and equations for MTF are basically of the same functional form as for Al-based systems. However, considerably less hard data is available for Au-based systems. Data that is available shows that the electromigration resistance of Au-based systems is superior to aluminum films [25]. Fig. 11 shows results for the electromigration MTF of some Au-based metalization systems and a large-grained, overcoated Al sample at 150°C air ambient [25]. Note that in this figure the evaporated Au films had W_n on the order of 1 or slightly greater while the sputtered films had $W_n \gg 1$. The electromigration resistance of Au-based systems is increased, similar to that of Al films, by use of an overcoat.

Contact characteristics of most concern for VLSI are contact resistance and contact stability. Contact resistance is of interest because in the best case the voltage drop across this resistance does not properly scale, and because difficulty is foreseen in properly opening and achieving good contact to submicron contacts. In either of the two scaling methods discussed in section 2 the contact voltage drop relative to critical voltages (e.g., supply and logic levels) scales <u>up</u> by K^2. This is not of critical concern as long as the "typical best" values of specific contact resistance obtainable with Al, PtSi (or Pd_2Si), or refractory metals are obtained. These values are in the low 10^{-6} ohm-cm^2 range, which means contact resistance in the range of a few hundred ohms for a 1 micron^2 contact [26,27]. However, obtaining good ohmic contact is often a problem and a random order of magnitude increase in this number across a slice or chip would not be at all surprising. An example of the importance of the contact voltage drop in scaling can be seen in Table 7, which gives some results from a scaling simulation [14] of a current

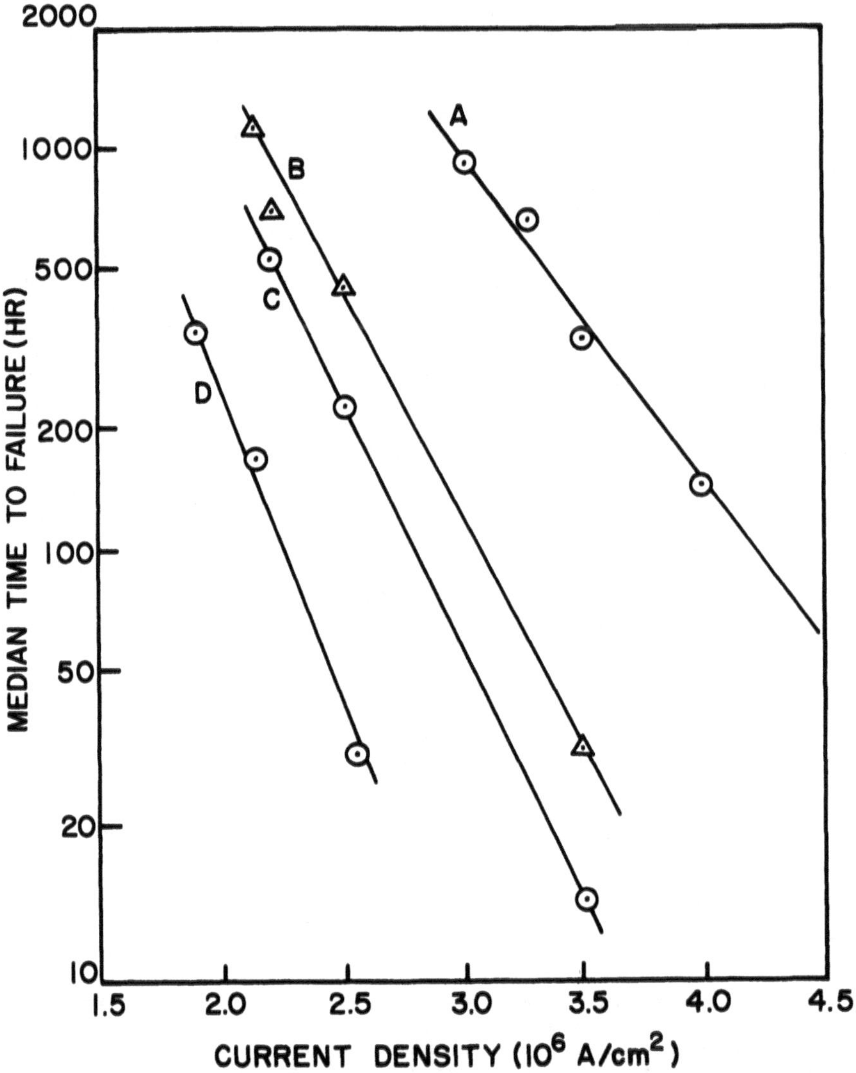

Fig. 11 Electromigration MTF for Au and Al stripes, w = 6.5 micron, t = 0.8 micron, 150°C air ambient. A: evaporated Au films (Mo-Au); B: annealed sputtered Au films (Ti:W-Au); C: un-annealed sputtered Au films (Ti:W-Au); D: large-grain Al films with protective coating (rf-silane SiO_2, silane SiO_2)

switch emitter follower bipolar logic gate. The circuit for which the scaling was performed had an unscaled voltage swing of 400mV which was intended to remain constant with scaling. The scaling was K $\simeq$ 10, and neither of the two

Table 7 Poorly scaling voltage drops in simulated scaling of a bipolar logic gate, after SOLOMON and TANG [14]

COMPONENT	VOLTAGE DROP(mV) BASIS	SCALED
METAL	7	36
CONTACTS	1	156
SUB-COLLECTOR	6	64
EPI-LAYER	260	182
TOTAL	274	438

sets of scaling "rules" discussed earlier was employed; instead all parameters were adjusted to optimize the performance. Notice that the scaled parasitic voltage drops reduce that swing by over 100mV. Note also that the specific contact resistance used in the simulation (2×10^{-7} ohm cm^2) is extremely optimistic, and a more realistic value would be in the range 5×10^{-7} ohm cm^2 (PtSi) to $1\text{-}2 \times 10^{-6}$ ohm cm^2 (Al) [24,25,26].

Contact stability in aluminum-based interconnection systems affects both yield and reliability of IC's through similar mechanisms. Future metalization systems may exhibit similar problems, and in fact gold-based systems do exhibit similar problems on failure of whatever barrier metal is used in the system (e.g., Ti:W, Pt). A fundamental problem is that during the contact sintering operation, or other high temperature operations subsequent to metal deposition, spikes or spears of aluminum metal penetrate vertically (spikes, 100 Si orientation) or laterally (spears, 111 Si orientation) into the silicon surrounding the contact window [28]. The formation of spikes or spears occurs for any Al-based system unless an effective barrier metal is interposed between aluminum and the substrate Si, or unless the solid solubility of Si in the aluminum thin film is satisfied by some technique other than dissolution of the single crystal Si surrounding the contact. The spikes may range in depth up to approximately 1 micron, and the lateral extension of the spears may range up to several microns in particular preferred directions from the contact window. The result of these can be junction degradation (at worst shorting) for shallow-junction devices. This is observable as a yield loss which can be quite high for pure Al metalization systems. Fig. 12 shows estimated yield loss due to junction penetration or degradation for pure aluminum, barrier metal-Al, and barrier metal-Au systems for an LSI complexity bar. Thus the mechanism impacts the feasible level of integration through increased process complexity and/or reduction of packing density.

The reliability problems posed by junction pitting are potentially as serious as the yield loss, for those circuits having very shallow junctions and using no barrier metal. Spiking and spearing is due to dissolution of the underlying single crystal silicon at the contact window, into the aluminum metal. As to whether this is due to an exothermic reaction between the native oxide in the contact window and aluminum, or whether it is due to solid state diffusion of silicon with aluminum, is not clear although evidence favors the former cause. In any case, silicon leaves the bulk and alloys with the aluminum at certain localized points. The volume of silicon that leaves the sub-

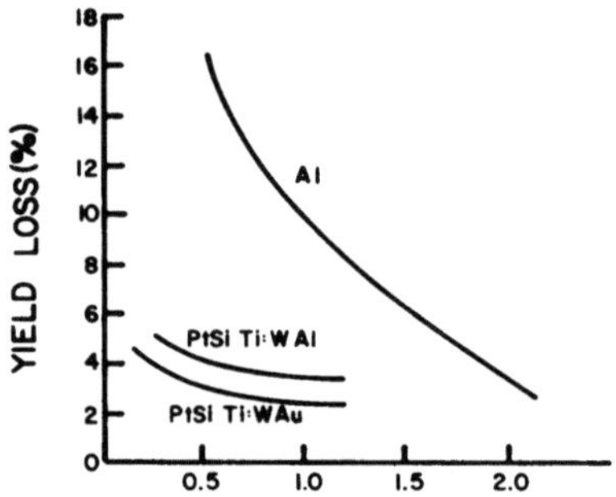

Fig. 12 Yield loss due to metal penetration at contacts, LSI bipolar chip

strate is replaced by Si-doped aluminum which is metallic. Two things can happen during subsequent operating life of the device. Silicon can diffuse further into the aluminum stripe, requiring further penetration of the Si-doped aluminum region into the substrate. Or, the silicon in the aluminum can be driven by electromigration forces, from a relatively negative silicon contact toward a relatively positive silicon contact down the aluminum metalization stripe. This results in localized depletion of silicon near existing Al-rich spikes or spears and concomitant growth of them since additional silicon will dissolve in the aluminum to satisfy the solid solubility at the temperature of operation. This growth can be a long-term reliability hazard particularly if spikes or spears are near the junctions immediately after device processing.

Practical solutions to contact pitting have been narrowed to three. One practical solution to contact pitting is to use a barrier metal such as TiPt or Ti:W alloy with PtSi contacts. The requirement for PtSi contacts and a barrier layer complicates the aluminum metalization process to the point that the complexity is approximately the same as gold-based metalization processes. Thus one of the primary motivations for using Al-based systems, i.e., process simplicity, does not exist if the PtSi-Ti:W-Al system is used. There have been attempts to use pure aluminum with a contact layer of Pd_2Si but without a barrier layer such as Ti:W [29,30]. Conflicting results have been reported on the effectiveness of such a metalization system in preventing contact pitting and it seems likely that a proven barrier layer will continue to be necessary to prevent contact degradation.

From the above discussion, it is clear that serious difficulty is to be expected from attempts to simply extend the LSI doped aluminum/doped polysilicon interconnection system to submicron line-widths. On the other hand, a clearly better alternative metalization system is not presently available. To sum up the requirements on the ideal interconnection system, in addition to obvious attributes such as compatibility with dry processing techniques, low resistivity, etc., a VLSI-compatible interconnection system must (1) have acceptable electromigration characteristics even for line widths of 0.5 micron and line thickness of 0.1 to 0.2 micron, (2) have low contact resistance and no tendency for contact spiking in spite of small contact areas (0.5 x 0.5 $micron^2$) and shallow junctions (1000Å - 2000Å), and (3) be compatible with techniques appropriate for multilevel (minimum two levels, probably three or four levels) interconnection schemes. This last requirement, although the least amenable to quantification, could turn out to be the hardest to satisfy for VLSI processes. It is one which interacts in some ways with the other two requirements, and one which interacts strongly with some other fabrication technology factors. For example, both adherence of metal to underlying dielectrics, and adherence of interlevel dielectrics to underlying metal, is required. In this last respect gold, which has desirable electromigration characteristics, is much poorer than aluminum for most known interlevel dielectric candidate materials. Compatibility of the lower-level metal system with material removal processes used to form interlevel vias is also required. In addition, consideration must be given, in the selection of interlevel dielectric materials, to their tendency toward charged species migration, polarization, charge spreading at dielectric-dielectric interfaces, and for military systems, to their sensitivity to ionizing radiation. Note in these respects the downward scaling of dielectric layer thicknesses and the resultant increase in circuit charge-induced instability sensitivity. In light of these several requirements on the interconnection system the optimum choice for VLSI technology is not clear.

2.2.2 Ionizing Particle Effects on Dynamic Logic Circuits

Limitations on minimum device size due to the effect of naturally occurring ionizing radiation have long been predicted [31]. Dynamic logic circuits are of particular sensitivity to charge generation by ionizing radiation due to the nature of the memory element, i.e., charge storage on a capacitive node. This is true whether the storage element is an oxide or depletion region capacitance, or a deep-depletion potential well as in complex MOS dynamic RAMs. Although only MOS types of dynamic logic/memory circuitry have existed in the past, VLSI power and packing density considerations may well lead to bipolar analogs of dynamic MOS circuits.

Recently reported results [32] lead to the conclusion that for practical purposes a mechanism has been observed which limits, or at least inhibits, the feasible complexity of dynamic logic circuitry including both bipolar and MOS dynamic RAMs. The mechanism involved is generation of minority carrier "packets" by incident alpha particles originating from naturally occurring contamination (U and Th) in the device package materials. This generation causes "soft" memory errors, which are defined as "random, non-recurring, single-bit errors". The influence of sea-level cosmic rays would be similar to that of the alpha particles, and as noted by MAY and WOODS the possibility of this contribution to their experimental observations [32] can not be ignored. Cosmic rays are of course an even more fundamental "fact of life" than alpha particles, and are independent of packaging materials. The limit on scaled complexity for either alpha particles or cosmic rays depends on the rate of soft errors which can be tolerated. For cosmic rays the estimated error rate versus critical charge (i.e., charge corresponding to a logical zero) jumps abruptly, by three orders of magnitude, at a critical charge corresponding to 3000 electrons [33] for a reasonable device size.

The steps involved in creation of a soft error in an MOS dynamic RAM are shown in Fig. 13 [32]. If alpha particles impinge on the chip surface, electron-hole pairs generated by the alpha particles as they come to a stop can be collected by depletion layers such that the generated electrons end up in storage wells as shown in the figure. If the fraction collected times the number generated exceeds a critical number which is somewhat less than the full-well capacity of the storage capacitor, a soft error will result. The alpha particles emitted by U and Th have energies up to 9MeV, with a peak in

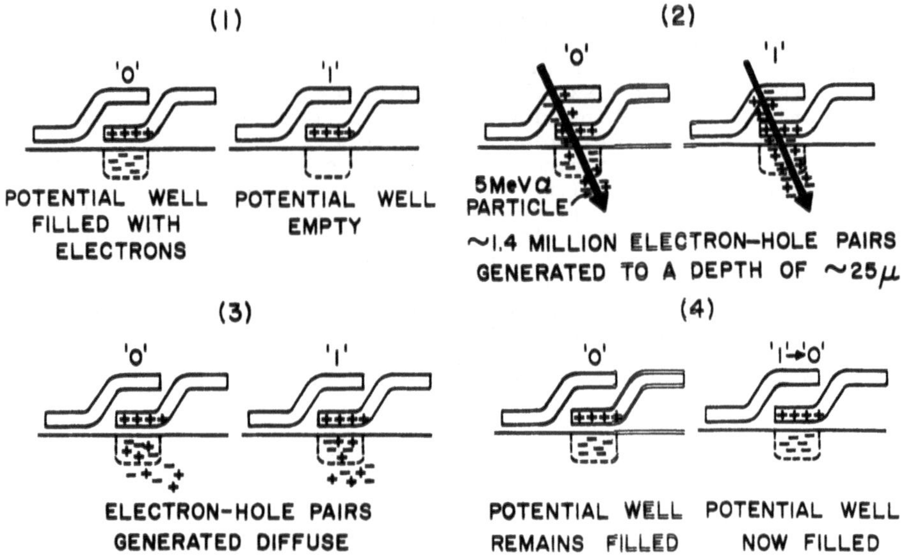

Fig. 13 Soft error creation in a one-transistor MOS dynamic RAM cell, after MAY and WOODS [32]

the distribution around 4MeV. They generate hole-electron pairs at the rate of 3.6 eV/pair up to their range in silicon of 5 - 50 microns, so that dynamic memory cells with critical charge greater than 2.5×10^6 electrons should experience a vanishingly small soft error rate due to these particles. Since the energy distribution for alpha particles from U and Th peaks at 4 - 5 MeV [range 25 micron, generated pairs $\simeq 1.4 \times 10^6$], memory cells with critical charge in the neighborhood of 10^6 electrons and smaller should show strong increase in soft error rate compared to less sensitive cells. This strong increase is present in the experimental data of Fig. 14 [32] when it is noted that the figure is plotted semilogarithmically.

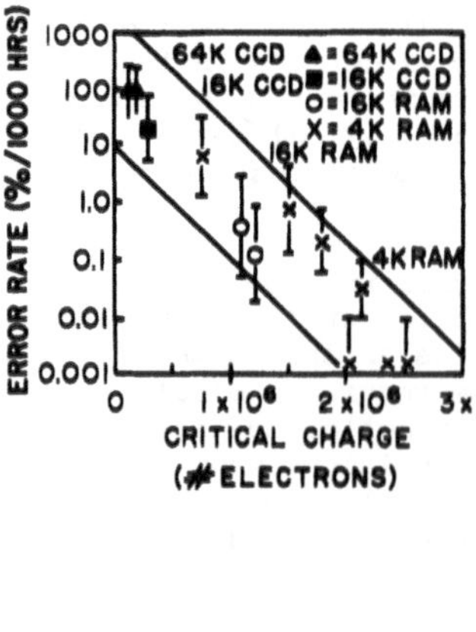

Fig. 14 Soft error rate versus critical charge, various commercial devices, after MAY and WOODS [32]

Examination of the scaling behavior of the full-well charge Q_{MAX} for MOS one-transistor RAM cells (or CCD's, for that matter) leads to the conclusion that

$$Q_{MAX} \rightarrow Q'_{MAX} = Q_{MAX}/k^2 \tag{14}$$

if the Dennard scaling rules are used. The strongest scaling factor in (14) is the scaling of area, $\sim 1/K^2$. The critical charge, which is reduced by such effects as incomplete charge transfer to sense amp bit lines, sense amp sensitivity, and thermal generation, should scale approximately the same as Q_{MAX}. Using LSI design rules, the current range of critical charge for available devices is 3×10^5 to 3×10^6 electrons. Thus the scaling behavior in (14) predicts serious problems due to alpha particles for necessary values of $K = 3 - 10$), even though the collection efficiency of charge storage nodes may be reduced due to sharing of the alpha-generated charge among several cells. The scaling behavior also predicts serious future problems due to cosmic ray-generated charge.

2.2.3 Active Scaling Parasitics

Parasitic device interactions in the SSI-MSI fabrication regime have been identified, and minimized or controlled. However, the scaling of minimum dimensions to the 3 - 5 micron region (LSI) and beyond to the 0.5 - 1 micron VLSI region opens up possibilities of new parasitic interactions caused by the narrow spacings, thinner oxides, etc. The fundamental problem lies in the invariance of some parameters, for example minority carrier diffusion length, under scaling as it is normally understood.

These new interactions are certainly not all predictable at this point in time, but two recently reported effects [34,35] give an indication of the sort of things which will be encountered as bipolar and MOS VLSI technology develops. As an example of the general tendency for parasitic interactions, note that it has been recently observed that the drain-substrate breakdown voltage for short-channel n-channel MOSFETs is lower than predicted and is a function of channel length [34,36]. In this case the drain-source breakdown voltage can be limited by parasitic interaction between the drain-substrate-source lateral bipolar transistor, the weak avalanche current from the drain-substrate junction, and the substrate spreading resistance [34]. The physical situation before breakdown and at breakdown for an n-channel MOSFET is shown in Fig. 15 [34]. Before breakdown, holes generated in the drain depletion region cause

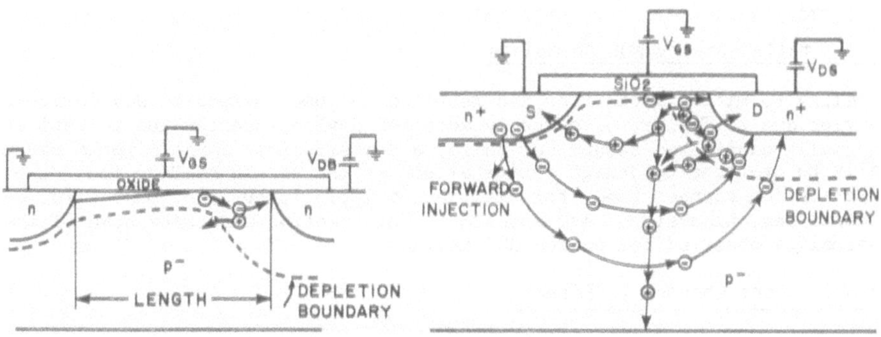

Fig. 15 N-channel MOSFET before and after breakdown due to parasitic lateral NPN transistor action, after SUN, et al [34]

substrate current, resulting in an increase in the substrate potential near the source. When the source-substrate potential reaches approximately 0.6V the junction is forward-biased and electron injection from source into substrate occurs. The lateral NPN transistor is thus in an active operating mode, as opposed to the cutoff mode, and drain-source breakdown occurs due to the BV_{CEO} of the lateral NPN transistor formed by the source, substrate, and drain. The α_{NPN} is given approximately by [34]

$$\alpha_{NPN} \cong 1 - (L/L_D)^2/2 \tag{15}$$

where L_D is the minority carrier diffusion length and L is the channel length. This formulation clearly exhibits the improper scaling behavior of α_{NPN} due to the assumed invariance of L_D. The resultant decrease in BV_{DS} as a function of L is shown in Fig. 16 [34]. This behavior fits other aspects of the parasitic transistor model developed by SUN, et al [34] down to the punchthrough

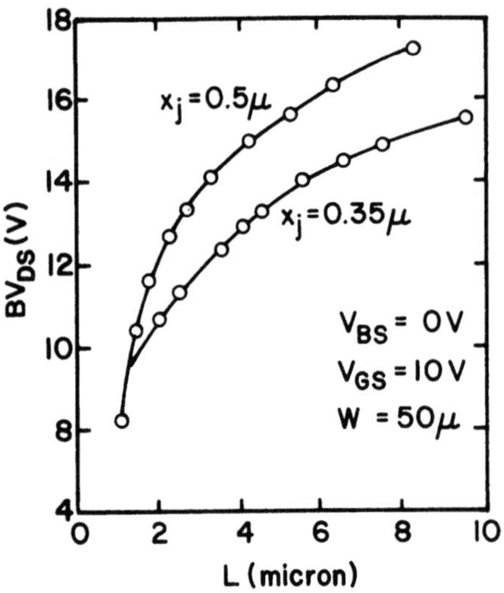

Fig. 16 Drain-substrate breakdown voltage as a function of channel length, n-channel MOSFET, after SUN, et al [34]

channel length limit for these devices of L $\simeq$ 0.8 micron.

2.3 Small-Geometry MOS Anomalies

Scaling to micron geometries has resulted in some unexpected MOS device behavior due to failure of commonly accepted device theories due to what are normally considered second (or third) order effects. These effects can usually be overcome by design perturbations on the simple scaling laws or by fabrication changes (see, for example, [6,13,15]), but they must first be identified, understood, and modelled. This section discusses some of the anomalies observed for scaled MOS devices.

2.3.1 Short Channel V_t Effect

For short-channel MOSFETs it is known that the magnitude of the threshold voltage decreases when the channel length is comparable to the depletion region widths of the source and drain diffusions [37,38]. A crossection of an n-channel device with this condition is shown in Fig. 17.

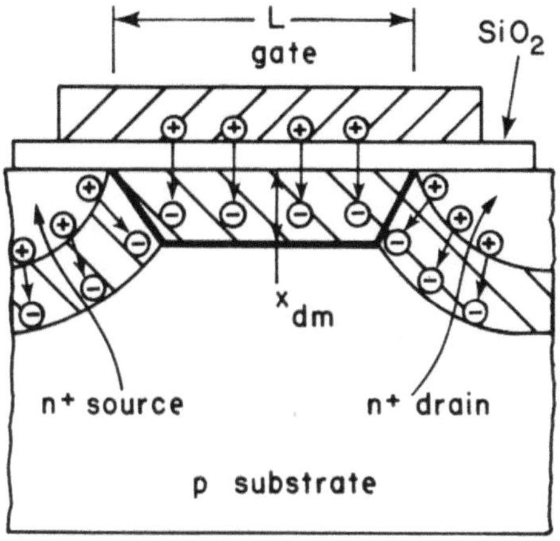

Fig. 17 Crossection of short channel MOSFET

For this case some of the electric field lines from the charge under the channel region terminate on ionized donor atoms in the source and drain depletion regions rather than on charge on the gate metal. The amount of charge Q_B contained in the trapezoidal region under the gate in Fig. 17 decreases with decreasing channel length. Since one of the terms in the equation for n-channel V_t is Q_B/C_{ox}, V_t decreases with decreasing channel length due to what is effectively a decrease in the average ("available") doping density under the channel. This also affects the substrate sensitivity of the device. Fig. 18 [38] shows the effect of both channel length and total back-gate potential on V_t for p-channel MOSFETs. The figure clearly illustrates the reduction in V_t with decreasing channel length and the reduction in substrate sensitivity (i.e., decrease in the effective average doping level under the gate) with decreasing channel length.

Although two-dimensional computer modeling is necessary to explain the details of the short channel effect, a simple model described by YAU [38] describes the effect well both qualitatively and quantitatively. In this model, at zero drain-source voltage, the field lines arising from the depletion charge in the gate region can be approximated as shown in Fig. 17. The field lines originating from the charges inside the trapezoidal depletion region are terminated on the gate electrode, whereas the field lines from the charges outside the trapezoidal region are terminated at the source and drain regions. Based on this geometrical approximation, the <u>effective</u> depletion charge inside the trapezoidal area is

$$Q_B(eff) = F \int_0^{X_1} C(x)\,dx = F\,Q_B \qquad (16)$$

where C(x) is the channel doping concentration, X_1 is the maximum depletion region thickness at inversion, $X_1 = f(V_{sx})$, and F is the form factor that can be obtained by the geometrical analysis,

$$F = 1 - \left(\sqrt{1 + 2X_1/X_J} - 1\right) X_J/L \qquad (17)$$

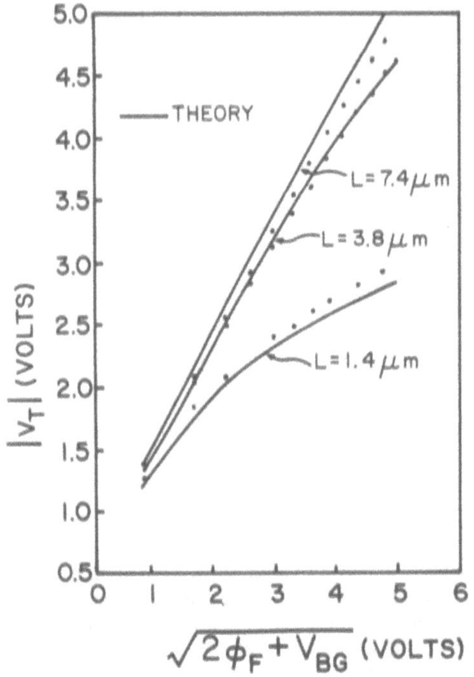

Fig. 18 Threshold voltage versus total back-gate potential for p-channel MOSFETs with various channel lengths, after YAU [38]

where X_J is the junction depth and L is the device channel length between source and drain. Thus the change in V_t for n-channel MOSFETs due to the short channel effect is

$$\Delta V_t = V_t\,(\text{short channel}) - V_t\,(\text{long channel}) = (F-1)\,Q_B/C_{ox} = (F-1)\,qN_a X_l/C_{ox} \tag{18}$$

for substrate doping N_a. The actual V_t is

$$V_t = V_{FB} + 2\,\psi_{si} + qN_a X_l\,F/C_{ox} + Q(\text{eff})/C_{ox} \tag{19}$$

The effect of drain voltage V_d on V_t for the short channel case can be accounted for by adding a term $0.5V_d$ to V_{sx} in calculating X_l for the correction factor F [39,40].

This simple model predicts most, if not all, of the significant short channel V_t effects. Fig. 19 shows predictions of the model for $|V_t - V_{FB}|$ versus channel length, for various substrate doping levels [38]. The solid lines of Fig. 18 are predictions using YAU's model, and clearly show a significant reduction in sensitivity to substrate bias. Note that for MOSFETs scaled as in Table 3, the upper two curves of Fig. 19 are the appropriate ones. However, the situation is less clear when ion-implanted, optimized devices [6] are modeled. Note that from the above model the short channel V_t correction factor should be only weakly temperature dependent, and the onset of the short channel V_t effect should generally be practically temperature independent. This has been verified experimentally [40].

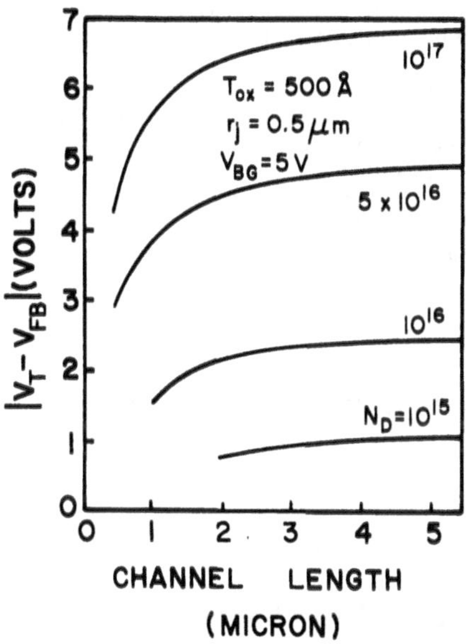

Fig. 19 Threshold voltage versus channel length predicted using the model of YAU [38]

2.3.2 Narrow Width V_t Effect

It has been established that a reduction of device channel width W, beyond a certain point, results in a narrow width effect on V_t, increasing the magnitude of V_t and increasing the substrate bias sensitivity of V_t [41,42]. An example of this effect showing both these characteristics is given in Fig. 20 [13]. Although there is some disagreement as to the details of the physical model involved [41,42], the effect is basically understood as being due to fringing of electric field lines in the substrate at the edges of the channel. Thus when the amount of charge, per unit channel length, uncovered in the substrate along the edge of the channel becomes comparable to the charge uncovered

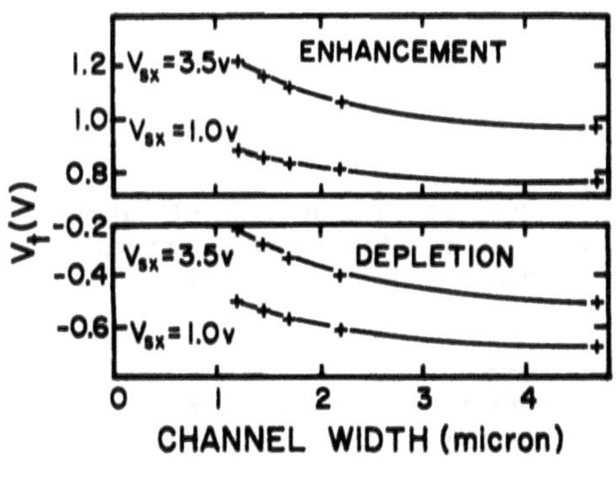

Fig. 20 Dependence of n-channel V_t on channel width for L = 15 micron, V_{ds} = 0.1V, implanted channel devices, after DENNARD, et al [13]

directly under the channel, the Q_B term in the expression for V_t increases and V_t is increased. The increase in the effective Q_B also causes the increase in substrate sensitivity. Note that the effect should not exist for open geometry Silicon On Insulator (e.g., Silicon On Sapphire) devices since there is no place for the "electric field lines in the substrate" to fringe to.

A simple geometrical model due to MERKEL [42] will be used to model the effect. Fig. 21 shows a schematic crossection, looking down the channel, of a narrow MOSFET with channel width W and maximum depletion region length at inversion X_1. Assuming that the space charge region spreads laterally in the substrate along the channel width and that the lateral extension of the space charge region is of a thickness X_1 and is cylindrical, the total charge in the substrate per unit channel length is given by

$$Q_{BT} = -qN_a W X_1 \left(1 + \frac{\pi X_1}{2W}\right) \tag{20}$$

Thus the contribution of Q_B to V_t is raised by the factor γ, where

$$\gamma = 1 + \frac{\pi X_1}{2W} \tag{21}$$

and the change in V_t for n-channel MOSFETs due to the effect is

$$\begin{aligned} \Delta V_t &= V_t\,(\text{narrow channel}) - V_t\,(\text{wide channel}) \\ &= \gamma\, Q_B / C_{ox} \end{aligned} \tag{22}$$

where of course Q_B is the substrate charge per unit gate area for long, wide devices, $Q_B = Q_B\,(V_{SX})$. The model thus predicts both the magnitude and at least one derivative of the effect, and shows that it is fundamentally decoupled from the short channel V_t effect.

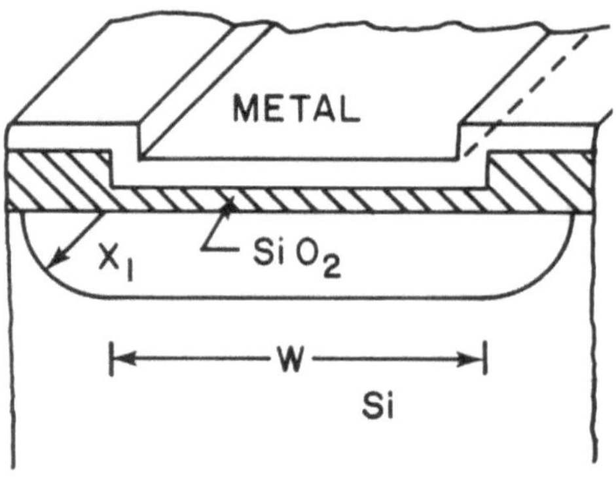

Fig. 21 Crossection of narrow MOSFET illustrating lateral spreading of channel depletion region

2.3.3 Combination Narrow Width/Short Channel (Minimum Size) V_t Effect

From the foregoing discussion it is clear that at least to first order the narrow width effect, which leads to a ΔV_t (NWE), and the short channel effect, which leads to a ΔV_t (SCE), are decoupled and that the V_t of any device including one of minimum size should be given by [41]

$$V_t = V_t\,(\text{large geometry}) + \Delta V_t\,(\text{SCE}) + \Delta V_t\,(\text{NWE}) \tag{23}$$

Here the size effects can be calculated using the simple models of the previous two sections, or more detailed models (see, for example, [41] or [43], although many others exist). Fig. 22 shows experimental results (solid line), and calculated results (points) obtained assuming decoupling of size effects [41].

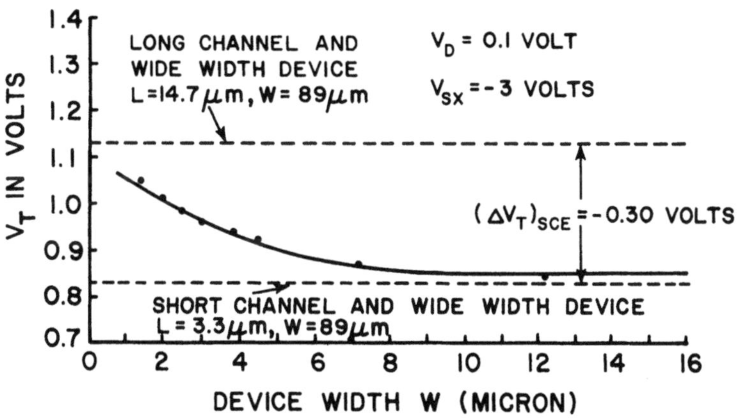

Fig. 22 Experimental and theoretical results for channel size effect on V_t, after WANG [41]

The particular device used in Fig. 22 was ion-implanted and more complicated (but conceptually simple) expressions were necessary for the calculation of the threshold voltage effects. However, results shown in the figure do bear out the principle of decoupled effects.

2.3.4 Short Channel Limit for Subthreshold Conduction Effects

MOSFETs fabricated with short channels without compensating adjustments in other fabrication parameters, show both strong increases in subthreshold drain current and dependence of subthreshold current on drain voltage [8,44]. That there is some effect is not unexpected since the subthreshold current is a sensitive function of the potential distribution in the channel and the nature of short channel phenomena revolves around the influence on the channel field of the source and drain depletion region fields. An emperical relationship for the limiting channel length, that length below which channel length effects on subthreshold current will occur, has been shown to be [8]

$$L_{min} = C_1\,(X_J\, t_{ox}\, W_1^2)^{1/3} \tag{24}$$

where C_1 is a constant (C_1 = 0.41 (angstrom)$^{1/3}$) and W_1 is the sum of drain and source depletion region widths, calculated for infinite planar junctions using the abrupt junction formulation [7]. Fig. 23 shows the behavior of L_{min} with variations in t_{ox} and doping level [8] as predicted from (24), for X_J = 1.46 micron. The scaling behavior of L_{min} is significant. For scaling by the prescriptions of Table 3, L_{min} scales as $K^{-4/3}$. Thus L_{min} scales faster than L if the Dennard rules are used. This conclusion relaxes at least one scaling constraint somewhat and gives greater latitude in scaling for optimized devices.

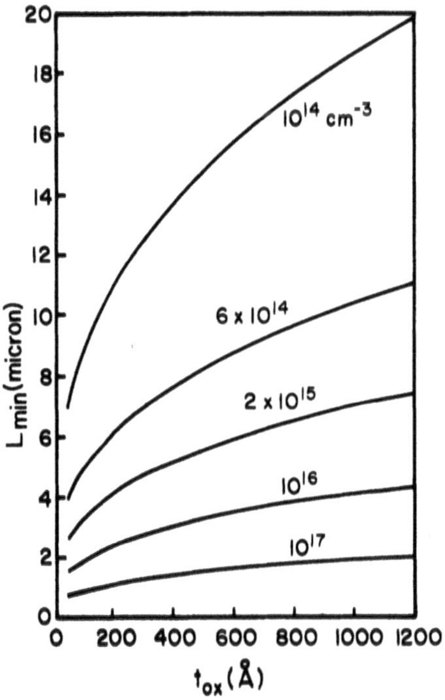

Fig. 23 Behavior of L_{min} with t_{ox} and Na variations, X_J = 1.46 micron, after BREWS, et al [8]

2.4 VLSIC Projections

The performance benefits in complexity and speed of scaling dimensions, discussed in section 2, were based on first-order scaling considerations which do not necessarily result in optimized circuits and which did not comprehend practical fabrication limits. Some realistic computer simulations of scaled circuit performance, taking into account many process considerations, have been done (see, for example, [11,45]) and some experimental data is available for realistically configured test vehicles [11,13,45-47] fabricated using advanced processes with VLSI potential. Figs. 24, 25 and 26 show results of some computer simulation work obtained in the course of a comprehensive evaluation of the scaling characteristics of various circuit technologies [11]. Also included in the evaluation were ECL and enhancement-mode load MOS. These two technologies exhibited unpromising packing densities and, except for some speed advantage for ECL for minimum dimensions greater than 1 micron, no advantages over the three technologies are shown in the figures.

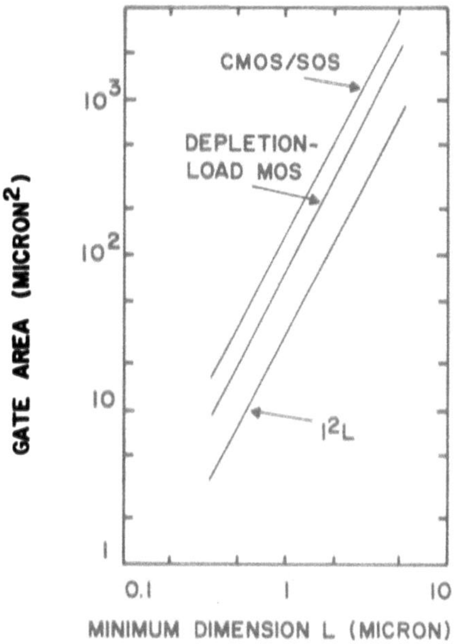

Fig. 24 Scaling of 4-input NOR gate area, after HART, et al [11]

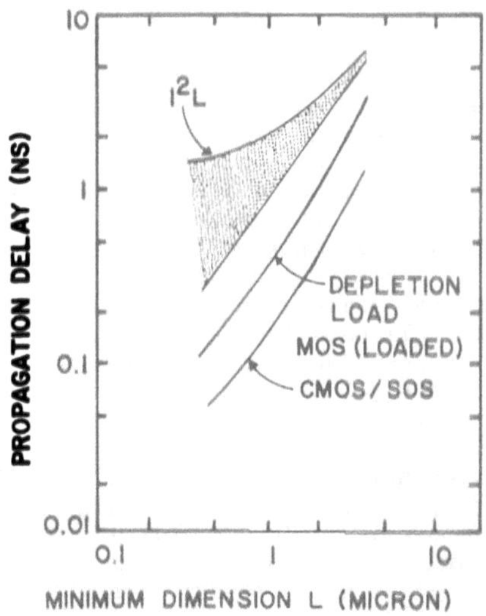

Fig. 25 Behavior of propagation delay of simple inverters, fan-out = 1, versus minimum dimension, partly after HART, et al [11]

For the MOS scaling the minimum dimension was taken to be the width of the polysilicon gate (somewhat greater than the real channel length L_{eff}), and shallow (0.25 micron) junctions, lower power supply voltages (20V for L = 3.3 micron, 3V for L = 0.65 micron) and thinner oxides (0.1 micron for L = 3.3 micron, 0.01 micron for L = 0.65 micron) were used in the modeling. Effects

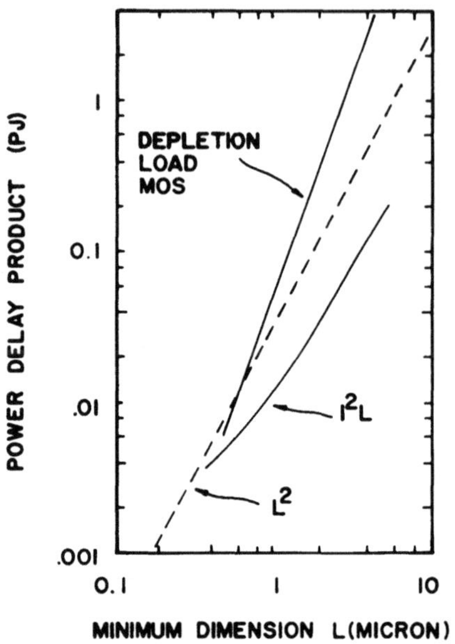

Fig. 26 Behavior of power-delay product of simple inverters, fan-out = 1, versus minimum dimension, after HART, et al [11]

such as carrier velocity saturation in the channel and overlap capacitances (e.g., gate-drain, particularly important for small L) were included [11]. For the I^2L scaling the minimum dimension was taken to be the interconnection width and the width of a contact opening. Thick oxide "isolation" was assumed, and the epitaxial layer thickness was taken to be 0.45 micron, constant with scaling, with approximately 0.23 micron intrinsic base width. This base width is well in excess of the limit, due to punchthrough and doping fluctuations, of approximately 0.08 micron [16]. For both technologies a specific contact resistance of 10^{-6} ohm cm^2 was used for contact to base diffusions, and 10^{-7} ohm cm^2 for contact to n^+ regions. From Fig. 26 it appears that the scaling behavior of the power-delay product is approximately as predicted by the simple theory of section 2, except for I^2L at small dimension which shows the increasing importance of collector-base sidewall capacitance [11]. This quantity scales as L scales, instead of as L^2 scales. Note that for the MOS devices the power-delay product should scale as L_{eff}^{-3}; this is somewhat stronger than L^{-3} due to the influence of the assumed fixed lateral diffusion distance of the source and drain region. Failure to scale as L_{eff}^{-3} is due to other effects (e.g., velocity saturation in the channel) coming into play at very small L_{eff}. The scaling behavior of MOS propagation delays in Fig. 25 are generally not in agreement with the behavior shown in Table 4. This is apparently because t_{ox} was scaled so strongly, and power dissipation per gate was allowed to rise. A wide range of I^2L propagation delay behavior can be predicted, as shown in Fig. 25, depending on exact assumptions made concerning fabrication technology capabilities. The upper limit of Fig. 25 is due to HART, et al [11].

For comparison to experimental VLSI-geometry results involving multiple fan-out - fan-in gates the results of Figs. 25 and 26 must be increased (by X2 for T_d, for fan-out = 4, for example). Experimental results [13] for scaled depletion load MOS gates with fan-out and fan-in of 3, loaded with 50×10^{-15} F of simulated wiring capacitance, indicate T_d = 1.9ns and T_d = 1.1ns for L_{eff} = 1.3 micron and L_{eff} = 1.05 micron respectively. Power-delay products were 0.247 pJ and 0.187 pJ for the two cases. These performances are in good agreement with the power-delay product of Fig. 26 (correcting for L_{eff}) and in reasonable agreement with the T_d behavior of Fig. 25, taking into account parameter differences. I^2L experimental results for fan-outs of 4 |45| are somewhat worse than the PT_d predictions of Fig. 26 and in reasonable agreement with the corrected T_d predictions of Fig. 25, for 1.5 micron minimum geometry.

2.5 Conclusions

It is clear that the driving forces exist for VLSI speed-complexity (FTR) increases of two orders of magnitude or more. Although technology problem areas exist, notably associated with the interconnection system, simple scaling theory predicts the potential for more than two orders of magnitude increase in FTR with scaling down the minimum dimension by a factor of 5 or so. Computer simulations and experimental data generally bear out this conclusion. Trends indicate that this degree of improvement is possible in five to ten years.

References

1. G. E. Moore, IEEE Spectrum 16, No. 4, 30, April 1979.

2. G. E. Moore, Proceedings of the International Electron Device Meeting, 11 (1975).

3. R. N. Gossen, IEEE Spectrum 16, No. 3, 42, March 1979.

4. H.-N. Yu, A. Reisman, C. M. Osburn, and D. L. Critchlow, IEEE J. Solid-State Circuits SC-14, 240 (1979).

5. L. R. Weisberg and L. W. Sumney, 1978 Digest of Papers of the Government Microcircuits Applications Conference, p. 18, Nov. 1978.

6. R. H. Dennard, F. H. Gaensslen, H. Yu, V. L. Rideout, E. Bossons, and A. R. Leblanc, IEEE J. Solid-State Circuits SC-9, 256 (1974).
A. N. Broers and R. H. Dennard, in Semiconductor Silicon, Electrochemical Society, H. R. Huff and R. R. Burgess, eds (1973).

7. S. M. Sze, Physics of Semiconductor Devices, John Wiley and Sons, New York (1969).

8. J. R. Brews, W. Fichtner, E. H. Nicollian, and S. M. Sze, IEEE Electron Device Lett. EDL-1, 2 (1980).

9. B. Hoeneisen and C. A. Mead, Solid-State Electron. 15, 819 (1972).

10. R. W. Keyes, Proc. IEEE 63, 740 (1975).

11. P. A. H. Hart, T. Van'T Hof, and F. M. Klaasen, IEEE J. Solid-State Circuits SC-14, 343 (1979).

12. G. H. Heilmeier, IEEE Spectrum 16, No. 3, 45, March 1979.

13. R. H. Dennard, F. H. Gaensslen, E. J. Walker and P. W. Cook, IEEE Trans. Electron Devices ED-26, 325 (1979).

14. P. M. Solomon and D. D. Tang, 1979 IEEE International Solid-State Circuits Conf. Digest of Technical Papers, p. 86, Feb. 1979.

15. F. M. Klaasen, Solid-State Electron. 21, 565 (1978).

16. B. Hoeneisen and C. A. Mead, Solid-State Electron. 15, 891 (1972).

17. C. Hart and A. Slob, IEEE J. Solid-State Circuits 7, 346 (1972).

18. F. M. Klaasen, IEEE Trans. Electron Devices ED-22, 145 (1975).

19. H. H. Berger and K. Helwig, IEEE J. Solid-State Circuits SC-14, 327 (1979).

20. S. A. Evans, IEEE J. Solid-State Circuits SC-14, 318 (1979).

21. B. L. Crowder and S. Zirinsky, IEEE Trans. Electron Devices ED-26, 369 (1979).

22. J. R. Black, Proc. 12th IEEE Reliab. Phys. Symp., 142 (1974).

23. G. A. Scoggan, B. N. Agarwala, P. O. Peressini, and A. Brouillard, Proc. 13th IEEE Rel. Phys. Symp., 151 (1975).

24. J. C. Blair, P. B. Ghate and C. T. Haywood, Appl. Phys. Letters 17, 281 (1970).

25. P. B. Ghate, J. C. Blair, and C. R. Fuller, Metalization in Microelectronics, International Conference on Metallurgical Coatings, 28 March - 1 April 1977, San Francisco, CA.

26. H. Sello, in Ohmic Contacts to Semiconductors, Electrochemical Society, B. Schwartz, ed. (1969).

27. P. L. Shah, IEEE Trans. Electron Devices ED-26, 631 (1979).

28. A. J. Learn, J. Electrochemical Soc. 123, 894 (1976).

29. H. Grinolds and G. Y. Robinson, J. Vac. Sci. Technol. 14, 75 (1977).

30. C. J. Kircher, J. Appl. Phys. 47, 5394 (1976).

31. J. T. Wallmark and S. M. Marcus, Proc. IRE 50, 286 (1962).

32. T. H. May and M. H. Woods, Proc. 16th IEEE Reliab. Phys. Symp., 33 (1978).

33. J. F. Ziegler and W. A. Lanford, 1980 Digest of Technical Papers, International Solid-State Circuits Conf., p. 70, Feb. 1980.

34. E. Sun, J. Moll, J. Berger and B. Alders, 1978 International Electron Devices Meeting Technical Digest, p. 478, Dec. 1978.

35. O. Kudoh, M. Tsurumi, H. Yamanaka, and T. Wada, IEEE J. Solid-State Circuits SC-13, 235 (1978).

36. T. Toyabe , K. Yamaguchi, S. Asai, and M. S. Mock, IEEE Trans. Electron Devices ED-25, 825 (1978).

37. H. C. Poon, L. D. Yau, and R. L. Johnston, 1973 International Electron Devices Meeting Technical Digest, p. 156, Dec. 1973.

38. L. D. Yau, Solid-State Electron. 17, 1059 (1974).

39. P. P. Wang, IEEE Trans. Electron Devices ED-24, 196 (1977).

40. F. H. Gaensslen, V. L. Rideout and E. J. Walker, 1975 International Electron Devices Meeting Technical Digest, p. 43, Dec. 1975.

41. P. P. Wang, IEEE Trans. Electron Devices ED-25, 779 (1978).

42. G. Merkel, "Short Channels - Scaled-Down MOSFET's" in "Processing and Device Modelling for Integrated Circuit Design", Noordhoff, Leyden, Netherlands, p. 705, (1977).

43. K. N. Ratnakumar, D. J. Bartelink and J. D. Meindl, 1980 IEEE International Solid-State Circuits Conf. Digest of Technical Papers, p. 72, Feb. 1980.

44. G. W. Taylor, IEEE Trans. Electron Devices ED-25, 337 (1978).

45. S. A. Evans, IEEE J. Solid-State Circuits SC-14, 318 (1979).

46. S. A. Evans, J. L. Bartelt, B. J. Sloan and G. L. Varnell, IEEE Trans. Electron Devices ED-25, 402 (1978).

47. J. M. Herman, S. A. Evans and B. J. Sloan, IEEE J. Sloan, IEEE J. Solid-State Circuits SC-12, 150 (1977).

3. Advanced Lithography

R. K. Watts

With 28 Figures

Optical lithography has made possible in integrated circuit manufacturing a unique combination of high volume, high precision, and low cost. Conventional semiconductor devices can probably function with dimensions very much smaller than they have at present [1,2] and too small for optical patterning. Development of electron and X-ray lithography has been underway for 14 and 8 years, respectively, both techniques offering increased resolution. Requirements for higher patterning precision, such as better level to level registration and better linewidth control, are often more difficult goals than smaller linewidth, leading to greater refinement and complexity of exposure tools.

3.1 Optical Lithography

Optical lithography comprises the formation of images in photoresist on wafers by contact, proximity, or projection printing with near ultraviolet light. These techniques have continued to be refined to satisfy needs for higher resolution and precision.

3.1.1 Optical Resists

Photoresists are of two types. A negative resist on exposure to light is made less soluble in a developer solution, while a positive resist becomes more soluble. Commercial negative resists, such as Kodak Microneg 747, consist of two parts: a chemically inert polyisoprene rubber, which is the film forming component, and a photoactive agent. The photoactive agent on exposure to light reacts with the rubber to form crosslinks between rubber molecules, making the rubber less soluble in an organic developer solvent. The reactive species formed during illumination can react with oxygen and be rendered ineffective for crosslinking. Therefore a nitrogen atmosphere is usually provided.

The developer solvent not only dissolves the unexposed resist. The exposed resist swells as the uncrosslinked molecules are dissolved away. The swelling distorts the pattern features and limits resolution to 2 to 3 times the initial film thickness.

Positive resists have two components: a resin and a photoactive compound dissolved in a solvent. The photoactive compound is a dissolution inhibitor. When it is destroyed by exposure to light, the resin becomes more soluble in an aqueous alkaline developer solution. The unexposed regions do not swell much in the developer solution, and so higher resolution is possible with positive resists. The development process of projection printed images in positive resists has been modeled theoretically. It is an isotropic etching process [3]. The sensitivity of most standard resists peaks in the 0.3-0.4μm spectral range. Two examples of commercially available positive resists are Shipley AZ-2400 and Hunt HPR-206.

The light intensity $I(\lambda,z)$ which is effective in exposing a volume element of resist at height $z<T$ above the substrate depends on the reflectivity of the substrate, the thickness T of the resist, and the convolution of the absorption spectrum of the resist with the spectrum of the incident light. The absorption spectra are fairly sharp and only a small part of the lamp spectrum is effective. The refractive index of most resists is about 1.6, leading to an optical mismatch at the air/resist interface. If the substrate is highly reflective, the standing wave pattern set up has a node at the substrate and a peak amplitude which depends on resist thickness T. This variation in peak intensity with resist thickness becomes less with decreasing substrate reflectivity and increasing absorption by the resist [4]. Since resist thickness varies at a step in substrate topology, the resulting difference in effective exposure leads to size variations in the resist image. Because of the high resolution of positive resists, the standing wave patterns are often well resolved, and the resulting corrugations are seen in the resist edge profile [5].

It seems clear that resists will be developed for use at shorter wavelengths where higher resolution is possible. A few such deep UV resists reported recently are polymethyl methacrylate (PMMA), sensitive for $\lambda<0.25\mu m$, polybutene sulfone (PBS), sensitive at $\lambda \lessapprox 0.2\mu m$, and poly methyl isopropenyl ketone (PMIPK), sensitive at $\lambda \approx 0.2\mu m$ [6]. At these shorter wavelengths the radiation quantum is large enough to lead to scission of the molecular chain. The resulting smaller molecules are more soluble in the developer. The scheme involving a dissolution inhibitor is not necessary.

Some work has been reported on inorganic negative resists consisting of a chalcogenide glass such as Se_xGe_y or As_2S_3 [7,8] coated with a thin Ag layer. Exposure leads to migration of Ag atoms into the glass, rendering it etch resistant to a developing alkaline solution or CF_4 plasma [9]. No swelling occurs, since the resist is a rigid inorganic material. Fig. 1 is an exposure curve plotted as fraction of initial thickness T_0 remaining after exposure to dose q and development [10]. The dose is the incident light flux times the exposure time. For this resist the slope γ of the linear portion ($\gamma=\Delta(T/T_0)/\Delta\log q$) is very large. The larger γ, the lower the contrast I_{max}/I_{min} required in the imaging light. Fig. 2 shows application of the inorganic resist in a two-level structure consisting of 2μm thick hard baked HPR-206 topped by a 2000 Å thick layer of $GeSe_2$ [11]. After development, the thin layer serves as a mask for reactive ion etching of the thick lower layer. A planar top surface is presented for patterning. The step coverage problem is solved, for the thick layer can be considerably over-etched without much change in lateral pattern dimensions.

Other properties of resists which are also quite important are good adhesion to the substrate and resistance to wet and dry etch processes. In general the commercially available optical resists are compatible with such processes. Adhesion of positive resists to certain types of substrate is less than for negative resists, and a thin silane adhesive layer is often first spun onto the substrate.

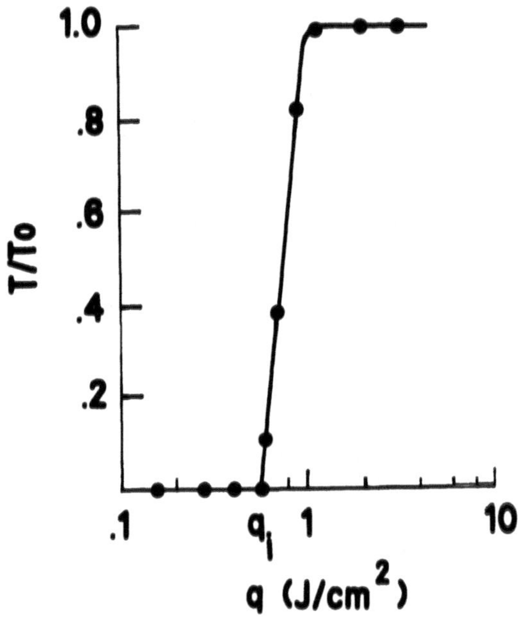

Fig. 1 Exposure curve of $GeSe_2$:Ag resist, [10]. q_1 is threshold dose. Slope of steep rise is γ.

The unwanted variation of feature size in the resist image is due to many effects, some related to resist properties and resist processing and others to the exposure tool. These contributions can be to some extent separated by writing the edge acuity or slope of a feature edge $\partial T/\partial X$ as $\partial T/\partial X = (\partial T/\partial q)(\partial q/\partial X)$. X is measured perpendicular to the pattern edge and parallel to the plane of the substrate. The first factor is proportional to γ/q. The second is proportional to the intensity gradient in the image, as shown for example, in Figs. 4 and 5. (The parameters Q, γ, and W are defined later).

3.1.2 Contact and Proximity Printing

In contact printing a photomask is pressed against the resist covered wafer with pressures typically in the range 0.05 atm to 0.3 atm, and exposure is by light of wavelength near 0.4µm. Very high resolution - $\leqslant$1µm linewidth - is possible, but because of spatial nonuniformity of the contact, resolution may vary considerable across the wafer. To provide better contact over the whole wafer a thin (0.2 mm) flexible mask has been used; 0.4 µm lines have been imaged in 0.98 µm resist [12]. The contact produces defects in mask and wafer so that the mask, whether thick or thin, may have to be discarded after a short period of use. For example, JONES [13] reports mask defect densities increasing to 37 defects/cm^2 after 15 exposures from 13 defects/cm^2 after 5 exposures. Defects produced are pinholes in the chromium film, scratches, intrusions, and star fractures [14].

Fig. 2 Two-level resist structure $GeSe_2$/HPR, [11]

Nevertheless contact printing continues in wide use. LIN [15] has imaged 0.25μm features in 1.8μm thick PMMA resist using 0.20-0.26μm radiation. Quartz or Al_2O_3 mask substrates must be used to pass these shorter wavelengths, since the usual borosilicate glass absorbs strongly wavelengths less than 0.3μm. The absorbing Cr film is more transparent at shorter wavelength and thickness must be increased to 0.16μm.

Proximity printing has the advantage of lack of contact between mask and wafer and thus longer mask life. Resolution is not as good as for projection printing, but printer cost is lower. Resolution is also less than for hard-contact printing.

Figure 3 shows the optical problem in the schematic form of a mask with a long slit of width W separated from a parallel image plane (wafer) by a gap g. We assume that g and W are larger than the wavelength λ of the imaging light and that $\lambda \ll g < W^2/\lambda$ - the region of Fresnel diffraction. Then the diffraction pattern which forms the image of the slit is a function

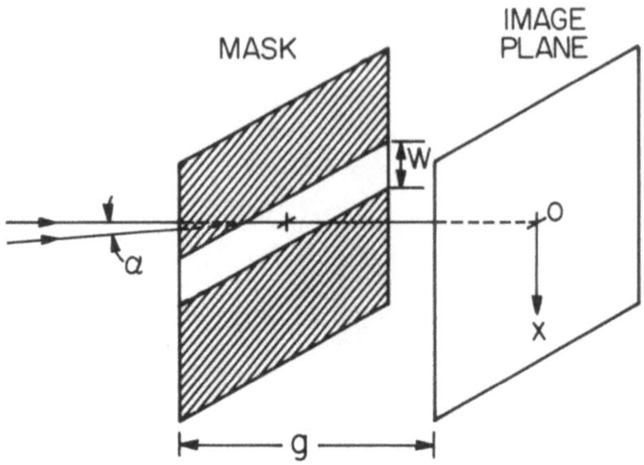

Fig. 3 Proximity printing idealization

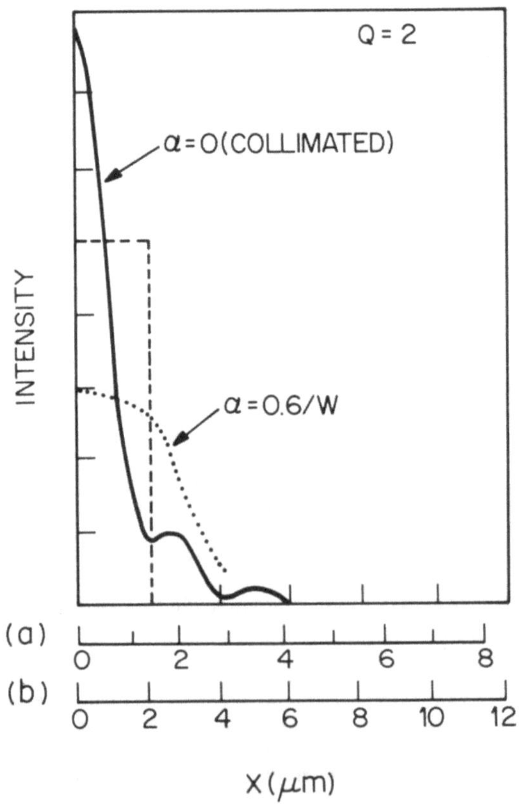

Fig. 4 Image of slit with Q = 2. Scale (a): g = 10μm; scale (b): g = 20μm.

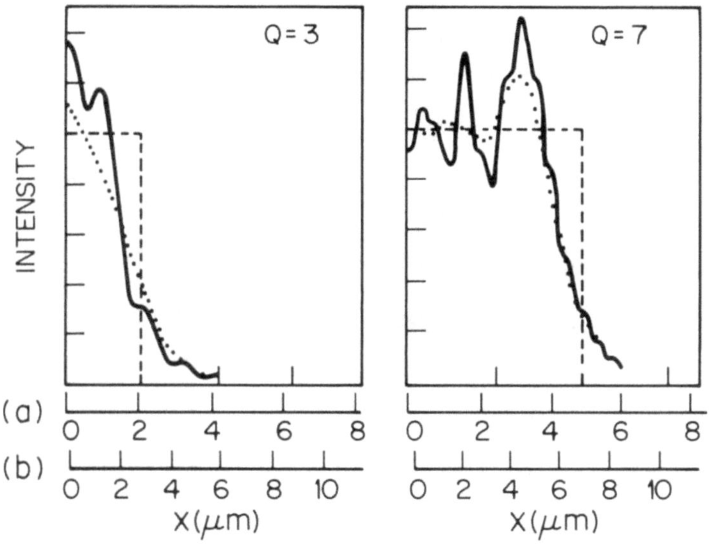

Fig. 5 Images of slit for Q = 3 and Q = 7. Scale (a): g = 10μm; scale (b): g = 20μm.

only of the particular combination of λ, W, and g which we shall call the parameter Q where

$$Q = W\sqrt{2/(\lambda g)} \tag{1}$$

The diffraction patterns are well known [16]. Some calculated by SKINNER [17] are shown in Figs. 4 and 5 for Q = 2, 3, and 7 (solid curves). The limit $g<W^2/\lambda$ corresponds to $Q>\sqrt{2}$. The dashed rectangle shows the light intensity at the mask. Since a particular value of Q can correspond to many possible combinations of W, λ, and g, we pick two combinations (a) λ = 0.4μm, g = 10μm; and (b) λ = 0.4μm, g = 20μm and show abscissas for each; W is then determined by (1). For Q = 3, for example, we have the same image pattern for a 4.2μm slit and 10μm gap (scale a) as for a 6μm line and a 20μm gap (scale b). The larger Q, the more faithful the image. We see that resolution becomes better at smaller gaps and shorter wavelengths.

Two other noteworthy features of the diffraction patterns are the ragged peaks and the slope near X = W/2. The peaks can be smoothed by use of diverging rather than collimated light. The dotted curves are for light with divergence half angle $\alpha = 1.5\lambda/W$ radians. If α becomes large, the edge slope is reduced further, increasing linewidth control problems. Some smoothing also occurs because of the spread of wavelengths in the source, but this is a smaller effect. The illumination system must provide an apparent source size large enought to give a value of α - typically a few degrees - optimized for the smallest features to be printed. This is illustrated in Fig. 6. The mercury arc used as source is too small to yield the required α, so a scheme like that illustrated at the right in the figure is used. The optical system must also minimize non-uniformity of intensity across the field; this may typically be 3%-15%. Fig. 7 shows the illumination system of a Canon proximity printer. The illumination is

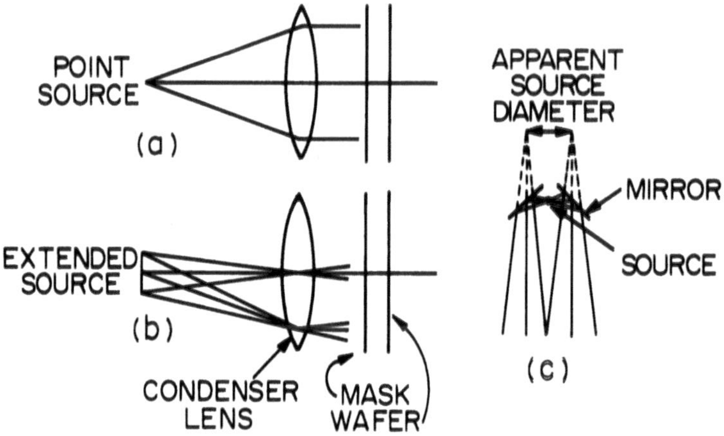

Fig. 6 Proximity printing with collimated and uncollimated light

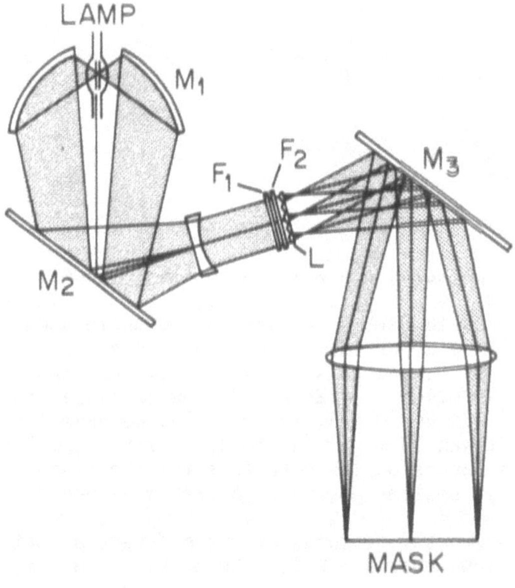

Fig. 7 Illumination system of Canon PLA 500F proximity printer. F = filter, M = mirror, L = fly's eye lens.

telecentric at the mask to prevent runout errors. With a Hg arc source the strong lines at 0.436µm, 0.405µm, and 0.365µm provide exposure flux. The same printer is available with Xe-Hg source for enhanced output in the 0.2µm-0.3µm spectral region. Using this source NAKANE et at. [6] report exposure times of 1 min. (a very long exposure compared with standard resists) with PMMA resist and 2µm resolution with g = 10µm to 20µm.

The edge of a feature in developed resist occurs at the position where the product of light intensity and exposure time It_e equals the resist threshold dose q_1. Linewidth control is in general more difficult in proximity printing than in contact printing. Thus linewidth is influenced by variations in exposure intensity, gap g, and resist threshold. MCGILLIS and FEHRS [18] have measured linewidth as a function of exposure intensity I and mask-wafer gap. They find that the linewidth variation defined as printed width minus width of the line on the mask is proportional to lnI and to $\sqrt{g}$, and that for each value of g there is an optimum exposure intensity which minimizes the effect on linewidth of gap variation. The mask-wafer gap can vary across a wafer and from one wafer to another because of wafer and mask bowing and dirt particles between wafer and wafer chuck. For exposure intensity near the optimum value, control of exposure to within 13% should provide linewidth control of ±0.5μm, even for large variations of the gap g. JONES [13] has reported linewidth control to within ±0.25μm with a 50μm gap.

Because of the sloping edges of the diffraction pattern of the slit, the image of a mask object consisting of equally spaced parallel lines will lose definition as the spacing between lines decreases and the edge tails begin to overlap. If I_M is the peak intensity in the image (on the lines) and I_m is the minimum intensity in the image (between lines, where this quantity should be zero), the modulation of the image M_i is given by

$$M_i = \frac{I_M - I_m}{I_M + I_m} \tag{2}$$

M_i is a function of linewidth, line spacing, gap, and wavelength. For 2.5μm lines and spaces it depends on mask-wafer gap in the following way, according to a calculation by HEIM [19].

g(μm)	M_i
5	0.62
10	0.50
20	0.25

For exposure of resist a ratio I_M/I_m of at least 4 is desirable. This corresponds to M_i = 0.6, a value which can be exceeded only at very small gaps. Further loss of modulation or contrast can be reduced by use of Fe_2O_3 or Cr/Cr_2O_3 masks which have much lower reflectivity than Cr, and reduce scattered light under the opaque parts of the mask.

3.1.3 Projection Printing

Projection printing offers higher resolution than proximity printing together with large separation between mask and wafer. Four important performance parameters of a printer are resolution, level to level alignment accuracy, throughput, and depth of focus. The resolution of an optical imaging system of numerical aperture sinα with light of wavelength λ is, according to Rayleigh's criterion, 0.61 λ/sinα - the separation of the barely resolved images of two point sources. The Rayleigh depth of focus is given by $\pm \lambda/(2 \sin^2\alpha)$. At the limiting resolution of the system, the contrast in the image is uselessly small, however. We shall see how optical performance can be specified in a more meaningful way by means of the system transfer function.

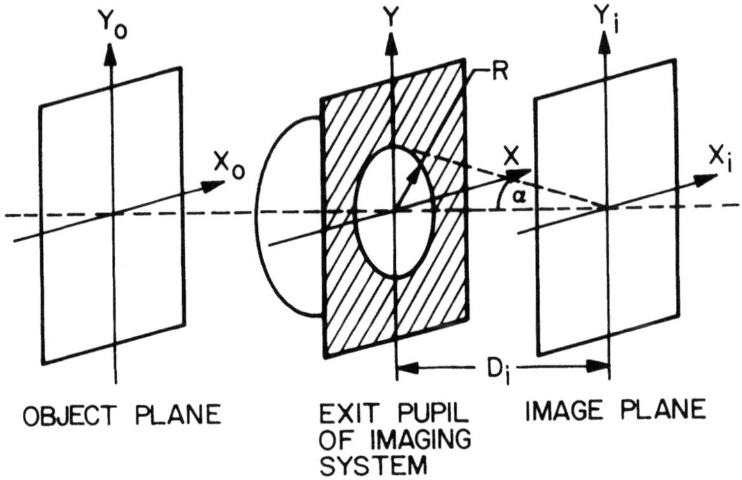

Fig. 8 Projection printing imaging system idealization

We consider a general optical system (Fig. 8) with no aberrations. That is, a spherical wave diverging from a point in the object plane is converted to a spherical wave converging to a point in the image plane. The f number of the system is $F = D_i/2R$, and the numerical aperture is $\sin\alpha = R/\sqrt{(D_i^2 + R^2)} \approx R/D_i$.

We want to recall briefly some results from the theory of image formation and to show how the transfer functions differ for coherent and incoherent imaging systems [20,21]. For coherent object illumination all points in the object have wave amplitudes with fixed phase relationships, and all phases have the same time dependence. For example in Fig. 6a the phase varies across the plane at right in a way simply determined by the path lengths from the point source. If there is no phase correlation, on the other hand, and the phases vary independently from point to point across the object (Fig. 6b), the illumination is spatially incoherent. We consider first the coherent case. It is not necessary that the radiation be monochromatic, only that the wavelength spread be small compared with the average wavelength. The wavelength λ can stand for the average wavelength. The field amplitude $U_i(x_i,y_i)$ at point x_i,y_i in the image plane due to an amplitude distribution $U_o(x_o,y_o)$ in the object plane is given by

$$U_i(x_i,y_i) = \iint_{-\infty}^{\infty} K(x_i-x_o', y_i-y_o')U_o(x_o',y_o')dx_o'dy_o' , \tag{3}$$

where $x_o' = mx_o$, and m is the magnification of the system. K is the amplitude at x_i,y_i due to a point source at x_o,y_o. K is the Fraunhofer diffraction pattern of the exit putil $P(x,y)$,

$$K = \left(\frac{1}{\lambda D_i}\right)^2 \iint_{-\infty}^{\infty} P(x,y)\exp\{-\frac{2\pi i}{\lambda D_i}[(x_i-x_o')x+(y_i-y_o')y]\}dxdy \tag{4}$$

We shall be concerned only with a circular pupil function, as in Fig. 8.

Then P = 1 for points x,y in the circle, and P = 0 outside the circle.

Let us consider a point source on axis, $x_o=y_o=0$. Then (4) can be evaluated to give the Bessel function,

$$K(x_i,y_i) = \frac{\pi}{(2\lambda F)^2}\,[2J_1(s)/s],\ s=(\tfrac{\pi}{\lambda F})\,\sqrt{x_i^2+y_i^2}. \qquad (5)$$

The quantity $[2J_1(s)/s]^2$, which is proportional to the intensity in the image of the point source, is the Airy pattern shown in Fig. 9. The first zero occurs at radius $x_i^{(0)} = 1.22\ \lambda F$. 84% of the light is contained in the central spot and 7% in the first ring. Thus (3) shows that the image is made up of a superposition of Airy patterns - one for each point in the object. The smaller the aperture R, the larger F, and the lower the resolution of the system, since each point is smeared into a circle of diameter 2.4 λF.

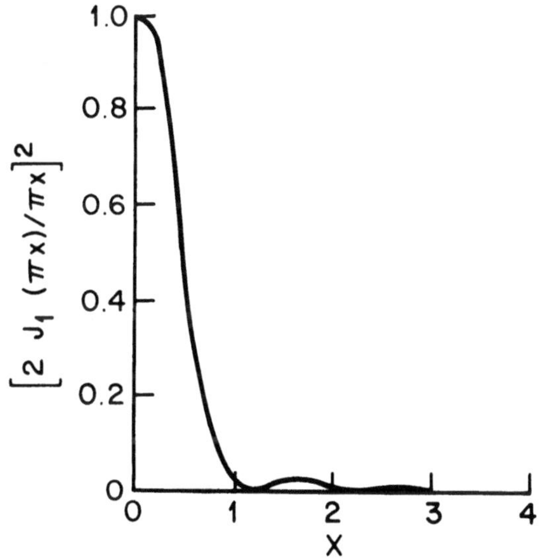

Fig. 9 Airy pattern. $X = (\pi/\lambda F)\sqrt{x_i^2+y_i^2}$

The Fourier transforms of U_o, U_i, and K will be called G_o, G_i, and H respectively:

$$G(u,v) = \iint_{-\infty}^{\infty} U(x,y)\exp\,[2\pi i(ux+vy)]\ dxdy \qquad (6a)$$

$$H(u,v) = \iint_{-\infty}^{\infty} K(x,y)\exp\,[2\pi i(ux+vy)]\ dxdy \qquad (6b)$$

In (6) G and U stand for G_o and U_o or G_i and U_i. G_i and G_o are related by

$$G_i(u,v) = H(u,v)G_o(u,v), \qquad (7)$$

and H(u,v) is the frequency response function of the optical system.

A component of the spatial frequency spectrum of the image is obtained by multiplying the corresponding component of the object spectrum by the value of H at that frequency. H(u,v) is called the coherent transfer function. Taking the inverse transform

$$K(x,y) = \iint_{-\infty}^{\infty} H(u,v)\exp\,[-2\pi i(ux+vy)]\;du\,dv \tag{8}$$

and comparing with (4) we see that

$$H(u,v) = P(u\lambda D_1,\; v\lambda D_1). \tag{9}$$

The coherent transfer function has the same shape as the pupil function. For the circular aperture

$$\begin{aligned} H(u,v) &= 1 \text{ if } \sqrt{u^2+v^2} \leq R/\lambda D_1 \\ &= 0 \text{ if } \sqrt{u^2+v^2} > R/\lambda D_1. \end{aligned} \tag{10}$$

This transfer function is shown in Fig. 10. It has the value 1 for frequencies up to the cutoff freauency $u_m/2$, where $u_m = 2R/\lambda D_1 = 1/\lambda F$.

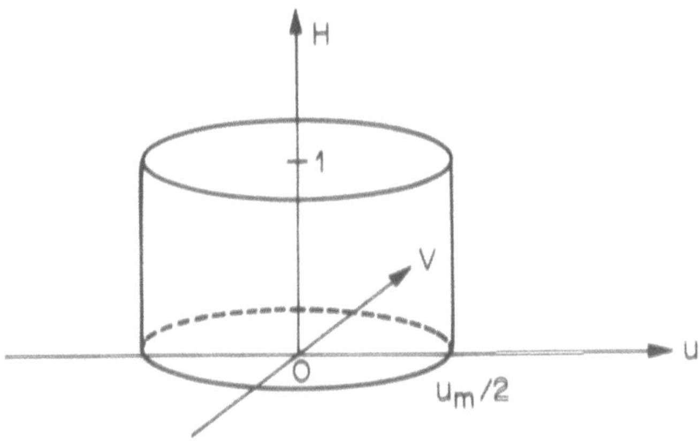

Fig. 10 Coherent or optical transfer function, round pupil

Now let us consider spatially incoherent illumination where the phases vary independently across the object. The intensity $I_1(x_1,y_1)$ is a time average $U_1(x_1,y_1,t)U_1^*(x_1,y_1,t)$ and for incoherent light is given by

$$I_1(x_1,y_1) = \iint_{-\infty}^{\infty} I_0(x_0',y_0')\,|K(x_1-x_0',y_1-y_0')|^2 dx_0'dy_0'. \tag{11}$$

(For the coherent case it is given by the squared magnitude of the right side of (3).) If $\mathcal{G}_0$, $\mathcal{G}_1$, and $\mathcal{H}_1$ are Fourier transforms of object and image intensities and of $|K|^2$,

$$\mathscr{I}(u,v) = \iint_{-\infty}^{\infty} I(x,y)\exp\,[2\pi i(ux+vy)]\; dxdy \tag{12}$$

$$\mathscr{H}(u,v) = \iint_{-\infty}^{\infty} |K(x,y)|^2 \exp\,[2\pi i\;(ux+vy)]\; dxdy,$$

then $\mathscr{I}_0$ and $\mathscr{I}_1$ are related linearly by

$$\mathscr{I}_1(u,v) = \mathscr{H}(u,v)\mathscr{I}_0(u,v). \tag{13}$$

The frequency response function $\mathscr{H}$ is called the optical transfer function and $|\mathscr{H}|$ is called the modulation transfer function or MTF. $|\mathscr{H}|$ is related to the pupil function in a more complicated way than before.

$$\mathscr{H}(u,v) = \left(\frac{1}{\lambda D_1}\right)^2 \iint_{-\infty}^{\infty} P(\xi+u\lambda D_1/2,\; \eta+v\lambda D_1/2)x \tag{14}$$

$$P^*(\xi-u\lambda D_1/2,\; \eta-v\lambda D_1/2)\; d\xi d\eta.$$

In this case the transfer function is given by the overlap of two pupils displaced from each other by the amount $u\lambda D_1$. Fig. 11 shows the case of the circular pupil, the line joining the pupil centers being taken as one axis. As u increases, the circles move apart and the transfer function decreases. $\mathscr{H}(u)$ is found by calculation of the shaded overlap area to be

$$\mathscr{H}(u) = \frac{2}{\pi}\,[\cos^{-1}(u/u_m)-(u/u_m)\sqrt{1-(u/u_m)^2}], \tag{15}$$

$$\text{for} \quad u \leq u_m$$

The limiting spatial frequency is twice as great as for coherent illumination, but image contrast or modulation is less at the lower frequencies. The transfer function describes how the system degrades the image of a sinusoidal object grating. In general there is a reduction in modulation and a phase shift. The MTF at frequency u is the ratio of image modulation (See (2)) to object modulation for an object grating of periodicity u^{-1}.

$$|\mathscr{H}(u)| = M_1(u)/M_0(u) \tag{16}$$

KING and GOLDRICK [22] describe MTF measurements for projection, proximity, and contact printers using a square wave bar target. The sine wave MTF can be calculated from the measured square wave response.

In modern printers the illumination is intermediate between the coherent and incoherent limits. The separation between points in the image which have correlated phases is neither infinite nor near zero. It can be sometimes varied by means of a diaphragm in the condenser illuminating the object. This is a familiar phenomenon to anyone who has used a microscope. As the diaphragm is closed, the image looks"sharper", but interference rings appear around edges, and the light level drops. This is the near-coherent case. Opening the diaphragm causes the rings to disappear and the image becomes brighter. The ratio S of the numerical aperture of the condenser to that of projection optics is often used as a measure of coherence. S = 0 implies near coherence; if S = 1 the apertures are matched and the entrance pupil of the projection optics is filled to give

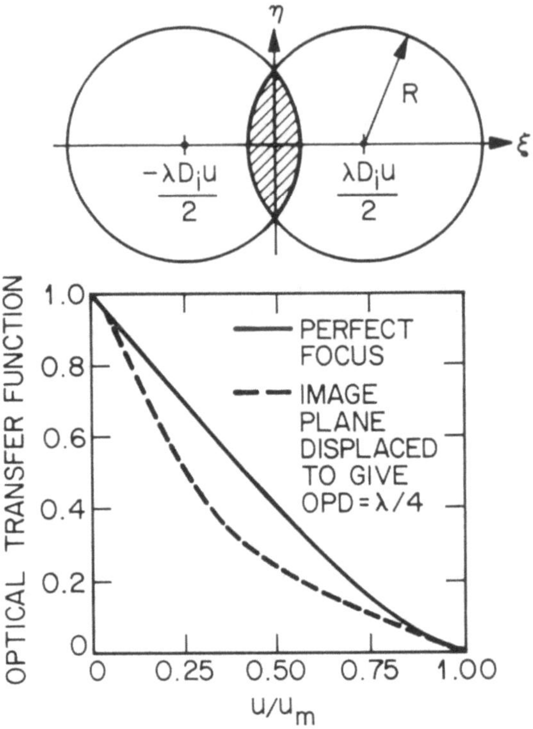

Fig. 11 Incoherent or optical transfer function, round pupil

nearly incoherent illumination. Partial coherence has some advantages over incoherent illumination. The useful range, MTF > 0.6, is extended to higher spatial frequencies; edge gradients in the image become steeper; and the image is a little less sensitive to defocus [23]. Fig. 12 compares the magnitude of the transfer functions for S = 0.75 and S = 1 for a particular F and λ rather than in terms of the normalized frequency u/u_m.

Since the transfer functions with round pupil are symmetrical under rotation about the vertical axis, they can be written and displayed as functions of a single variable u as in (15) and (16) and Fig. 10 and 11. But it is understood that u stands for $\sqrt{u^2+v^2}$. This two-dimensional nature explains why a small contact hole with predominant spatial frequency components $u=v=u_o$ requires a different exposure from a long line of the same width with $u=u_o$, $v\approx 0$, for $|H(\sqrt{2}\,u_o)|$ is less than $|H(u_o)|$. If both types of feature occur on the same mask, both types of resist image will not have correct dimensions. In general for very small features representing high spatial frequencies, the required exposure depends on shape of the feature and proximity to other features, setting a practical resolution limit much less than u_m.

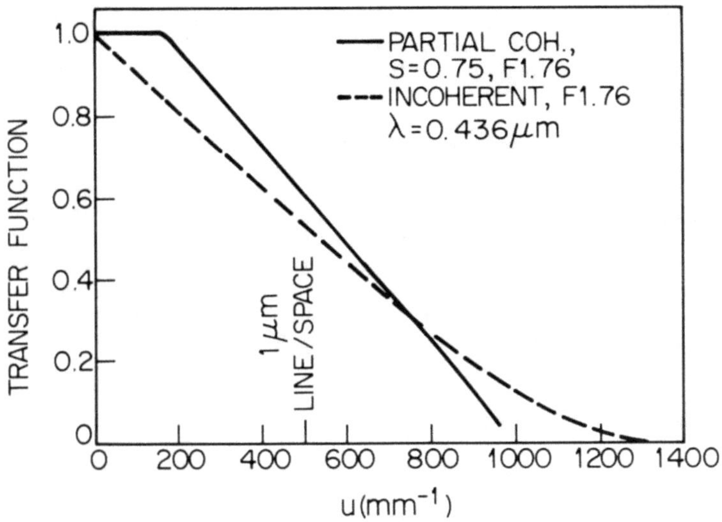

Fig. 12 Comparison of incoherent and partially (S = 0.75) coherent transfer functions with F = 1.76

Aberrations can be treated by including a phase error $\phi(x,y)$ in the pupil function, $P(x,y) \rightarrow P(x,y) \exp[i\phi(x,y)]$. This reduces the magnitude of the transfer function, except in the coherent case as can be seen from (9). Most printers have very nearly diffraction limited optics. Fig. 13 shows yield of meander patterns as a function of meander linewidth as obtained by

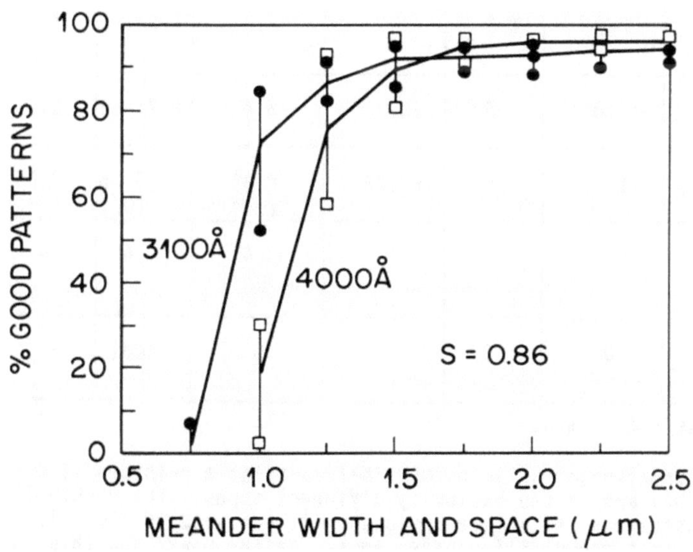

Fig. 13 Meander pattern yield as function of meander linewidth, Perkin Elmer printer with λ = 0.31μm and 0.4μm. [24]

BRUNING [24] with a modified Perkin Elmer printer at $\lambda=0.31\mu m$ and $\lambda=0.4\mu m$. The feature sizes at which the breakpoints in yield occur are roughly in the ratio of the two wavelengths, indicating diffraction limited behavior at both wavelengths.

The focus error is a simple but important aberration. In Fig. 11 the dashed curve shows the effect of a displacement of the image plane or wafer from the focal plane by one Rayleigh unit $w = 2\lambda F^2$, corresponding to a phase error of $\pi/2$ at the edge of the pupil. Defocus affects different linewidths differently. NARASIMHAM and CARTER [25], using a Perkin Elmer printer, find that defocus of 3w changes the width of 7μm lines by 2μm and reduces edge slope of the positive resist profile by 28°. MORITZ [26] , using a similar printer and AZ1350J resist, finds that a 3σ linewidth variation of ≤0.5μm is achieved for defocus ≤w, some of the variation being due to illumination non-uniformity.

Table 1 Some projection printers

	Perkin Elmer 140	Cobilt 3000A	Canon FPA141	Canon FPA112FA	GCA 4800		Electro mask 700SLR	Censor SRA100
m	1	1	0.25	1	0.1	0.2	0.1	0.1
F	3.0	2.63	1.75	2.50	1.76	2.2	1.76	1.47
Field [mm^2]	82x100	10(dia)	10x10	22x22	10x10	20x20	10x10	10x10
$L_{0.6}$[μm]	2.0	1.8	1.2	1.7	1.2	1.5	1.2	1.0
w [μm]	±7.8	±6	±2.7	±5.4	±2.7	±4.2	±2.7	±1.8
Align [μm]	±1		±0.25	±0.3* (2σ)	±0.35 (3σ)		±0.25	±0.2* (2σ)
Wafer Dia. [in.]	4	4	3	4	4		4	4
Throughput [Wafers/hr.]	~60	~60		30	≤44	≤61		≤60

*Automatic Alignment

Some information on a few commercially available printers is collected in Table 1. They are of two basically different types. The Perkin Elmer and Cobilt printers use reflective optics, as illustrated in Fig. 14. A curved lamp source (not shown) illuminates an arc on the mask, and this is imaged by the mirror optics onto the wafer with unity magnification (m=1). In this way only a small zone of the spherical primary mirror is used, providing

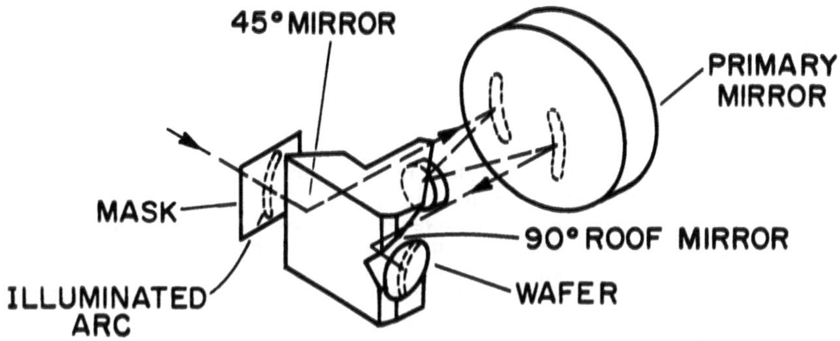

Fig. 14 Projection optics, Perkin Elmer printer

nearly diffraction-limited (more nearly so for the Perkin Elmer optics) imaging. Mask and wafer are swept through the arc to form an image of the whole mask [27]. The other printers use refractive optics - highest quality lenses containing more than ten elements [28]. All but the Canon FPA112FA project a reduced image of the mask with reductions of 10:1, 5:1, or 4:1. The smaller fields are stepped to cover the whole wafer.

As rough measures of resolution and depth of focus $l_{0.6}$, the linewidth of the equal line/space pattern for which the MTF of an incoherent system would have value 0.6, and w, the Rayleigh unit of defocus for an incoherent system, are listed. Alignment tolerances and throughputs are those given by the manufacturer. The Canon FPA112FA and Censor SRA100 have automatic alignment of the image to the pattern on the wafer before each exposure. If the whole image field cannot be filled with a large chip or several smaller ones and must be reduced in size, then more steps are needed to cover the wafer, and throughput decreases. The very small depth of focus of the high resolution optics requires close control of lens to wafer separation. Automatic optical (GCA, Censor) or air gauge (Electromask) sensors are employed on some printers, providing claimed control tolerances of ±0.25µm to ±0.5µm. Image distortion/magnification errors are generally small - ±0.25µm for the 10 cm field of the Perkin Elmer 200 printer, a more advanced version of the model 140, and ±0.2µm for the GCA4800 with 1 cm field.

Since the size of the smallest feature resolvable with an optical system is proportional to λF, higher resolution is obtained by working at shorter wavelengths. There is more room for improvement here in projection printing than in proximity printing where resolution is proportional to $\lambda^{1/2}$. Or F can be reduced, but useful field size drops with decreasing F. In addition depth of focus $\sim\lambda F^2$. Thus it is better to reduce λ than F. The reduced depth of focus and degree of linewidth control accompanying higher resolution seems to represent an important practical limit. Some surmountable problems encountered in the λ=0.2µm-0.3µm wavelength range are increased Rayleigh scattering ($\sim\lambda^{-4}$), greater difficulty in reducing aberrations, and a need for new resists. Reflective optics are more suitable; there are few optical glasses available for use in this wavelength range. Step-repeat exposure offers level to level registration precision independent of wafer size by separate alignment of each exposure field. The mask pattern dimensions are more convenient in the reduction projection systems than in those imaging with unity magnification. For chips larger than an exposure field it may prove possible to abut fields sufficiently accurately that two or more different reticle images can be joined at the edges [29].

3.2 Electron Lithography

Electron lithography offers higher resolution because of the small wavelength of the 10-25 keV electrons used. The resolution of electron systems is not limited by diffraction, however, but by electron scattering and by limitations of the resists. Scanning electron beam systems have been under development for more than a decade, and several commercial systems are available. The EBES machine [30] has proved to be the best photomask pattern generator; it is becoming more widely used in mask shops. Because of the serial nature of the pattern generation, throughput is much less than for optical systems. However, some special products such as microwave transistors have for many years been manufactured by direct wafer patterning [31]. In the first applications to low volume integrated circuits some levels have been patterned optically and some by electron beam [32].

3.2.1 Resists

Electron exposure of resists occurs through formation of bonds or crosslinks between polymer chains (negative resist) or bond breaking (positive resist). The incident electrons have energies far greater than bond energies in the resist molecules. Thus there is no resonant absorption of the energy. Both bond scission and bond formation occur. Which predominates determines whether the resist is positive or negative.

In a negative resist, electron beam induced crosslinks between molecules make the polymer less soluble in the developer solution. One crosslink per molecule is sufficient to make the polymer insoluble. Sensitivity increases with increasing molecular weight. If the molecules are larger, then fewer crosslinks are required per unit volume for insolubility. The polymer molecules in the unexposed resist will have a distribution of lengths or molecular weights and thus a distribution of sensitivities to radiation. The narrower the distribution, the higher will be the contrast γ. Sensitivity and contrast are defined as in section 2.1. The exposure dose q has units of charge deposited by the beam per unit area - C/cm^2. As in the case of optical negative resists, there may be present after irradiation long lived active species which produce crosslinking and are quenched on exposure to air.

In a positive resist the scission process predominates, the exposure leading to lower molecular weights and greater solubility. Again high molecular weight and narrow distribution are advantageous [33].

Two factors are of major importance in limiting resist resolution: swelling of the resist in the developer - more important for negative resists - and electron scattering. Swelling of negative resists, whether optical, electron, or X-ray, has two deleterious effects. Two adjacent lines of resist may swell enough that they touch. On shrinking in the rinse they may not completely separate, leaving a "bridge" here and there. This expansion and contraction weakens adhesion of very small resist features to the substrate and can cause small undulations in narrow (0.5μm) lines. Both problems become less severe as resist thickness is reduced.

When electrons are incident on a resist or other material they enter the material and lose energy by scattering, producing secondary electrons and X-rays. This fundamental process limits resolution of electron resist to an extent depending on resist thickness, beam energy, and substrate composition. More electrons are scattered back into the resist from a heavy metal substrate than from a silicon substrate, for example. The envelope of the electron cloud in the material can be pictured as an onion bulb which is pulled closer to the surface as beam voltage decreases. At higher beam voltage the electrons penetrate farther before being scattered over large lateral distances.

For an electron beam of Gaussian shape $\exp(-r^2/\beta^2)$ incident on the resist covered substrate the distribution of energy deposited in the resist at depth z is closely approximated by [34]

$$f(r,z) = a_1\exp(-r^2/\beta_1^2) + a_2\exp(-r^2/\beta_2^2). \quad (17)$$

$$\beta_1^2 = \beta^2 + \beta_f^2(z)$$

$$\beta_2^2 = \beta^2 + \beta_b^2(z)$$

$\beta_f(z)$ represents the forward scattering of the beam in the resist, and $\beta_b(z)$ represents the back scattering from the substrate with $\beta_b >> \beta_f$. Generally the value of z of interest is that corresponding to the resist substrate interface z_i. For a 25 keV electron beam penetrating 0.5μm thick resist on a silicon substrate, for example, $\beta_f(z_i) = 0.06\mu m$, $\beta_b(z_i) = 2.6\mu m$, and $a_2/a_1 = 2.7 \times 10^{-4}$. As separation between lines decreases, the back scattered electrons contribute a greater dose between the lines where the dose should be zero. This is somewhat similar to the reduced modulation in an optical image at higher spatial frequencies. An exposed pattern element adjacent to another element receives exposure not only from the incident electron beam but also from scattered electrons from the adjacent element. This is called the proximity effect and is, of course, more pronounced the smaller the gap between elements. For example 0.5μm lines require 20-30% more exposure when isolated than when separated from each other by 0.5μm. Thus as pattern density increases, it becomes necessary to adjust exposure for various classes of elements, or even in the extreme case for different parts of elements.

Resist resolution is better in thinner resist layers. Minimum thickness is set by the need to keep defect density sufficiently low and by resistance to etching in device processing. For photomasks where the surface is flat and only a thin layer of chrome must be etched, resist thicknesses in the range 0.2 - 0.4μm are used. For device processing where topographic steps must be covered and dry etching conditions are more severe, thicknesses of 0.5μm to 2μm are required. Most electron resists are not as resistant to dry etching as optical resists. One way of alleviating the problems of proximity effect, step coverage, and process worthiness is through use of a multi-level resist structure in which the thick bottom layer consists of a process resistant polymer (Fig. 2) In one realization of a three-level structure the upper most layer of electron resist is used to pattern a thin intermediate layer, such as 1200 Å of SiO_2, which serves as mask for etching the thick polymer below [35]. For electron lithography a conducting layer can be substituted for the SiO_2 to prevent charge buildup which can lead to beam placement errors. In another two layer resist structure both thin upper layer and thick lower layer are positive electron resists, but they are developed in different solvents [36]. The thick layer can be over-developed to provide the undercut profile which is ideal for lift off processing.

Table 2 Some electron resists

Resist	Sensitivity [C/cm^2]	Resolution [μm]	γ
COP (Mead Chem.)(-)	$3x10^{-7}$ @ 10 kV	1	0.95
PBS (Mead Chem.) (+)	$8x10^{-7}$ @ 10 kV	0.5	1.3
Waycoat (Hunt Chem.) (-)	$3x10^{-7}$ @ 10 kV	1	1
OBER-100 (Tokyo Ohka) (-)	$5x10^{-7}$ @ 20 kV	1	0.77
SEL-N (Somaru) (-)	$4x10^{-7}$ @ 27 kV	0.5	1.3
FMR-E101 (Fuji Chem.) (+)	$5x10^{-7}$ @ 20 kV	0.5	3.3
PMMA (+)	$5x10^{-5}$ @ 15 kV	0.1	
PMIPK (+)	$6x10^{-6}$ @ 10 kV	0.5	
AZ2400 (Shipley) (+)	$2x10^{-5}$ @ 25 kV		

Table 2 lists a few readily available electron resists. There are many other resists under development also [37]. PMMA stands for poly methyl methacrylate, the highest resolution resist known, and PMIPK for poly methyl isopropenyl ketone. AZ2400 is an example of a commercial optical resist which is also electron sensitive [38]. The sensitivity used in practice depends on the application; the values indicated are nominal values. Since faster electrons penetrate more deeply, more current must be used at higher voltages. A resist is about 2x less sensitive at 20 kV than at 10 kV. The values given for resolution are meant to serve for rough comparison. Resolution obtained in practice depends on such factors as resist thickness, substrate, and resist development technique.

3.2.2 Mask Generation

Since the first widespread use of electron beam pattern generators has been in photomask making, let us see what advantages they have in this application. Fig. 15 shows two methods of making a photomask. On the left a reticle is patterned by means of an optical pattern generator - a machine which under computer control exposes and places pattern elements to form the chip image at 10x scale, making 60 to 100 exposures per minute [39]. The pattern element is the image of an illuminated aperture of variable size. The reticle is then used as object in a step repeat camera, which steps the reduced image to fill the desired mask area. Stepping accuracy is

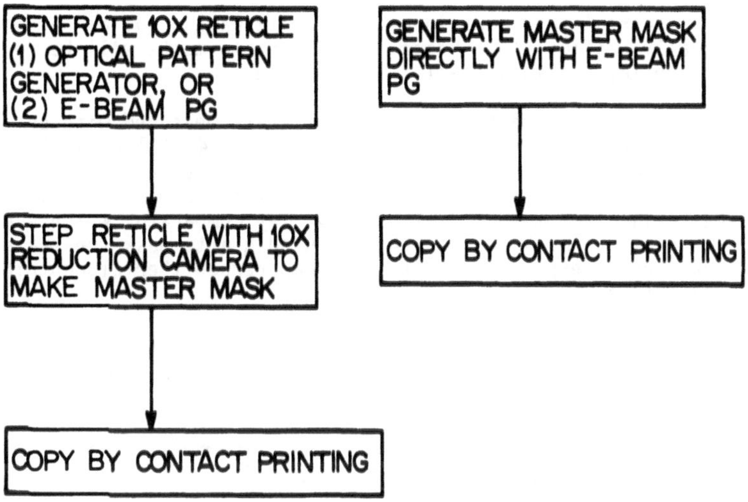

Fig. 15 Two paths for generation of a photomask

interferometrically controlled. The master mask produced may be used in a projection printer or copied and the copies used in a contact printer. If an e-beam pattern generator is used to make the reticle, the main advantage is speed in the case of complex chips. A large dense chip can require 20 hours or more of optical pattern generator time, but between one and two orders of magnitude less time with electron beam.

Table 3 Specifications of master masks, 10 cm Array

Optically Generated*		MEBES Generated [40]
Stacking [μm]	±0.75	±0.2
CD Tolerance [μm]	±0.3	
CD Uniformity [μm]	±0.12	±0.1
Defects [cm^{-2}]	≤0.16	≤0.23

*Master Images, Inc., Santa Clara, California

At right in the figure the more efficient method of e-beam patterning at final size is indicated. Table 3 shows specifications of the two types of mask. The optically generated mask has tolerances which the vendor guarantees will not be exceeded. The results for the masks generated with a MEBES electron beam pattern generator are typical results from the National Semiconductor mask shop [40]. CD stands for critical dimension. Nevertheless it can be seen that stacking errors - the overlay misregistration between two masks of a set - are considerably less for the e-beam masks. (The term registration will be used for overlay precision of two levels on a wafer and stacking, for two mask patterns.) Since a mask set for a circuit may contain as many as 12 masks, not all of which have equally critical tolerances, more than two masks will be required to stack or overlay within a specified tolerance.

Pattern placement errors on a mask are measured with scanning, computer controlled X-Y optical measuring machines [41,42] or with the e-beam pattern generator [43]. Fig. 16 shows measurements at 49 points of a 76 mm x 76 mm array. For this mask maximum placement error was 0.11μm. Stacking of two masks is often measured with an overlay comparator which superimposes the image of one mask on the other. Fused quartz mask substrates with their low thermal expansion coefficient can reduce thermal contributions to stacking errors.

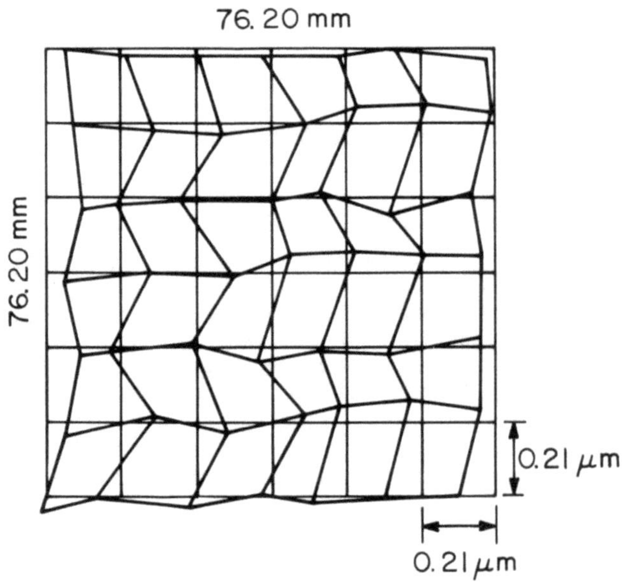

Fig. 16 Photomask placement errors measured by EBES at 49 points. Maximum error = 0.11μm. RMS error = 0.07μm.

Misregistration of one circuit level to another on the wafer can be an important source of yield reduction. There are many contributions Δ_i to the total registration error Δ_t. If the contributions are mutually independent, then

$$\Delta_t = (\Sigma_i \Delta_i^2)^{1/2}. \tag{18}$$

If the two masks used to image the level patterns do not overlay perfectly, there will be a registration error on the wafer. Thus the mask stacking errors contribute to Δ_t. The alignment tolerance of the printer forming the image of the mask also makes a contribution (Table 1). Another contribution comes from wafer processing (etch tolerances, process-induced wafer distortion [44]). If each of these three errors - mask stacking, alignment, and processing - has the rather optimistically small value 0.2μm, then the total registration error would be $\Delta_t \geq 0.35\mu m$. Since the ratio of minimum feature size to registration tolerance is usually in the range 3 to 5, registration of this precision would be barely adequate for circuits with 1μm minimum features. There may be other contributions to Δ_t than the three mentioned [45]. For example, a change in adjustment of the printer transferring the mask image can lead to a change in image distortion with resulting placement errors.

Mask defects are either opaque spots in areas which should be transparent or pinholes where the chromium layer should be continuous. Sometimes critical masks are repaired to reduce the number of defects. At least one vendor offers "zero defect" masks. If a master mask is copied by contact printing, the defect density of the copy will exceed that of the master. Defect density is generally measured by an optical instrument which compares one chip with a similar one on the mask and records differences or defects greater than a minimum size (1μm-2.5μm) [46,47]. Not every mask defect will result in an inoperable chip.

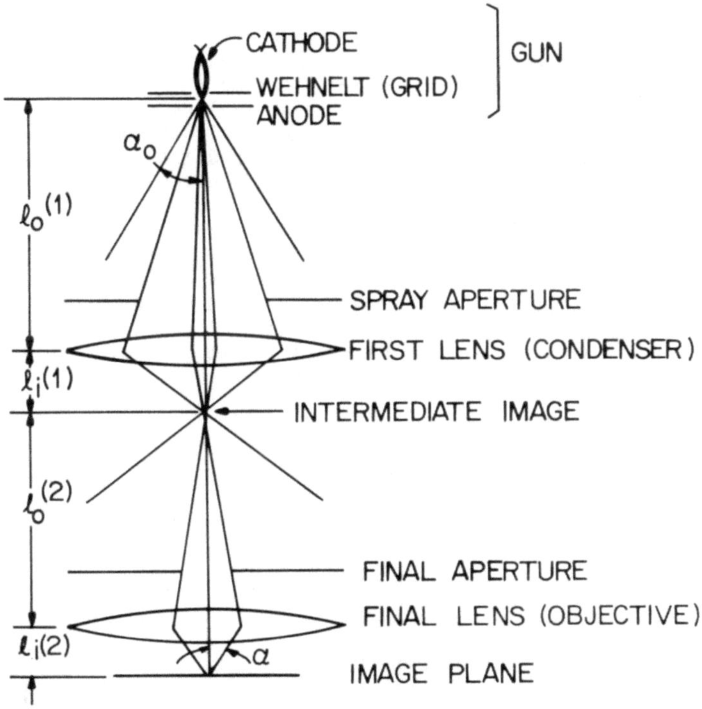

Fig. 17 A simple two-lens probe-forming electron optical system

3.2.3 Electron Optics

Scanning electron beam pattern generators are similar to scanning electron microscopes, from which they are derived. Fig. 17 shows the basic probe forming electron optical system. Two or more magnetic lenses form a demagnified image of the source on the wafer image plane. Provisions for scanning the image and blanking the beam are not included in the figure.

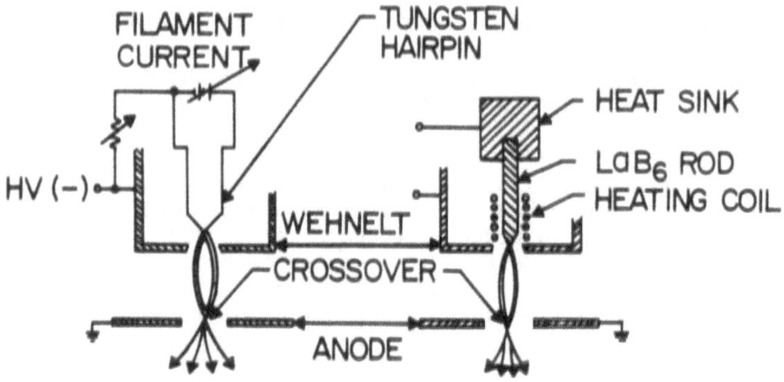

Fig. 18 W gun (left) and LaB_6 gun

The electron gun is shown in more detail in Fig. 18. The cathode is generally a thermionic emitter - either a tungsten hairpin or a pointed LaB_6 rod. Field emitters are also being explored. The rod may be sintered material or a crystal; it may be directly heated, as at left, or indirectly heated by radiation and electron bombardment, as at right. Emission current density from the cathode J_c is given by

$$J_c = AT^2\exp(-E_W/kT) \tag{19}$$

LaB_6 has a lower work function E_W than W and about the same value of A (Table 4). The electrons are accelerated through the voltage $V \approx 10$ to 25 kV and focused by the gun to a spot called the "crossover" of diameter $d_o = 10\mu m$ to $100\mu m$. If I_b is the beam current at the crossover, the brightness - current density per unit solid angle - is

$$\beta = 4I_b/(\pi d_o \alpha_o)^2 \tag{20}$$

Table 4 Thermionic Emitters

	E_W [eV]	A [A/cm^2K^2]	T* [K]	J_c [A/cm^2]	Ref.
Sintered LaB_6	2.4	40	1923	65	[48]
LaB_6 xtl. (100)	2.47	14	1923	17	[49]
W	4.5	60	2700	1.75	[48]

*Typical operating temp.

The maximum value of the brightness is

$$\beta = J_c eV/\pi kT, \tag{21}$$

and the maximum current density at the crossover is

$$J_b = \beta\pi\alpha_o^2, \tag{22}$$

The thermal field emission source consists of a tungsten tip of radius 0.5μm to 1μm heated sufficiently (1100 C - 1400 C) to provide annealing of sputtering damage for relaxed vacuum requirements and better stability compared with a cold field emitter [50]. Eqs. (20) and (22) do not apply, and there is no gun crossover. The source in this case is a small (~200 Å) virtual one located inside the tip. Thus the gun and the imaging optics are very different from those we consider. Thermal field emitters are of interest because for small image spots they promise higher currents than thermionic sources [51].

The lenses in the electron optical column (Fig. 17) are magnetic. The field of such a lens, $\vec{B} = \hat{e}_r B_r(r,z) + \hat{e}_z B_z(r,z)$, has cylindrical symmetry about the optical axis z. If a parallel beam of radius r_o enters the field of the lens, the electrons experience a force which causes those not on axis to rotate about the axis and turn toward it. For a thin lens the electron path beyond the lens is given by

$$dr/dz \approx -(r_o e/8mV) \int_{-\infty}^{\infty} B_z^2 dz. \tag{23}$$

The constant $dr/dz = -r_o/f$, where f is the focal length of the lens. Many rules of light optics apply here also, such as the thin lens law relating object distance, image distance, and focal length,

$$1/l_o + 1/l_i = 1/f. \tag{24}$$

The magnification of a lens is $M = l_i/l_o$. An intermediate image is formed by the first lens with magnification M_1 and this is the object which is further demagnified by the second lens to form a spot in the image plane of diameter

$$d_i = Md_o, \qquad M = M_1M_2 \tag{25}$$

Typically $M_2 \approx 10^{-3}$-10^{-1}. Although the lengths $l_o^{(1)}$ and $l_i^{(2)}$ are fixed, $l_i^{(1)}$ and $l_o^{(2)}$ are variable. If current through the windings of the first lens of Fig. 17 is increased, $l_i^{(1)}$ decreases, M_1 is reduced, and the beam current passing through the final aperture is reduced, because of the increased divergence of the beam at the intermediate image. The current density and current in the image plane are

$$J \le \pi\beta\alpha^2 \quad , \qquad I = J(\pi d_i^2/4). \tag{26}$$

Spot sizes of interest are in the range 0.1μm - 2μm. This is far from the diffraction limit. From section 2.3 the diameter of the central spot of the Airy pattern is $1.2\lambda/\alpha$. α is the numerical aperture. For 15 kV electrons the wavelength λ = 0.1 A. Taking $\alpha = 10^{-2}$ rad, we have a diffraction spot width

$$d_{diff} = 1.2\lambda/\alpha = 10^{-3}\mu m. \tag{27}$$

Thus diffraction can be ignored. However, aberrations of the final lens and of the deflection system will increase the size of the spot and can change its shape as well. Aberrations of the other lenses are not as important since the intermediate images are larger. Fig. 19 shows a typical double deflection system before the final lens, arranged so that the deflected beam always passes through the center of the lens. In some designs the deflected beam is incident normal to the wafer surface (telecentric) to minimize runout error for warped wafers. The deflection coils provide a magnetic field perpendicular to the beam axis. There will be another set of deflection coils to provide deflection perpendicular to the plane of the figure also.

The aberrations are treated as independent contributions to spot broadening, each contributing a circle of confusion to the spot. The actual spot size in the presence of aberrations is then the square root of the sum of the squares of the independent contributions. The aberrations are of two types: aberrations of the undeflected beam and aberrations which are functions of the distance r in the image plane from the axis to the position of the deflected beam. The aberrations of the first type are

Spherical aberration , $d_s = (\frac{1}{2}C_s)\alpha^3$

Chromatic aberration , $d_c = C_c(\Delta V/V)\alpha$

Spherical aberration is the focusing at different distances of rays passing through different parts of the lens; a ray near the axis has a longer focal length than one farther from the axis. $e\Delta V$ is the spread of electron energies, usually a few eV. C_s and C_c are constants characterizing the aberrations. Astigmatism, a third possibility, results from breaking of cylindrical symmetry in the column and can be removed by introducing compensating fields with stigmator coils (Fig. 19). Deflection aberrations, such as coma and field curvature, are proportional to $r^2\alpha$ or $r\alpha^2$. We lump them all together and call their contribution to the spot spread d_{df}. In addition to the deflection aberrations there is field distortion, but distortion is a shift of the image point, not a broadening of the spot. Table 5 shows these aberrations for the IBM VS1 machine [52]. We note that deflection aberrations dominate. Distortion and some of the deflection aberrations can be reduced by addition of correction signals to the deflection fields [53]. Distortion with this machine is ±2.5μm at the edges of a 5 mm x 5 mm field, or $\pm 5 \times 10^{-4}$ expressed as a fraction of the field. For optical printers this ratio is typically much smaller (section 2.3).

Table 5 Aberrations at the corners of 2mm x 2 mm and 4 mm x 4mm exposure fields for VS1

	2mm fld.	4mm fld.
Spherical, d_s	0.01μm	0.01μm
Chromatic, d_c	0.04	0.04
Deflection, d_{df}	0.087	0.35
Total, no dynamic correction	0.095	0.35
Total with dynamic correction	0.043	0.050

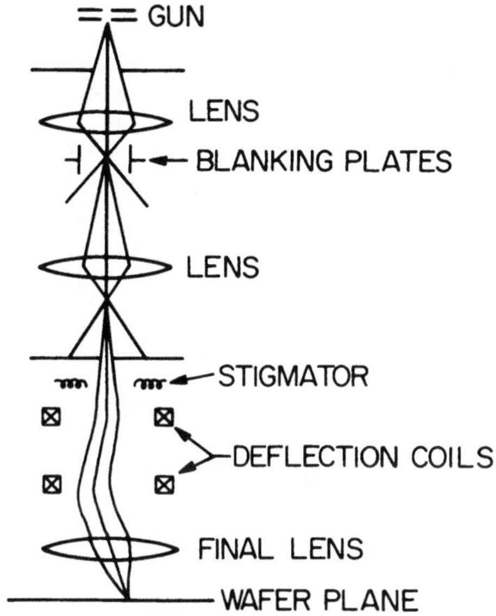

Fig. 19 Electron optical system with double deflection

Another source of spot broadening is the mutual Coulomb repulsion of the electrons as they traverse the column. For total column length L this contribution is approximately given by [54]

$$d_{ee} \approx [LI/(\alpha V^{3/2})] \times 10^8 \mu m. \quad (28)$$

This spreading only becomes important at large currents when small spot sizes or sharp edges of large spots are sought. For example if $I = 500$ nA, $L = 70$ cm, $V = 20$ kV, and $\alpha = 10^{-2}$ rad., then (28) gives $d_{ee} \approx 0.12\mu m$. Another effect of the inter-electronic interactions is an increase of energy spread ΔV of the electrons. Through chromatic aberration of the optics this leads also to broadening of the spot [55]. In the presence of aberrations (26) becomes

$$I = (\pi^2/4)\beta\alpha^2[d^2-d_s^2-d_c^2-d_{df}^2-\ldots] . \quad (29)$$

d is the beam diameter in the presence of aberrations,

$$d^2 = d_i^2+d_s^2+d_c^2+d_{df}^2+\ldots \quad (30)$$

For a given set of aberrations an optimum α can be found which gives minimum spot size for a particular current. However, in practice the final aperture is not changed for each current, but optimized for some small spot size. Then the current is less and is given by (29),

$$I \sim (d^2 - \text{constant}). \quad (31)$$

3.2.4 Vector Scan

In a vector scan pattern generator the beam is directed sequentially to the parts of the chip pattern to be exposed. The pattern is decomposed into a number of elements (rectangles, triangles, etc.) and each one is filled in by the writing beam. Vector scan machines generally expose in step repeat fashion. Fig. 20 shows an exposure field of dimensions F x F. The alignment marks in the field corners are first scanned to set deflection amplitude, offset, rotation, and X-Y scan axis orthogonality. Marks may be etched trenches in silicon or they may be metal layers or other features which give sufficient contrast. Mark position is sensed by the change produced in the number of back-scattered and secondary electrons on scanning over the mark edges. Marks may be initially etched into a wafer and several circuit levels overlayed by reference to the marks. Accuracy of overlay depends on many variables such as field size, mark condition, resist thickness, and stability of field distortion from one level to the next. Reported values lie in the range ±0.1μm to ±0.6μm.

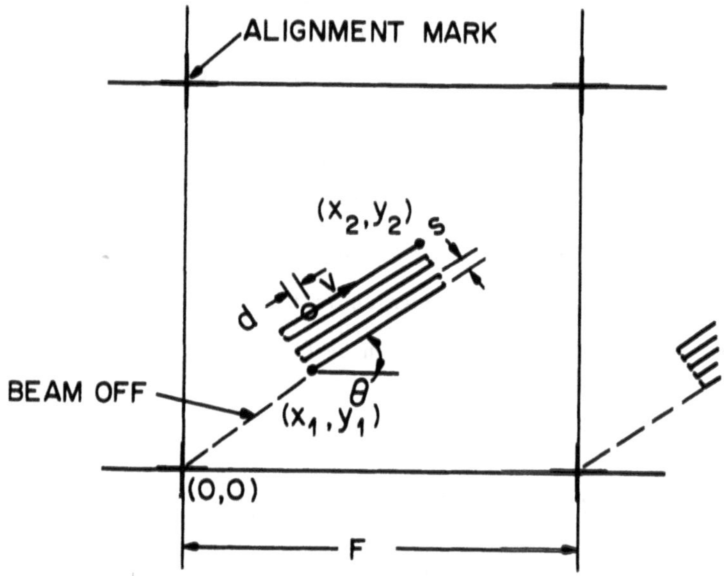

Fig. 20 Vector scan writing of a rectangular pattern element

The beam is scanned over each pattern element in turn. Dotted lines indicate where the beam is turned off, that is, deflected by voltages applied to the blanking plates (Fig. 19) and prevented from passing through an aperture. For example the rectangle shown might be specified in the program by an instruction such as

$$(x_1,y_1)\mathrm{RECT}(x_2,y_2)R(\theta). \qquad (32)$$

Software, as in optical pattern generators, consists of an operating system and a high level system in which pattern elements and commands are simply expressed. In one such machine the point (x_1,y_1) is determined by 16 bit digital to analog converters (65536 points resolution in x and y) and scan line lengths are determined by other DAC's [56].

After all elements in the field are exposed, the stage is stepped to bring the next field under the beam and the exposure process is repeated.

The stage need not provide highly precise motion. Stage position is generally monitored by a laser interferometer, and small differences from desired stage location are compensated by small offsets of the beam. If a chip is larger than an exposure field, then several fields can be used to expose the chip, but scan distortion at the field edges must be small enough so that patterns in adjacent fields are properly joined. If the machine is used to pattern masks rather than wafers, no alignment marks are used on the mask. In this case the deflection parameters are set by periodic reference to a mark fixed on the stage.

Once the beam current is set, exposure is controlled by varying the scan speed v and scan spacing s. These are set to desired values for the whole pattern, or v might be set to different values for some pattern elements to compensate for proximity effects. If, for example, we wished to give the rectangle 20% more exposure than the general level set for the chip, the instruction used might be

EXPOSE, 1.2 (33)

$(x_1,y_1)\mathrm{RECT}(x_2,y_2)R(\theta)$.

From (32) and (33) it is obvious that few pieces of data are needed to trace out a pattern element which may contain very many beam diameters or "pixels". The proper exposure is found by exposing a test pattern at a series of scan speeds.

The scan field must be as large as allowed by deflection aberration and distortion because stage stepping time affects throughput. Reported exposure fields lie in the range F = 1 mm to 6 mm - much smaller than for optical step-repeat machines. Between exposure fields the stage must accelerate and decelerate and then settle so that the motion is damped to a small smplitude before writing can begin.

Let us see what other factors affect throughput of a vector scan machine. The minimum time necessary for exposure of the fraction η of the field area F^2 with current I and resist sensitivity q is

$$T_e = \eta F^2 q/I \; . \tag{34}$$

q and I are related to scan speed v and scan spacing s by

$$vs = I/q \; . \tag{35}$$

In order to write faithfully a rectangle or other feature of minimum dimension ℓ, the scan spacing and beam diameter should be no larger than

$$s = \ell/5 \; , \quad d = \ell/4 \quad , \tag{36}$$

with some dependence on the way in which beam diameter is defined. Combining (35) and (36) gives

$$v = 5\ell J\pi/(64q) \; . \tag{37}$$

There will be some upper limit on the scan speed v set by the deflection system. To increase the maximum scan speed small low-inductance coils driven with large currents are employed in magnetic deflection systems.

Or faster electrostatic deflection may be used. If the desired scan speed exceeds the maximum, the current density must be reduced or a less sensitive resist used. In the region of small ℓ where $v \ll v_{max}$ a more sensitive resist might be used to increase throughput. However, another limit can be encountered here. Generally the more sensitive resists developed to date have had less resolution and less etch resistance than less sensitive ones. Theoretically we should want the number of electrons deposited by the beam in a pattern element of very small size to be large enough that proper exposure is assured despite random fluctuations which will change this number from one small element to another [57]. For example 10^{-6}C/cm^2 corresponds to 491 electrons into a 0.1μm x 0.1μm element of area. The standard deviation characterizing the fluctuations is $\sqrt{491} = 22$.

The total time to pattern a wafer by stepping through N fields is

$$T_{tot} = N(T_e+T_{sr}+T_{oh})+T_o \ . \tag{38}$$

T_{sr} is the time required for registration and stage movement. Let us assume an average stage speed 5 cm/sec, a settling time of 0.1 sec., and an alignment time of 0.2 sec. T_{oh} is the pattern overhead time which includes computer dead time and DAC settling time. This might vary from 3μs to 100μs per pattern element. Let us assume 10^5 pattern elements per 6 mm x 6 mm field and 3μs per element. T_o can be made negligible compared with the other contributions. We take a 100 cm diameter wafer containing N = 190 fields or chips and plot throughput, T_{tot}^{-1}, as a function of current in Fig. 21. In using (34) for T_e we are assuming that beam current, not scan speed, is the limiting factor. The parameters used in the figure are

$\eta = 0.2$ $\qquad$ $T_{sr} = 0.4$ sec.
$q = 3\mu$ C/cm^2 $\qquad$ $T_{oh} = 0.3$ sec.
$F = 6$ mm $\qquad$ $T_o = 0$
$N = 190$

At high beam currents T_e becomes small compared with time required for stage motion and overhead, and the curve saturates. With current density J = 300 A/cm^2 the current available in a 0.1μm diameter spot is 24 nA. The IBM VS1 machine provides a current density this great, but resists less sensitive by an order of magnitude than 3μ C/cm^2 are used because of limited deflection speed [58].

Commercially available vector scan machines are EBMF-2 of Cambridge Scienfitic Instruments, Beamwriter of Philips, and JBX5-A of JEOL. Vector scan systems have also been developed by Hughes Research [59], IBM [58], and Texas Instruments [60]. Table 6 lists some specifications of the Philips and IBM machines.

3.2.5 Raster Scan

In raster scan the beam is deflected repetitively over the exposure field, as in a television raster. The beam is turned on at various points in the scan to expose the desired pattern. The EBES machine [30], developed by Bell Laboratories and available commercially from ETEC Corp., and Extrion, uses beam deflection in one dimension. The writing scheme is shown in Fig. 22. The stage moves continuously in a direction perpendicular to the writing beam. The pattern data are decomposed into a number of stripes, and one stripe is written over all chips before the next stripe is begun.

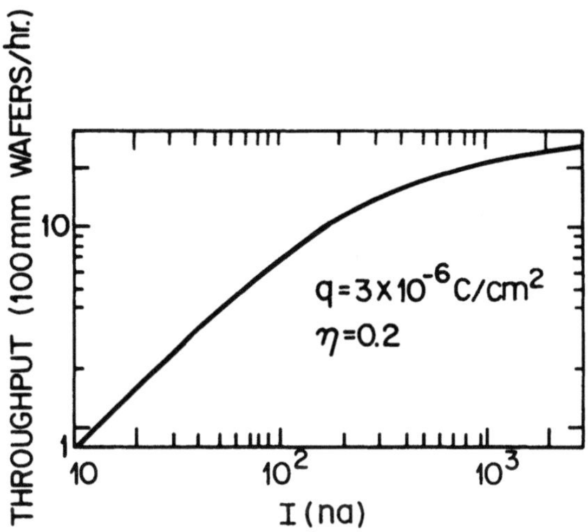

Fig. 21 Throughput as function of current for step-repeat exposure under conditions given in text

Table 6 Some specifications of two vector scan pattern generators

	IBM VS1	Philips Beamwriter
Cathode	LaB_6	W
Image Current Density [A/cm^2]	300	100
Exposure Field [mm^2]	4 x 4	1.6 x 1.6
Beam Diameter [μm]	≥0.05	0.025-1 *
Accelerating Voltage [kV]	25	20
Max. Defl. Rate [MHz]	10	10
Typ. Overlay Error [μm]	±0.1	±0.1

*Automatic Focus, Each Field

The stripe is 512 addresses wide, an address corresponding to 0.5μm or 0.25μm. Beam diameter may be varied from 0.2μm to 1μm. The pattern information comes to the blanking plates from a shift register at a 20 MHz rate. (A new version will operate at 40 MHz.) Since total time to write the 512 address scan is 31.6μs - 50 ns per spot + 6μs for flyback -, the writing time is approximately 2 cm^2/min. with the 0.5μm spot and 0.6 cm^2/min. with the 0.25μm spot. Sensitive resists are required because spot current density is at most 30 A/cm^2.

Stage motion is monitored by laser interferometer, and small position corrections are made by beam deflection, larger ones by drive motor speed variation. Beam position errors are checked and corrected periodically by scanning over a fiducial mark on the stage.

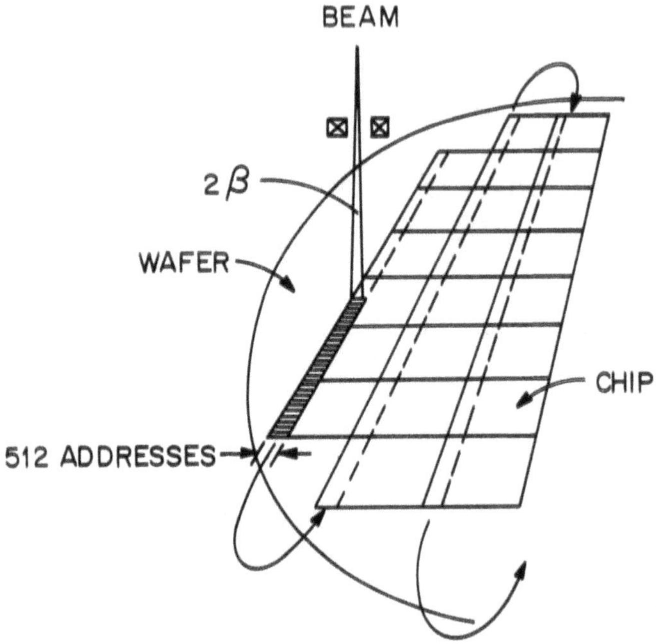

Fig. 22 EBES writing scheme. Curved arrows indicate serpentine stage motion

The EBES machine is the most widely used electron beam photomask pattern generator. The small beam deflection angle $\beta = 9$ mrad and the repetitive nature of the scan provide very accurate pattern placement. For writing on wafers three or more alignment marks on the wafer are scanned to provide level-to-level registration. In raster scan the whole field is scanned. This is less efficient than vector scan, but has the advantage that polarity of the pattern is easily changed. The exposure is determined by the beam current and the (fixed) 50 ns dwell time. It is not changed during writing. Thus corrections for proximity effect, which are easily made with vector scan, cannot be made. However, the variable scan speed of vector scan places a severe burden on the deflection system, and eddy current effects must be compensated or eliminated over a wider frequency range of the deflection system.

The IBM EL1 raster scan system exposes 5 mm x 5 mm fields in step-repeat fashion with a 2.5μm x 2.5μm square beam [61, 62]. The square spot is the image of a square aperture which is illuminated by a tungsten gun. Current in the image spot is 3μA. To reduce the effects of electron-electron interactions at this high beam current, a deflection system contained in the final lens rather than the double deflection system of Fig. 19 is used, allowing a shorter interaction path. The deflection distortion over the large exposure field is reduced from 1000 ppm to 30 ppm by scanning a reference grid to generate deflection error correction signals. Reduction of distortion is especially important if some circuit levels are to be printed with, for example, projection printers and others, such as metal levels in custom logic chips, are patterned by e-beam. Such a distortion correction procedure is not practical with vector scan because of the non-repetitive scan. Level-to-level registration is accomplished as in the vector-scan case by reference to four alignment marks in the field corners; resulting 3σ overlay errors are $\leq 0.5\mu m$. A number of these systems have been direct writing on wafers in a production environment for several years [31]. They are highly automated and throughput is 22 wafer levels (57 mm dia. wafer) per hour.

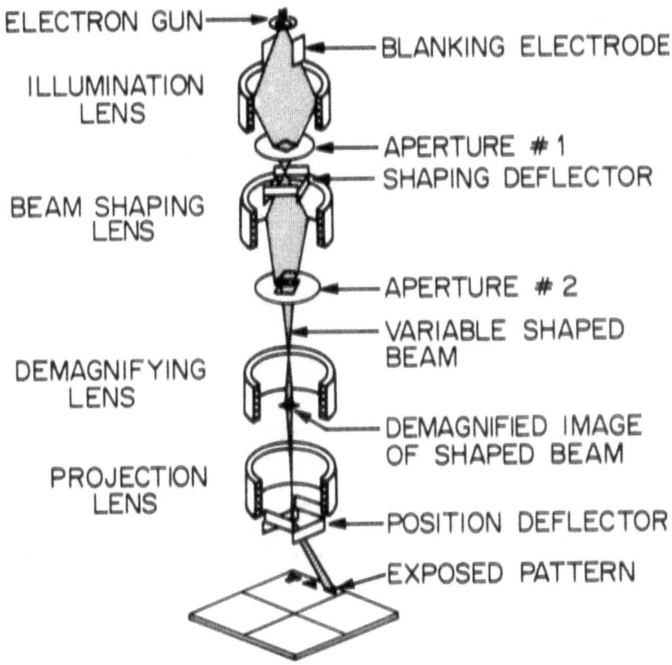

Fig. 23 JEOL JBX6A electron optical system

3.2.6 Variable Beam Shape

In the JEOL JBX6-A machine shown schematically in Fig. 23, the spot is rectangular of variable size and shape. The image of the $125\mu m \times 125\mu m$ square aperture #1 may be shifted in two dimensions to cover various portions of aperture #2, which is the object imaged on the mask or wafer. Minimum rectangle width is $1\mu m$; maximum dimension is $25\mu m$. Maximum current density is only 0.1 A/cm^2, so high sensitivity resists must be used. Electrostatic beam deflection is employed. Proximity effect corrections are easily made by variation of the exposure time of individual rectangles. Exposure field is 2 mm x 2 mm. Thus many fields must be stepped to expose a mask or wafer. For a chip containing 8.5×10^4 rectangles with a $1\mu\ C/cm^2$ resist the patterning time would be

T_e = 1.0 sec.

T_{oh} = 0.09 sec.

T_{sr} = 5.5 sec. (including 1 sec. align time)

6.6 sec.

Total time to pattern a 100 mm wafer containing 282 chips would be 31 min.

The Zeiss-Jena ZBA-10 is another example of this type of pattern generator. Exposure field is very small - 1 mm x 1 mm. Current density in the image is 1.3 a/cm^2, and rectangle edge length is variable from 0.1μm to 10μm [63].

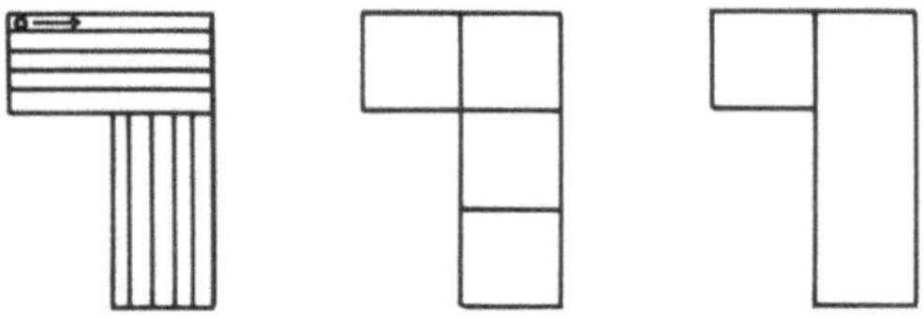

Fig. 24 Exposing a pattern element in three different ways. Minimum element dimension is equal to beam width in the case of the square beam.

Fig. 24 compares patterning an L-shaped element with a scanned round beam, a square beam, and a rectangular beam. In a proposed extension of the variable beam shape the second aperture plate would contain a number of apertures with the shapes of pattern elements [64]. The whole L would be imaged at once. These shaped-beam methods offer a way to project many image points in parallel and to achieve faster exposure without the need for the fast deflection system required with a small scanned beam.

3.2.7 Electron Projection

Electron projection systems are another method of achieving high resolution over a large field with high throughput. Rather than using a small beam writing the pattern in serial fashion, a large beam provides parallel exposure of a large area pattern. The systems are of two types.

In the 1:1 projection system, invented at Westinghouse, parallel electric and magnetic fields image electrons from a full size mask onto the wafer. The "mask" is of quartz patterned with CsI on the side facing the wafer. Photoelectrons are generated on the mask/cathode by backside UV illumination. Westinghouse, Thomson CSF, Mullard, and Radiant Energy Systems have developed systems. Only Mullard and RES now work in this area. In Table 7 are specifications of the Mullard machine [65]. The advantages of the system include: stable mask, good resolution, fast exposure with low sensitivity electron resists, large field, and fast alignment. The disadvantages largely arise from the fact that the wafer is the anode. The edge of the wafer creates field inhomogeneities which make the outer pattern area useless. Tantalum alignment marks are needed on the wafer for the alignment scheme in which X-rays produced by electrons incident on the tantalum are detected behind the wafer. The electric field causes backscattered electrons to fall back on the wafer, reducing contrast. Proximity effects can only be compensated by undersizing or oversizing the features on the mask. An electrostatic chuck to hold the wafer flat in the vacuum environment is necessary but difficult to develop because it interferes with the imaging fields.

Table 7 Electron projection systems

Mullard 1:1 System	
Minimum Linewidth [μm]	∼0.5
Field Size [cm. dia.]	5
Depth of Field [μm]	60
Expose Time (PMMA) [min.]	1.5
IBM 10:1 System	
Minimum Linewidth [μm]	0.25
Field Size [mm^2]	3 x 3
Registration Tolerance [μm]	±0.05
Depth of Field [mm]	1
Expose Time per Field [sec.]	0.1

In the projection reduction systems invented at Tübingen University and developed at IBM a reduced image of a foil stencil mask is stepped to fill the wafer. Specifications of the IBM system [66] are also included in the table. Special problems include long alignment and stepping times and a difficult mask-making technology.

3.3 X-Ray Lithography

Since the proposal for X-ray lithography by SMITH and SPEARS [67] in 1972 the technique has been under development in many laboratories. We saw how diffraction effects are reduced and resolution improved by reducing the wavelength in optical lithography. If the wavelength is reduced further, all optical materials become opaque because of the fundamental absorption, but transmission increases again in the X-ray region. In X-ray lithography an X-ray source illuminates a mask which casts shadows onto a resist-covered wafer. Materials useful for the absorptive and transmissive parts of the mask, the atmosphere in the exposure chamber, and the resist are in large part determined by the absorption spectra of these materials in the X-ray region.

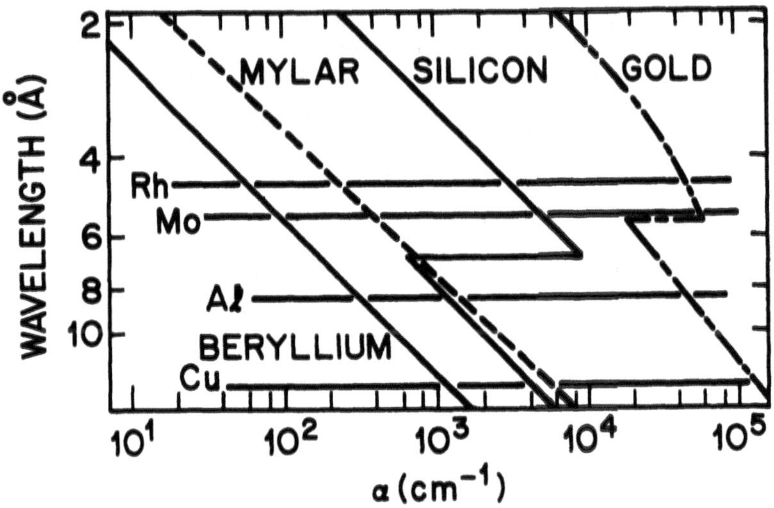

Fig. 25 Absorption coefficients of several materials in the soft X-ray region. Sharp lines in spectra of four sources are indicated.

The X-ray absorption of several materials is shown in Fig. 25. Over wide ranges of wavelength the absorption coefficient α of an elemental material of density ρ and atomic number Z is proportional to $\rho Z^4 \lambda^3$. As λ increases, the proportionality constant decreases in step function fashion at the "absorption-edge" wavelengths corresponding to the ionization energies of the inner electrons of the K, L, and other shells. We note the large differences in α observed for different materials at the same wavelength.

3.3.1 Resists

Any electron resist is also an X-ray resist [68], since an X-ray resist is exposed largely by the photoelectrons produced during X-ray absorption. The energies of these photoelectrons are much smaller - 0.3 keV to 3 keV - than the 10 keV to 30 keV energies used in electron lithography, making proximity effects negligible in the X-ray case and promising higher ultimate resolution. Measured photoelectron path lengths in resist increase with energy from 50 Å for 0.28 keV (λ = 45 Å) to 650 Å for 2.7 keV (λ = 4.6 Å). Diffraction effects are in spatial extent roughly equal to the wavelength. Thus highest resolution should be obtained at an X-ray wavelength ~50 Å, where the photoelectron range is comparable with the wavelength. Such high resolution has been demonstrated in PMMA. However, system considerations favor the wavelength range 4 Å to 10 Å, and most work has been done in this range.

On traversing a path of length z in resist or any other material an X-ray flux is attenuated by the factor exp($-\alpha z$). For most polymer resists containing only H, C, and O and with density $\rho \approx 1$ gm/cm^3 a 1µm thick resist film absorbs about 10% of the incident X-ray flux at the AlKα X-ray wavelength λ = 8.3 Å. This small absorption has the advantage of providing uniform exposure throughout the resist thickness z and the disadvantage of reduced sensitivity. As in the optical case, X-ray resist sensitivity is generally quoted in terms of incident dose q(J/cm^2) required for exposure; sometimes absorbed dose αq(J/cm^3) is used. As in optical and electron lithography there is some dependence of sensitivity on substrate, although the effect is smaller: a heavy metal substrate will produce more photoelectrons at the resist/substrate interface than a silicon substrate, for example.

For the shorter X-ray wavelengths the λ^3 dependence of the absorption coefficient leads to low sensitivity. This can be offset by incorporation of heavier elements to increase absorption. For PdLα radiation, λ = 4.37 Å, the DCOPA resist - a mixture of poly (3,3-dichloro-1-propyl acrylate) and PGMA/EA - incorporates chlorine. The ClK absorption edge at 4.40 Å provides higher absorption, and chlorine is a chemically reactive species; thus two effects contribute to higher sensitivity [69].

Table 8 Some X-ray resists

Resist	Sensitivity [mJ/cm^2]	γ	λ [Å]	Ref.
PMMA (+)	500	3	8.3	[68]
PGMA/EA (-)	5	1.07	8.3	[68]
PBS (+)	10		8.3	[68]
DCOPA (-)	10	1.2	4.4	[69]
XXR-102 (-)	11	1.1	8.3	[70]
XR-7 (-)	2	2.1	8.3	[71]

Some X-ray resists are listed in Table 8. The negative resists are limited in resolution by swelling during development, as in optical and electron lithography. Thus 1μm minimum features can be resolved in DCOPA of final thickness 0.5μm. Improved resolution can be obtained using thinner resist in a multilevel structure [35], as in optical and electron lithography. Conventional X-ray sources produce flux at the wafer which is typically less than that available in optical lithography by a factor of $\sim 10^3$. Resist sensitivity must be $\lesssim$10 mJ/cm^2 to avoid long exposure times.

3.3.2 Proximity Printing

Because the X-ray wavelength is small, diffraction effects can be ignored and simple geometrical considerations used in relating the image to the pattern on the mask. The opaque parts of the mask cast shadows onto the wafer below. The edge of the shadow is not absolutely sharp because of the finite extent S of the X-ray source (diameter of focal spot of the electrons on the anode) at distance D from the mask. If the gap between mask and wafer is called g, this blur δ is given by (Fig. 26).

$$\delta = g(S/D) \tag{39}$$

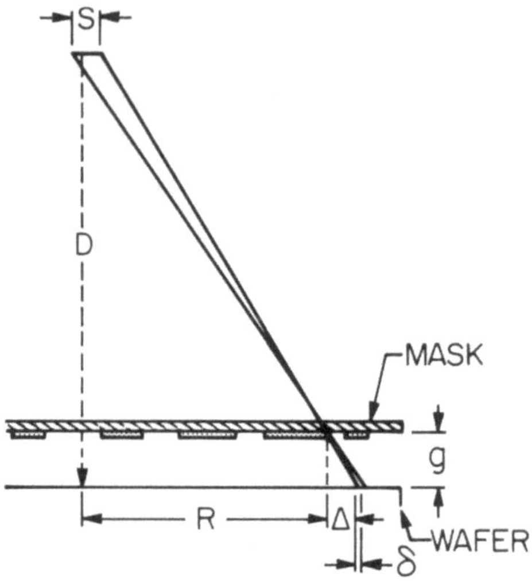

Fig. 26 X-ray proximity printing

Typical values are g=20μm, S=3 mm, D=40 cm, δ=0.15μm. Resolution is determined by δ, by the minimum linewidth achievable in mask fabrication, and by properties of the resist used.

The angle of incidence of the X-rays on the wafer varies from 90° at the center of the wafer to $\tan^{-1}(D/R)$ at the edge of a wafer of radius R. X-ray optics exist, but numerical apertures are much too small for the lithography case where flux cannot be wasted. No telecentric design like that of Fig. 7 is possible. The shadows are slightly longer at the edge by the amount

$$\Delta = g(R/D). \tag{40}$$

This small magnification is generally of no concern. In the special cases where it may be undesirable, it can be compensated when the mask is patterned. For multilevel devices it must have the same value for each level; or at least its variation must be within the registration tolerance. This implies stringent control of the gap g. Wafer warping in processing depends on wafer thickness and orientation and on the nature of the process; it can be largely removed with a proper vacuum chuck. It is not necessary that the gap have the same value at all points on the wafer, only that the spatial variation be the same, within close tolerance, for all levels.

Let us suppose that the mask-wafer gap is controlled within a tolerance of 1μm, a tolerance difficult to achieve. Then if R/D = 0.1, the variation Δ will be within 0.1μm. We fix δ at 0.1μm and want to calculate exposure time as a function of wafer diameter 2R and gap g. X-ray flux ϕ at the wafer is proportional to D^{-2}. For resist of sensitivity q the exposure time is

$$T = (q/C)D^2. \tag{41}$$

C is the proportionality constant, $\phi = C/D^2$. If we think of increasing D to accommodate larger wafers, keeping R/D = 0.1, we can then let the spot size S and the input power increase also. The flux can increase roughly as $S^{1.5}$ for an elliptical spot. Taking $C = 788S^{1.5}$ mw cm^2 and q = 2mJ/cm^2 or 20mJ/cm^2, a range within which recently reported negative X-ray resists lie [69-72], we obtain the curves of Fig. 27. The figure illustrates the speed advantage of whole wafer exposure over step-repeat exposure. For example, from the lower set of curves, exposure of a 10 cm wafer takes 56 sec. with g = 30μm. If instead the wafer were exposed as four 4 cm x 4 cm fields (5.7 cm. diagonal), total exposure time would be 92 sec. with the same 30μm gap. The minimum gap which can be used depends on the type of gap control mechanism and cleanliness and flatness of mask and wafer surfaces. The smallest controllable gap is used. The shortest exposure times reported are 1 to 3 min. for 2 in. and 3 in. diameter wafers [69-72].

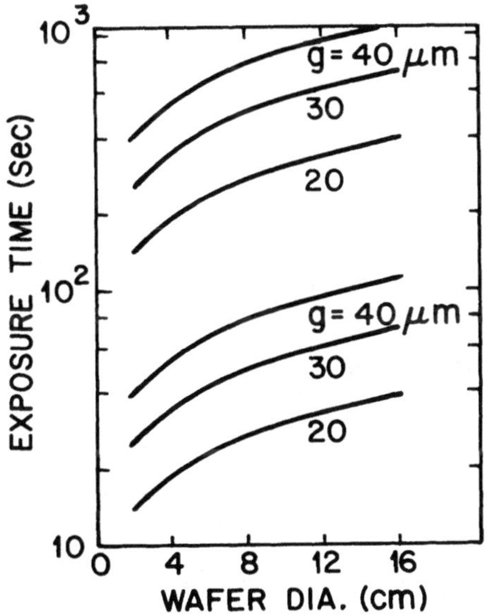

Fig. 27 Exposure time as function of wafer diameter under conditions discussed in text. Upper set of curves are for a resist of sensitivity 20mJ/cm^2; lower set are for 2mJ/cm^2.

X-ray exposure systems have been undergoing a transformation, becoming more similar in general layout to optical proximity printers with the source above the mask and wafer. The larger source-mask separations allowed by higher power generators and the helium atmosphere (instead of vacuum) possible with the shorter wavelengths have made development of registration systems simpler. Most manual systems in use employ high power - 200x to 700x - microscopes with provision for focusing simultaneously on the separated mask and wafer object planes [71, 73].

Although it may at first consideration not seem worthwhile to align with optical radiation and expose with wavelengths smaller by a factor of 500, registration precision is not limited by diffraction to the extent that resolution is limited. In fact diffraction effects can provide a leverage for high alignment precision [74]. The principle of the photoelectric microscope has also been applied with success [72]. Automatic registration is desirable. Two automatic systems with precision better than 0.1μm have been reported [72, 75]. Such precision places stringent requirements on mechanical design of the alignment stage; the mechanisms found in optical proximity printers are in general not adequate.

Machine imprecision is only one component of the total registration error. Others such as mask stacking errors due to pattern generator placement errors, mask distortions, and wafer process-related contributions also play a role.

3.3.3 X-Ray Sources

X-rays are produced by the interaction of incident electrons and a target material. The maximum X-ray energy is the energy E of the incident electrons. If E is greater than the excitation energy E_c of characteristic line radiation of the atoms of the material, the X-ray spectrum will contain these lines.

In early X-ray lithography experiments the X-ray source was often an electron beam evaporator with the chamber modified to accept a mask and wafer. The target metal could be easily changed to modify the X-ray spectrum. X-ray generation by electron bombardment is a very inefficient process, most of the input power being converted into heat in the target. The X-ray flux which can be produced is generally limited by the heat dissipation possible in the target. With electrons focused to a spot of diameter 1 mm on an aluminum target on a water-cooled stem, 400-500 W is a typical upper limit for the input power. The X-ray power produced is then only of the order of 10 mW, and this is distributed over a hemisphere. The X-ray power is proportional to the electron current. The power in the line radiation is also proportional to $(E-E_c)^{1.63}$ for a target thin enough that absorption of X-rays by the target itself can be neglected [76]. However, as E increases, the electrons penetrate a thick target more deeply, the characteristic X-rays produced must pass on the average through more material on the way out, and are absorbed more strongly. This effect and the extension of the high energy cutoff of the X-ray spectrum set an upper limit for E.

Much higher X-ray fluxes are available from generators which have high speed rotating targets. The heat dissipation is spread over a larger area; the electron focal spot is on the rim of a spinning, water cooled wheel. The wheel may be tilted or the rim beveled so that an elliptical focal spot appears round when viewed from the direction of interest. Source brightness can be increased in this way, because more power can be dissipated in the larger spot, at the expense of a relative enhancement of the higher energy portion of the spectrum and some increase in the flux spatial non-uniformity. High power X-ray generators with up to 60 kW electron beams are available for crystallographic work. For the softer (longer wavelength) X-rays of interest for lithography a 25 kW generator is the most powerful yet reported. The Rigaku SX-5 generator has two alternate electron guns - a fine focus gun which supplies at most 0.2 A. at 20 kV into a 1 mm spot and a coarse focus gun which supplies up to 0.9 A. into a 3 mm spot on an aluminum target. Although target loading can be increased by increasing rotation speed and target wheel diameter, practical limits are set by longevity of seals and effects of thermal stress on the wheel. We may expect to see the limits slowly rise with application of new techniques. YOSHIMATSU and KOZAKI [77] have reviewed the status of high power X-ray generator development. Table 9 lists some high power generators reported in the literature recently. The Bell Laboratories machine achieves fairly high power without a rotating anode by letting the electrons impinge not on a flat surface but on a conical depression which allows more efficient cooling because of the greater surface area. This machine requires less maintenance because of the lack of high speed motion with the attendant sealing problems [78].

Table 9 High power X-ray sources

Power [kW]	Target	Reference
25	Al, 6000 rpm	[75]
20	Si	[72]
18	Al, 8000 rpm	[71]
12	Al, 2500 rpm	[78]
8	Pd, Stationary	[79]
2.5	Al, 4000 rpm	[80]

3.3.4 Masks

An X-ray mask consists of a patterned metal absorber on a transmissive membrane substrate. In contrast with a photomask the ratio of metal thickness to substrate thickness is not so small, making fabrication more difficult. These thicknesses are determined by the transmission of the materials for the X-ray wavelength of interest.

Of the heavy metals with large ρZ^4 suitable for absorbers gold has been widely used because it is easily patterned. The thickness of gold necessary for absorption of 90% of the incident X-ray flux is 0.5µm, 0.7µm, 0.5µm, 0.2µm, or 0.08µm for the X-ray wavelengths 4.4 Å(PdL), 7.1 Å(SiK), 8.3 Å(AlK), 13.3 Å(CuL), or 44.8 Å(CK), respectively. Thus in general the metal is considerably thicker than the chromium layer on a photomask. Methods for patterning the metal with high resolution include ion milling, sputter etching, and electroplating. Electroplating [81] produces excellent definition with vertical walls, but requires a vertical wall primary pattern in resist of thickness equal to that of the metal to be plated, unless a complex process of mask replication is employed. More often a subtractive process has been employed in which a thinner resist layer is used to pattern a thin layer of a refractive metal; the refractive metal serves as a mask for ion milling the underlying gold [71]. With this method it has been possible to form walls which depart from the vertical by 20° or less. The minimum linewidth attainable by ion milling in 0.5µm thick gold is ∿0.4µm.

For higher resolution work longer wavelengths such as the 13 Å CuL radiation may be used where gold thickness can be reduced. FLANDERS and SMITH [82] have replicated 0.16μm lines with this radiation. There are several problems associated with longer wavelength radiation in a system, however. A high power electron gun is more difficult to design because accelerating voltage must be kept below 8.9 kV to avoid exciting the short wavelength CuK radiation. Either mask and wafer must be in vacuum where mask heating may be a problem or a very thin vacuum window could be required. A smaller mask-wafer gap is also necessary to maintain resolution. In addition sensitive resists capable of the higher resolution will be required.

The membrane forming the mask substrate should be as transparent to the X-rays as possible, smooth, flat, dimensionally stable, reasonably rugged, and transparent to visible light if an optical registration scheme is used. Materials which have been used include polymers such as polyimide [83] and polyethylene terephthalate [73], silicon [67] SiC [71] Si_3N_4 [84], Al_2O_3 [85], and a $Si_3N_4/SiO_2/Si_3N_4$ sandwich [72] structure. Although different mask substrates are appropriate for different portions of the soft X-ray spectrum, there is not yet general agreement on the best material for any particular wavelength.

The major questions remaining about X-ray masks concern their dimensional stability, minimum defect densities attainable, and ease of handling. Polymer membrane substrates can be distorted locally by the absorber metallization. For greater dimensional integrity a stiffer substrate material is favored. In Fig. 28 the dimensional stability of a mask [86] with substrate consisting of a BN membrane covered with a polyimide layer is shown. The figure shows the displacement due to mask processing of a set of fiducial marks written originally on the mask with an EBES pattern generator and read after processing. In this typical result for a 3 in. diameter array maximum pattern displacement is 0.38μm and average placement error is 0.18μm, about three times worse than the photomask of Fig. 16.

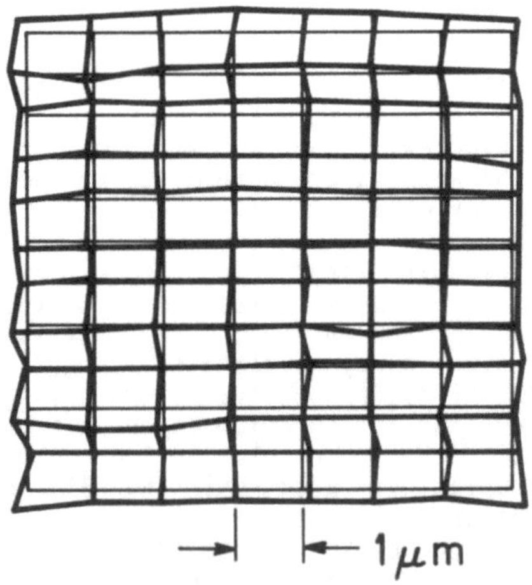

Fig. 28 EBES measurement of pattern placement + mask distortion errors at 64 points of 3 in. x 3 in. array on X-ray mask, from ref. [86]. Max. error = 0.38μm. RMS error = 0.18μm.

For these masks defect density is typically $\sim 2/cm^2$. Approximately half of these mask defects do not lead to defects in the resist image on the wafer. Ease of handling implies not only avoidance of mask breakage but also compatibility with cleaning procedures. Although X-ray printing is insensitive to common dust, reference planes on the mask used in precision setting of the mask-wafer gap must be kept clean.

3.3.5 Synchrotron Radiation

Some experiments have been reported in which the synchrotron radiation from electron synchrotrons and storage rings was used for X-ray lithography [87, 88]. The small angular divergence of the radiation simplifies mask-wafer registration, and the high intensity of the radiation leads to short exposure times. A single storage ring could provide radiation to a large number of exposure stations.

In synchrotrons and storage rings, high energy electrons are forced into closed curved paths by magnetic fields. An electron moving through a perpendicular magnetic field has an acceleration directed toward the center of the orbit and emits radiation. For the high energy electrons of interest, which have velocities very nearly equal to that of light, the radiation is emitted in a narrow cone in the forward direction of motion of the electron. An observer looking along a tangent to the orbit sees a bright spot. The radiation is very different from that of a point source because of the narrow beam from each electron. The radiation from a circular machine would come from all tangents and have the shape of a disc. The angular divergence of the radiation in the vertical direction is $\psi \approx (1957E)^{-1}$, where E is the electron energy in GeV, and ψ is in radians.

High energy electrons are provided to the storage ring by a microtron, a small synchrotron, or a small linear accelerator. The ring is briefly operated as a synchrotron to boost the electron energy to the final value. Then the electrons may circulate for many hours in a stable orbit, the loss due to the power radiated as synchrotron radiation being compensated by one or more acceleration cavities around the ring.

The peak of the power spectrum of the synchrotron radiation occurs at wavelength λ_p. This is related to the electron energy E (in GeV) and magnet bending radius R (in m) by

$$\lambda_p = 2.35\ R/E^3. \tag{42}$$

A 0.83 GeV machine with R = 2.1m would have a power spectrum peaked at λ_p = 8.4 Å.

There are two spectral regions of interest for X-ray lithography. In the 10 Å region mask substrates may be thicker and mask and wafer may be used at atmospheric pressure. In the 40 Å region limiting resolution is best because the path length of secondary electrons produced in resist exposure is ∼40 Å. At a port 10 m from the source spot on the circumference the radiation flux in the 7-12Å range from a 0.9 GeV ring would be 400 mW/cm^2 for 500 mA circulating current. This is 3 orders of magnitude larger than for conventional X-ray sources. The irradiated area is a rectangle which must be scanned across the wafer.

Table 10 illustrates how the better collimation of the radiation makes the registration problem simpler. We see that for a change in mask-wafer gap of 2μm, for example, the resulting alignment error for a 3 in. wafer is much smaller than obtained with a standard X-ray generator. The resolution is also better (smaller δ), even with a larger mask-wafer gap of 100-400μm.

Table 10 Comparison of X-ray generator and storage ring with regard to geometrical projection parameters

	Shadow δ [μm]	Runout Δ [μm]	Misregistration for 2μm Gap Change [μm]
X-ray generator, g=20μm D=50 cm	0.12	1.5	0.15
Storage ring, g=100μm, D=10 m	0.01	0.38	0.008
Storage ring, g=400μm, D=10 m	0.04	1.5	0.02

3.4 Ion Lithography

Ion lithography is in a very early stage of development. A finely focused ion beam or a collimated beam incident through a stencil mask could be used in several ways. When an ion beam is used to expose resist, higher resolution than with electron beams is potentially possible because of less scattering. In addition resists are more sensitive to ions than to electrons. And there is the possibility of wafer processing without resists if the ion beam is used to implant or sputter in selected areas.

The sensitivity of PMMA resist has been measured for 30 keV, 60 keV [89], and 200 keV[90] He^+ ions and for 100 keV and 150 keV Ar^+ ions [89], and the dose required for exposure is nearly two orders of magnitude less than with 20 keV electrons. Of course the statistical limitation imposed by fluctuations (section 3.4) implies a resolution/sensitivity tradeoff for ions also. The perpendicular straggle of the path of an ion penetrating material and the range of low energy secondary electrons produced are less than the range of backscattered electrons produced in electron lithography. The ion energies of interest for exposing resist depend on the ion. If the ion must penetrate 2500 Å of resist, then a proton would need 14 keV and a Au ion, 600 keV for the projected range to be 3000 Å.

For a beam used in fabrication by sputtering, lower energies are generally of interest. When an ion beam is incident on a material the sputtering yield increases with beam energy for beam energy larger than some small threshold value as more damage occurs near the surface. Beyond some energy the yield decreases as the ions penetrate more deeply and fewer surface atoms receive enough energy to leave the surface. For example the peak in the sputtering curve for Ar^+ ions incident on Cu occurs at 23 keV. For ion implantation, energies from 30 keV to 500 keV are used, and doses range up to 10^{15} ions/cm^2 (or 1.6×10^{-4} C/cm^2 for monovalent ions, representing a much larger dose than for resist exposure.)

The problems of ion optics for a scanning ion system are very much more severe than for electron optics. The brightest sources are the two types of field ionization source in which ions are produced in the strong field near a pointed tungsten tip. The source of ionized material is a gas surrounding the tip or a liquid metal which flows to the tip from a reservoir. The largest current densities obtained in the focused image of such a source are 1.5 A/cm^2 for Ga^+ in a 0.1μm spot [91] and 15 mA/cm^2 for H^+ ions in a 0.65μm spot [92]. Total beam current is severely limited. There are no bright sources for such useful implant species as B and P. Electrostatic lenses rather than magnetic must be used for focusing ion beams. If a magnetic lens were used to focus an ion beam the field would have to be much larger than in the electron optics case since from (23) the required field is proportional to $(mV/f)^{1/2}$, where m is the particle mass. Different isotopes would be focused to different points. Similarly magnetic deflection is much less practical than electrostatic. Electrostatic optical systems generally have higher aberrations, necessitating small aperture α and small scan fields.

SELIGER et al. [91] have reported a prototype scanning system in which a beam of 57 keV Ga^+ ions is focused to a 0.1μm diameter spot with current density 1.5 A/cm^2. Spot size is apparently limited by chromatic aberration of the electrostatic lens and the large 14 eV energy spread of the source. There is also work on use of collimated ion beams with stencil masks. FREE and MEADOWS [93] have employed a collimated 500 eV Hg^+ beam to transfer a pattern by sputter etching.

3.5 Conclusion

A number of techniques are under development to satisfy future patterning requirements. Optical lithography with step-repeat exposure is being extended to the regime of one micron resolution. For smaller features other methods are of interest - the first applications being to laboratory devices and special products where circuit performance is a more important consideration than low fabrication cost.

Addendum (see p. 295)

References

1. B. Honeisen, C. A. Mead: Solid-State Electron 15, 819 (1972)
2. B. Honeisen, C. A. Mead: Solid-State Electron 15, 891 (1972)
3. F. H. Dill, W. P. Hornberger, P. S. Hauge, J. M. Shaw: IEEE Trans. El. Dev. ED22, 445 (1975)
4. J. D. Cuthbert: Solid-State Technol., August, p. 59 (1977)
5. G. R. M. Thomas, H. L. Coleman, M. Lanahan: Proc. SPIE 174. 15 (1979)
6. Y. Nakane, T. Tsumori, T. Mifune: Proc. Kodak Interface '78, p. 32 (1978)
7. H. Nagai, A. Yoshikawa, Y. Toyoshima, O. Ochi, Y. Mizushima: Appl. Phys. Lett. 28, 145 (1976)
8. A. Yoshikawa, O. Ochi, H. Nagai, Y. Mizushima: Appl. Phys. Lett. 31, 161 (1977)
9. M. S. Chang, J. T. Chen: Appl. Phys. Lett. 33, 892 (1978)
10. K. L. Tai, L. F. Johnson, D. W. Murphy, M. S. C. Chung: 155th Electrochem. Soc. Mtg., Boston, Mass., May 6, 1979
11. K. L. Tai, W. R. Sinclair, R. G. Vadimsky, J. M. Moran, M. J. Rand: 15th Symp. on Electron, Ion, and Photon Beam Technol., Boston, Mass., May 29, 1979
12. H. I. Smith: J. Electrochem. Soc. 121, 1503 (1974)
13. W. N. Jones: Proc. Kodak Interface '75, p. 49 (1975)
14. W. R. Pratt, M. P. Risen: Proc. SPIE 80, 126 (1976)
15. B. J. Lin: J. Vac. Sci. Technol., 12, 1317 (1975)
16. M. V. Klein: Optics (Wiley, New York 1970) pp. 374-376
17. J. G. Skinner: Proc. Kodak Interface '73, p. 53 (1973)
18. D. A. McGillis, D. L. Fehrs: IEEE Trans. El. Dev. ED22, 471 (1975)
19. R. C. Heim: Proc. SPIE 100, 104 (1977)
20. M. Born, E. Wolf: Principles of Optics, 5th Ed. (Pergamon Press, New York 1975) pp. 459-532
21. J. W. Goodman: Introduction to Fourier Optics (McGraw-Hill, New York 1968) ch. 6
22. M. C. King, M. R. Goldrick: Solid-State Technol., Feb. p. 37 (1977)
23. M. M. O'Toole, A. R. Neureuther: Proc. SPIE 174, 22 (1979)
24. J. H. Bruning: 15th Symp. on Electron, Ion, and Photon Beam Technol., Boston, Mass., May 6, 1979
25. M. A. Narasimham, J. H. Carter, Jr.: Proc. SPIE 135, 2 (1978)
26. H. Moritz: IEEE Trans. El. Dev. ED26, 705 (1979)
27. D. A. Markle: Solid-State Technol., June (1974) p. 50
28. G. Ittner: Proc. SPIE 100, 115 (1977)
29. G. Bouwhuis, S. Wittekoek: IEEE Trans. El. Dev. ED26, 723 (1979)
30. D. R. Herriott, R. J. Collier, D. S. Alles, J. W. Stafford: IEEE Trans. El. Dev. ED22, 385 (1975)
31. G. L. Varnell: Proc. Kodak Interface '72, p. 56 (1973)
32. E. V. Weber, H. S. Yourke: Electronics, Nov. 10 (1977) p. 96
33. L. F. Thompson, R. E. Kerwin: Ann. Rev. Mat. Sci. 6, 267 (1976)
34. M. Parikh, D. F. Kyser: J. Appl. Phys. 50, 1104 (1979)
35. J. M. Moran, D. Maydan: 15th Symp. on Electron, Ion, and Photon Beam Technol., Boston, Mass., May 29, 1979
36. W. D. Grobman, H. E. Luhn, T. P. Donohue, A. J. Speth, A. Wilson, M. Hatzakis, T. H. P. Chang: IEEE Trans. El. Dev. ED26, 360 (1979)
37. M. J. Bowden, L. F. Thompson: Solid-State Technol., April (1979) p. 72
38. J. M. Shaw, M. Hatzakis: IEEE Trans. El. Dev. ED25, 425 (1978)
39. G. M. Henriksen: Solid-State Technol., Aug. (1977) p. 81
40. K. Wishnuff: Proc. SPIE 174, 84 (1979)
41. M. Kallmeyer, K. Kosanke, F. Schedewie, B. Solf, D. Wagner: IBM J, Res. Dev. 17, 490 (1973)
42. H. D. Jacoby: Proc. SPIE 80, 130 (1976)

43. J. G. Skinner: Proc. Kodak Interface '76, p. 92 (1976)
44. L. D. Yau: IEEE Trans. El. Dev. ED26, 1299 (1979)
45. J. A. Underhill: Proc. SPIE 80, 85 (1976)
46. J. G. Skinner: Proc. SPIE 100, 20 (1977)
47. P. Sandland: Proc. SPIE 100, 26 (1977)
48. J. I. Goldstein: Practical Scanning Electron Microscopy, J. I. Goldstein and Y. Yakowitz, eds. (Plenum Press, New York 1975) ch. 2
49. L. W. Swanson, T. Dickinson: Appl. Phys. Lett. 28, 578 (1976)
50. L. W. Swanson, N. A. Martin: J. Appl. Phys. 46, 2029 (1975)
51. J. E. Wolfe: 15th Symp. on Electron, Ion, and Photon Beam Technol., Boston, Mass., May 29, 1979
52. T. H. P. Chang, A. J. Speth, C. H. Ting, R. Viswanathan, M. Parikh, E. Munro: Proc. 8th Int. Conf. on Electron and Ion Beam Sci. and Technol. (Electrochem. Soc., Princeton, 1976) p. 377
53. E. Munro: J. Vac. Sci. Technol. 12, 1146 (1975)
54. A. V. Crewe: Optik 52, 337 (1978)
55. K. H. Loeffler: Z. Angew. Phys. 27, 145 (1969)
56. G. L. Varnell: SPE Conf. on Photopolymers - Principles, Processes, and Materials, Ellenville, N. Y., Oct., 1976
57. I. E. Sutherland, C. A. Mead, T. E. Everhart: Rand Corp. Report #WN-9404-ARPA (1976)
58. T. H. P. Chang, A. D. Wilson, A. J. Speth, C. H. Ting: Proc. 8th Int. Conf. on Electron and Ion Beam Sci. and Technol. (Electrochem. Soc., Princeton 1976) p. 392
59. F. S. Ozdemir, W. E. Perkings, R. Yim, E. D. Wolf: J. Vac. Sci. Technol. 10, 1008 (1973)
60. G. L. Varnell, D. F. Spicer, R. Robbins, C. Carpenter, M. Malone: 15th Symp. on Electron, Ion, and Photon Beam Technol., Boston, Mass., May 29, 1979
61. J. L. Mauer, H. C. Pfeiffer, W. Stickel: IBM J. Res. Dev. 21, 514 (1977)
62. D. E. Davis, R. D. Moore, M. C. Williams, O. C. Woodard: IBM J. Res. Dev. 21, 498 (1977)
63. H. J. Binder, P. Hahnmann, E. Christ, H. G. Eichorn, K. Sonnefeld: Jena Rev. 23, 211 (1978)
64. H. C. Pfeiffer: IEEE Trans. El. Dev. ED26, 663 (1979)
65. J. P. Scott: J. Vac. Sci. Technol. 15, 1016 (1978)
66. M. B. Heritage: J. Vac. Sci. Technol., 12, 1135 (1975)
67. D. L. Spears, H. I. Smith: Electronics Lett. 8, 102 (1972)
68. L. F. Thompson, E. D. Feit, M. J. Bowden, P. V. Lenzo, E. G. Spencer: J. Electrochem. Soc. 121, 1500 (1974)
69. J. M. Moran, G. N. Taylor: 15th Symp. on Electron, Ion, and Photon Beam Technol., Boston, Mass., May 29, 1979
70. D. W. Johnson: Proc. SPIE 174, 108 (1979)
71. R. K. Watts, K. E. Bean, T. L. Brewer: 8th Int. Conf. on Electron and Ion Beam Sci. and Technol., Seattle, Wash., May 21, 1978
72. T. Hayashi: 8th Int. Conf. on Electron and Ion Beam Sci. and Technol., Seattle, Wash., May 21, 1978
73. D. Maydan, G. A. Coquin, J. R. Maldonado, S. Somekh, D. Y. Lou, G. N. Taylor: IEEE Trans. El. Dev. ED22, 429 (1975)
74. D. C. Flanders, H. I. Smith, S. Austin: IEEE Trans. El. Dev. ED23, 412 (1976)
75. S. Yamazaki, S. Nakayama, T. Hayasaka, S. Ishihara: J. Vac. Sci. Technol. 15, 987 (1978)
76. M. Green, V. E. Cosslett: Brit. J. Appl. Phys. (J. Phys. D) 1, 425 (1968)
77. M. Yoshimatsu, S. Kozaki: X-Ray Optics, H. J. Queisser, Ed. (Springer, Berlin 1977) ch. 2

78. J. H. McCoy: Proc. SPIE 100, 162 (1977)
79. J. R. Maldonado, M. E. Poulsen, T. E. Saunders, F. Vratny: 15th Symp. on Electron, Ion, and Photon Beam Technol., Boston, Mass., May 29, 1979
80. G. A. Wardley, R. Feder, D. Hofer, E. E. Castellani, R. Scott, J. Topalian: Circuits Mfg., Jan (1978) p. 30
81. E. Spiller, R. Feder, J. Topalian, E. Castellani, L. Romankiw, M. Heritage: Solid-State Technol., April (1976) p. 62
82. D. C. Flanders, H. I. Smith: J. Vac. Sci. Technol. 15, 1001 (1978)
83. D. C. Flanders, H. I. Smith: J. Vac. Sci. Technol., 15, 995 (1978)
84. E. Bassous, R. Feder, E. Spiller, J. Topalian: Solid-State Technol., Sept. (1976) p. 55
85. P. A. Sullivan, J. H. McCoy: IEEE Trans. El. Dev. ED23, 412 (1976)
86. D. Maydan, G. A. Coquin, H. J. Levinstein, A. K. Sinha, D. N. K. Wang: 15th Symp. on Electron, Ion, and Photon Beam Technol., Boston, Mass., May 29, 1979
87. E. Spiller, D. E. Eastman, R. Feder, W. D. Grobman, W. Gudat, J. Topalian: J. Appl. Phys. 47, 5450 (1976)
88. H. Aritome, T. Nishimura, H. Kotani, S. Matsui, O. Nakagawa, S. Namba: J. Vac. Sci. Technol. 15, 992 (1978)
89. M. Komuro, N. Atoda, H. Kawakatsu: J. Electrochem. Soc. 126, 483 (1979)
90. Y. Tarui: Proc. Int. Electron. Dev. Mtg., Washington, D. C., Dec. 3, 1977, p. 2
91. R. L. Seliger, J. W. Ward, V. Wang, R. L. Kubena: Appl. Phys. Letters 34, 310 (1979)
92. J. Orloff, L. W. Swanson: Scanning Electron Microscopy (ITT Res. Inst., Chicago 1977) p. 57
93. B. A. Free, G. A. Meadows: J. Vac. Sci. Technol. 15, 1028 (1978)

Additional references (on p. 295)

4. Computer Aided Design for VLSI

P. Losleben

Over the past three decades, we have witnessed the remarkable growth of the semiconductor industry from the invention of the transistor to an industry that is all-pervasive in its effect on modern life. LSI technology is visible everywhere; in our watches, calculators, automobiles, appliances, telephones, televisions, weapons, computers, and even our toys. Each time geometries are reduced by half, circuit complexity has potentially increased by four. MOORE's law shows a resultant doubling of circuit complexity each year. The profusion of new products using this technology are the direct result of a designer's ability to economically mold his creative ideas to the latest semiconductor process. The success of microprocessors is an excellent example of an approach which broadens the number of potential applications while reducing design cost.

We all know that square law growth is ultimately bounded. We could, for example, ultimately deplete all the silicon in the universe! Practically speaking, there are much more immediate limits. Earlier chapters in this book addressed the limits of lithography and fabrication. Here we see considerable room for growth as geometries potentially shrink to the sub-micron range. Yet, even MOORE [1] has modified his law to project a potential doubling of circuit complexity only every two years in the early eighties. This reduction is not driven by the technology itself, but by our ability to apply it.

As we approach VLSI[1] complexity circuits we are confronted with rapidly increasing design costs and design time. Of the recently developed 16 bit micro-processors, it has been reported that the M68000 required 52 manyears of effort to design [2]. The Intel 8086 required 13 manyears of effort merely for layout [3].

It is easy to see that few custom applications can support design costs of this magnitude. Expressed differently, the number of applications for VLSI will depend largely on future ability to reduce design cost and time.

---

1 The author defines VLSI as 10,000 gates on a single chip where a gate is defined as a single level of logic, ie NAND, NOR.

4.1 What is Computer Aided Design?

Just as we have seen the proliferation of semiconductor technology across industry, we see it increasingly applied in its own evolution. Computers abound in the modern semiconductor facility. In this chapter we are concerned with the use of computers as design tools. This is most obvious as the interactive graphics system used for layout or the remote terminal to a large computer used for ciruit simulation.

"In the broadest sense, Computer Aided Design includes the use of computers to assist in everything from specifying the characteristics of a product to meet a marketing objective to enumerating the details of how it is to be designed, manufactured, tested, and field supported" [4].

As with any automation, the primary goal of Computer Aided Design is cost reduction. In this chapter, we place primary emphasis on the reduction of non-recurring costs; the costs associated with performing design.

4.2 History

Computer Aided Design is not a new field. It is rumored that the developers of the first software programmable digital computer discussed how they might use that machine to help design the next. It is significant to note that the earliest known published paper on CAD by CRAY and KISCH in 1956 [5] began with the observation: "The subject of computers designing other computers has been a popular one for several years." The term Computer Aided Design actually became recognized following the work in the late 1950's on interactive graphics at M.I.T. [6].

Whatever the origins, the field was well established by the 1960's and shared the unbridled optimism of that period. There appeared to be no limit to what computers could conceptually do. New concepts were seriously underestimated and oversold. Enormous investments were made, especially in the defense, aerospace, and semiconductor industries. Most of these early systems failed.

While the 1960's were characterized by many failures in large software systems, those in CAD were particularly vexing because they directly affected the productivity of the electronics industry. In those cases where design engineers were reluctant to use these new tools, management directives were sometimes invoked to force usage. The results were predictable. In many companies a severe backlash resulted. Design engineers and their management avoided the field for more than a decade. In many companies, pockets of resistance still exist, especially among managers who still recall their experience as engineers during that period.

The 1970's brought about a revolution in hardware. While major emphasis in the 1960's was on time-sharing large computer systems, the turn of the decade saw the introduction of small, low cost, turn-key systems. The reduced cost of hardware, notably large capacity disc, minicomputers, and storage displays gave rise to the successful introduction of commercial CAD systems, a trend that is continuing.

This revolution came at an opportune time. The use of manual drafting and preparation of Rubilith artwork had sufficed as the primary design tools of the semiconductor industry well into the 1970's. Pressure for increasingly complex circuits at finer geometries brought about the development and use of photo plotters. The line drawing plotters of the sixties, never widely accepted for integrated circuits, gave way to pattern generators capable of producing 10X reticles at high precision. These machines at once eliminated the need for one step of photo reduction and permitted the finer tolerances needed for the small geometries introduced toward the end of the decade.

It was immediately obvious that manual preparation of data for these machines was impossible. The introduction of CAD in the form of low cost, stand alone digitizers and interactive editors provided the first clearly successful application of CAD to the semiconductor industry.

The 1970's saw major advances in CAD beyond that evident in the graphics systems. Most of these advances occurred in a few large corporations. Most were not applied in the semiconductor industry. It was in this context that BERNHARD in writing about progress in the 1970's stated: "CAD did not measure up back then, and LSI evolved in another direction; toward modular and distributed systems, highlighted today by microprocessors coupled to RAM's, ROMs, and other peripheral devices, and heavily reliant on software for systems integration." He goes on to state: "Now, with the semiconductor industry in sight of VLSI, the question is whether CAD will finally develop into the least expensive way to use mass-produced VLSI in arbitrarily complex semiconductor designs." [7]

4.3 State-of-the-Art

This section is intended to provide a brief description of the various sub-disciplines which are applicable to CAD for integrated circuits. There are several excellent references which should be owned by anyone working in the field who is interested in a more detailed treatment of the subject.

First and foremost, VANCLEEMPUT's, Bibliography [8], now in four volumes and handily cross referenced, is an indispensable guide to most relevant publications of the past decade. BREUER's two reference books [9,10] provide a good overview of the field, even though they are not directed primarily toward integrated circuit design. The Proceedings of the IEEE Computer Society/ACM Design Automation Conference [11] are probably the best single source of information although most copies can now only be obtained from good technical libraries. The IEEE Computer Society Tutorial on Design Automation is a good overview of high level simulation [12]. Finally, MEAD and CONWAY build a good framework for future design approaches in VLSI. [13].

The examples given in the remainder of this section are intended to be representative of the class of tool rather than be exhaustive.

4.3.1 Specification

Specification is the formal definition of the design to be implemented, its parts, and the relationships between its parts. Specification is seldom recognized as a legitimate design automation activity, yet it is a vital part of getting a design into an automation system. In a very general sense, all data input, by whatever means, could be considered here. Overall, most current systems use rather unremarkable means to communicate with a user. While a great deal has been published on the man-machine interface, little has found its way into CAD systems.

For the purpose of this paper two methodologies will be addressed:

Language

Language is used here in the context of written communication. Numerous design languages have been developed, especially for specifying behavior at the register transfer level of design. (SDL, SCALD, CDL, DDL [12]) Likewise, languages have been used for specification of physical attributes of the design. (XYMASK [14], HDL [15], PLOTS [16]). The latter were most widely used and have come to be standard interfaces in many automation systems (although most are unique to a given company). With the increased use of graphics equipment, the physical description languages have become much less desirable as a means of manual data preparation.

Graph

Graphical specification of functions through approaches such as state diagrams, and the most familiar, logic diagrams and circuit schematics, are becoming much more popular as interactive graphic equipments mature. Numerous systems now exist, but interface with any larger CAD system is usually left as an exercise for each installation.

4.3.2 Synthesis

Synthesis is the expansion of a functional specification at one level into a set of simpler functions at a lower level. Synthesis is almost always done manually, although a few automated approaches have been developed and have seen limited use.

Logic Synthesis

In the most general sense, synthesis of detailed logic from a higher level functional specification received a great deal of attention in the 60's and early 70's. (There are over a thousand references in VANCLEEMPUT's Bibliography [8]). Most of this work sought to minimize logic gates and found little application in LSI design due to the more complex goals encountered in this type of design.

PLA Minimization

The use of PLAs, first as standard products, and then as building blocks in

LSI chips brought about the development of minimization software in the 70's. (PLATO [17], MINI [18]). Most of this software was unique to a given physical structure, but, nonetheless, proved useful at a few installations.

Macro Expansion

The most primitive form of synthesis (in fact, many would not even qualify it as synthesis) is the simple pre-design of a function in terms of simpler functions with its expansion each time it is invoked. Most gate level simulators provide this capability. A few layout systems do also. (CALMOS [19], GAELIC [20]).

Compilers

Compilers have long been used for software, and the close parallels to simulators made this approach popular in early programs. Compiler simulators are generally quite fast, but most consume extensive memory and have been difficult to maintain. The approach remains viable, however, with new developments in hardware and software. While the concept of hardware compilers (Silicon Compilers) [21] is now claimed to be possible, some disagreement is evident as to whether current approaches should be considered to be macro assemblers. Whatever the academic argument, current work is in its infancy and appears to hold considerable promise.

4.3.3 Simulation

Simulation, in the context used here, is the use of the computer to predict circuit operation and performance. All simulation is an approximation of the actual circuit performance. More work has been done in this area of CAD than any other. As a result, a great deal has been published about simulation and quite a few programs are now commercially available.

Simulation is normally thought of as a spectrum of tools:

Functional Simulation - High Level Modeling of System Behavior

(GPSS [22], SIMSCRIPT [23]). This level of simulation is most applicable to the conceptual design of a system and is most useful in modeling the behavior of a proposed function or set of functions. It has rarely been used for LSI design.

Register Transfer Simulation - High Level Modeling of a System in Terms of Registers and the Data Paths and Functions Connecting Them

(DDLSIM [24], CDL [25]). This level of simulation begins to bear close resemblance to the desired physical architecture. A great deal of work has been done in the area, but its use for LSI design is only recently gaining in popularity.

Gate Level Logic Simulation - Modeling of a System or Subsystem in Terms of Individual Logic Gates

(LOGSIM [26], SALOGS [27], FLOGIC [28]). This level of simulation closely approximates the design to be implemented in terms of primitive functions such as NAND gates, NOR gates, FLIP-FLOPs, etc. There has probably been more development in this class of simulator than any other and many programs exist. Most were originally developed for PC board applications using standard components at the SSI or MSI level of integration, but are widely applied to the design of LSI circuits. They are currently the most valuable tool for verifying the correctness of the functional design.

Fault Simulation - Modeling of a Circuit in a Faulted Mode, Normally for Evaluating the Completeness of a Test Sequence or for Fault Isolation

(TEGAS [29], LAMP [30], LOGIC V [31]). Most of these programs are an outgrowth of Gate Level Simulators and are intended to improve confidence in the tests developed for acceptance or maintenance of systems, subsystems, or PC boards. They have seen some use for LSI designs, but all suffer from excessive run time at current levels of integration. Their application to VLSI is doubtful, at least in the manner that they are currently used.

Circuit Analysis - Modeling of the Circuit Wave-Forms in Terms of Primitive Components

(SPICE [32], ECAP [33], SCEPTRE [34], MOSTRAN [35]) These programs model basic transistor operation and are the primary design tool used by most integrated circuit designers. Their widespread availability and ease of use has played a major role in most semiconductor component designs.

Device Simulation - Modeling of the Characteristics of a Single Transistor

(SPICE [32], TANDEM [36]). These programs are normally used during process development. In many cases, this level of modeling is done in custom software which is abandoned after the immediate need is resolved.

Process Simulation - Modeling of the Individual Semiconductor Process Steps

(SUPREM [37]). This work is fairly recent and the SUPREM program developed at Stanford is the most widely used. As the semiconductor processes have become more complex, this tool has become increasingly more valuable and popular.

4.3.4 Testability

As components become more complex, the question of determining whether a component is operating correctly, first initially, and then operationally, becomes increasingly expensive to answer. Design for testability became accepted in the 70's emphasizing the need to address testability throughout the design process. Numerous tools were developed or adapted from earlier work:

Automatic Test Generation

The fully automatic development of test sequences from the design specification received a great deal of attention in the seventies and many programs were implemented. (CDALGO [38], D-LASAR [39], TITAN [40]). Virtually none were successful at LSI complexity levels due to excessive compute times.

Testability Prediction

The excessive compute times encountered in both automatic test generation and in fault simulation (discussed earlier under simulation) brought about a new set of tools in the late 70's intended for rapid and economical evaluation of how difficult a given design will be to test (SCOAP [41] Most combine algorithmic approaches with heuristics to assist a designer in producing a testable design. Since these tools are relatively new, use has been limited but promising.

Testable Design Rule Enforcement

In order to predictably and economically obtain testable designs a few companies have restricted the logic design to primitives and structures which permit simplified test development. (LSSD is the most widely publicized [42]). While penalties of this approach are claimed to be small, it is predictably controversial as well as expensive to implement. The approach appears most applicable to custom designs performed at systems developers and is being used extensively at least at one company. There is no known current use by the semiconductor industry.

4.3.5 Partitioning

Partitioning is the physical division of electrical components into progressively higher levels of assembly. Most work in this area was done in the early 70's and was abandoned when it became obvious that the goals of partitioning were either too complex or too poorly understood to permit automation. While some work continues on developing aids (PSL/PSA [43]) there is currently no known practical application to LSI design.

4.3.6 Layout

Physical layout includes automated and computer assisted means of developing the physical geometries of a design. Layout has been the single essential task in integrated circuit design, and as such it has emerged as a heavily used area of CAD. Many approaches have been attempted and only the most significant are described here.

Interactive Graphics

Interactive graphics is the most widely used form of physical layout in the industry today. While the goal of interactive design has proved to be elusive, improved hardware and software now available seems to indicate a gradual move in that direction. Three approaches are most significant:

Manual Digitize and Edit

Turn-key graphics systems based on digitizers and interactive edit stations proved to be the most practical systems for evolving from a manual drafting mode of layout to a computer assisted mode. While most layout is still done manually in the semiconductor industry, these systems are used almost exclusively for data entry and error correction. (CALMA and APPLICON systems are the most widely used for integrated circuit design).

Dynamic Graphics

While most early work was done on dynamic refresh graphics, this approach proved too expensive and provided performance that rapidly degraded with design complexity. Most early attempts to apply these equipments to integrated circuit design in the late 60's and early 70's failed. The availability of low cost, high performance semiconductor memory and high resolution raster displays is now re-emerging as a very promising replacement. Color display makes this approach even more appealing.

Symbolic

Symbolic layout, in the most general sense, substitutes a set of symbols for the mask layers to assist the user in conceptually displaying his design. The earliest work was done at Rockwell in the early 70's and has been used extensively since [44]. This work, and SLIC (a deriviative approach developed at AMI) [45] substitutes alpha-numeric characters for unique mask overlays. This permits use of low cost displays. The display is somewhat awkward to read for the uninitiated and only permits display at one scale. Both are now using a color display which greatly enhances readability. The approach appears to be very cost effective for some applications.

A second approach (STICKS [46], CABBAGE [47]) uses dynamic color graphic display to enable a designer to manipulate lines and shapes on a floating grid. The shapes, being symbolic representations of actual geometric shapes or collections of geometric shapes, require less compute resources to display. The one-to-one mapping of a symbol to a component (e.g., transistor, via, piece of interconnect) and the uncluttered display are claimed to provide valuable additional information to the CAD system as well as permit the user to better conceptualize his design. This approach is very new and is just beginning to be used at a few companies.

Automated Layout

Automated approaches to layout use the computer to locate (place) functions on a chip and then interconnect (wire) them. Extensive work has been done in this area since the mid 60's although it has seldom found use in the semiconductor industry due to less efficient use of silicon area than for manual layout. The approach is more widely used by system developers. Four basic approaches are most common:

Standard Cell

Originated in the mid 60's at TI and RCA, variations on this approach have appeared many places and it is the most broadly used automatic layout approach (PRF [48], PR2D [49], MP2D [50], LTX [51], CALMOS [19]). The approach arranges cells (functions to be interconnected) in rows with areas for making interconnect between the rows. Usually the cells are constrained to be of uniform height but variable width. Earlier programs arranged cells back-to-back sharing a common power (and sometimes clock) distribution while newer programs use a single row with connections made on top and bottom. The latter approach permits a higher degree of freedom in placement and wiring resulting in smaller chip sizes. Most programs do both automatic placement and wiring and allow the chip to expand to assure complete or near complete wiring.

The approach is most applicable to applications where rapid design and ease of making changes dominate costs. Approximately 500 cells are reasonable for most programs, although some may go higher. It is almost exclusively used for MOS technologies.

Master Slice

Also called Gate Array, Universal Array, and Standard Array, this approach is based on a standard array of components where wafers are processed up to first layer of metalization and then stockpiled for future customization. The approach was inspired by the numerous "breadboard" circuits that existed in the 60's where customization was achieved through wire bonds. Numerous approaches have been made to automating the placement and wiring dating back to a total wafer approach attempted by TI in the 60's. The most successful to date has been by IBM at a level of about 400 gates per chip.

Because the under structure is rigidly defined, the most serious impediment to automation is wiring. 100% wiring completion is seldom achieved and IBM's experience indicates that a limit of 75% population of the possible number of circuits on the chip is desirable to optimize the amount of wiring to be completed automatically. Most other systems use manual placement and wiring.

Macro Cell

In an effort to increase circuit complexity beyond that attainable for standard cell and master slice, a few attempts have been made at predefining higher level functions for subsequent use in standard cell or master slice arrays. This approach has been successfully used with manual intervention (PRF, MP2D, CALMOS) but presents difficulties which have yet to be resolved in automated layout. Most automated placement algorithms begin with an assumption of unit cell size and where large disparities exist in actual sizes, the algorithms become inefficient. Further, local placement within these macrocells will usually not be compatible with global placement goals over the whole chip resulting in considerable difficulty in wiring. Manual intervention is almost always required.

Custom Cell

The placement and wiring of arbitrary shapes has always been an elusive goal of automated layout. Most current approaches (SICLOPS [52], AVESTA [53]) restrict the shapes to arbitrary rectangles. Work in this area is still very limited and there are no known successful applications at this time.

Other Layout Aids

Because of the immediate importance of layout, numerous approaches have been developed to assist the designer. A few of the most significant are included here:

Wirability Prediction

Originally developed for use on PC boards, these programs are used to determine if a circuit is wirable before expending the computer resources to actually do the wiring. (AGRAWAL [54], FOSTER [55]). They are applicable to Master Slice layout approaches but few applications are reported to date.

Compaction

Compaction programs are intended to squeeze layouts down to minimum geometries. Some success is reported in FLOSS [56], but the most important use is in conjunction with symbolic layout (CABBAGE [47]). This work is very new and is severely handicapped by difficulty in accommodating complex layout rules, especially where many exceptions exist.

Compensation

This is a very important new capability to expand or contract polygons in integrated circuit artwork. It has only recently become available, but is already extensively used in conjunction with scaling of artwork to adapt existing designs to new semiconductor processes.

4.3.7 Topological Analysis

This area of CAD includes those tools which analyze the physical layout. Most people who are familiar with integrated circuit design will recognize the expression, "eyeball hours" as a rather descriptive although painful measure of the effort involved in checking physical layouts. The requirement for absolute accuracy on highly complex physical layouts has resulted in literally thousands of eyeball hours of meticulous manual checking for conformance to design rules, implementation of the desired function, and calculation of critical electrical parameters. As with most manual operations, this has proved to be highly error-prone and as automated aids have become available, their value has become obvious through improved yield and reduced costs.

Design Rule Checks

The most heavily used type of topological analysis examines the physical geometries to measure distances, area, and other geometric relationships to determine conformance to layout rules. Initial approaches were heavily compute bound, generated many extraneous false error messages, or severely constrained the geometries. The late 70's produced a great deal of progress in this area, and while most programs still exhibit problems, the area is well established as critical to the development of high complexity circuits.

Connectivity Checks

Connectivity checks are automated techniques for assuring that the physical layout implements the desired function. They typically consist of two parts: a trace of the layout to determine the connectivity that has been implemented and a comparison of that connectivity against an independent description. For example, the connectivity of a layout and the connectivity of a logic diagram could each be extracted and then checked for agreement. In most cases the designer must communicate with the program in terms of primitive elements for which the nodes are predefined. For standard cell or symbolic layout approaches, this constraint is easily accommodated, but for custom cell layouts, more common to the semiconductor industry, the constraint has limited application of this class of program.

Device Recognition

A few connectivity checking programs attempt to recognize transistors from the unique features of the artwork in order to accommodate custom cell layouts. No general solution to this problem is known, but usable approaches have been developed for specific semiconductor technologies. The larger problem of assembling transistors into logic gates is even more difficult and little work has been done.

Electrical Rule Checks

Electrical Rule Checks are a very recent development which searches for illegal connections of transistors. For example, connection of an MOS transistor across the power supply, or connection of a source or drain of a depletion NMOS transistor to ground might be considered to be illegal.

Electrical Parameter Extraction

Calculation of load capacitances, coupling capacitance, transistor parameters, or resistance values has been done since at least the early 70's. Most approaches were very primitive until combined with connectivity checks and mask compensation. Association of a capacitance value with a net enabled ready interface with simulators while mask compensation permitted the modification of "as drawn" artwork to approximate the changes which occur in geometries during semiconductor processing. While work was reported in this area in the mid 70's, few programs exist.

4.3.8 Timing Analysis

Timing analysis is the prediction of network delay times without simulation. Basically more powerful than simulation, this approach evaluates timing of a circuit without having to synthesize a data path to insure propagation of the signal. The designer merely identifies the start and end of a data path and the program provides the propagation time for a signal through that path. The more powerful programs calculate propagation time as a function of load capacitances and compensate for rise and fall times of signals. These programs have not seen much use for LSI circuits due to the need for detailed characterization of the circuit elements ahead of time.

4.3.9 Documentation

Many of the tasks of documenting integrated circuit designs are now possible on current interactive graphics equipment. Numerous systems have been developed for logic diagram and circuit schematic preparation. In a few cases connectivity may be extracted from this for subsequent use in the CAD system. While these systems are now readily accepted for PC design, use for integrated circuit design is rare.

4.3.10 Data Management

The management of data in CAD systems in terms of data protection, backup, archive, configuration control, and design procedure management is accepted, but still not extensively implemented for LSI design. Few installations claim to have a data base more sophisticated than a collection of files owned by the user. This whole area remains largely undeveloped except for a few installations.

4.3.11 Summary of Current Use of CAD

Most integrated circuit design is still highly dependent on manual operations. The most successful applications of CAD have been circuit simulation, graphical digitizing and editing, and design rule checks. These applications now consume more computer resources than all the others combined. Systems manufacturers make fairly extensive use of automated layout and logic simulation as well. Beyond that, most CAD tools are too expensive, too difficult to use, have limited availability, or are simply too new to have more extensive use.

4.4 Perspective

If one assumes that circuit feature size will ultimately be reduced to the sub-micron range (.2 microns by some estimates), it is reasonable to expect chip complexities of 10^5 gates and memories of 10^6 bits. These numbers are so much larger than numbers reflecting current complexities that they are difficult to comprehend. We have passed the threshold of human comprehension. Like the national debt, who can relate to the total problem?

It is useful to cite simple analogies to give these numbers meaning. What are the implications of a two or three order of magnitude increase in circuit complexity?

4.4.1 The Connectivity Problem

In the early 70's metal interconnect of 10μ width and 7.5μ spacing was fairly common. This minimum width and spacing determines the minimum wiring grid size (17.5μ in this case) that can be used to route interconnect. If one were to use a 1/4 inch grid (like a standard engineering quadrille pad) to represent this wiring grid a 250 mil chip could be drawn (at a scale of 357X) on a sheet of paper less than 7 1/2 feet square. That's small enough to hang on your office wall.

Using the same 1/4 inch grid, the following table projects the size sheet of paper that would be required to draw a 250 mil chip as geometries are reduced:

Year	Grid Size	Scale	Paper Size	Equivalent Area
'70	17.5μ	357X	7.4 ft.	Wall Hanging
'80	8μ	800X	17 ft.	Large Living Room Floor
Mid 80's	2μ	3200X	67 ft.	Slightly Less Area than a Basketball Court (50 x 94)
Ultimate	.4μ	16000X	333 ft.	Slightly Less Area than Two Football Fields Including End Zones (360 x 160 x 2)

Table 5.1 Analogy of Wiring Grid on 1/4 Inch Quadrille Paper

The spectre of a battalion of draftsmen crawling about on two football fields vividly illustrates the need to approach layout in some other way. SEITZ [57] draws a very similar analogy in comparing a wiring grid to a city block. When extrapolated to ultimate geometries he envisions a map of the entire North American continent at urban densities. That spectre would give even a Los Angeles city planner cause for concern!

The Time Constant Problem

SEITZ goes on to point out that while transistor parameters scale nicely (speed goes up, power goes down, while geometries are reduced), interconnect does not scale. Interconnect resistance remains constant across one grid and as oxides are thinned, capacitance is also approximately constant. One can expect, at most, an order of magnitude reduction in sheet resistance so that even potential process changes do not help much. Even at current geometries circuit performance is dominated by the RC time constant of the interconnect paths. Interconnection of functions in the future will obviously be much more difficult and Mead's observation [58] that logic is cheap while interconnect is expensive will dramatically change the manner in which design is accomplished in the future.

The Noise Immunity Problem

Of course, for many applications, speed is not important. Unfortunately there is still no escape from the increased resistance of interconnect. The minimum geometry transistors associated with high density, low performance applications represent an increasingly small proportion of the capacitive load on a net. Meanwhile reduced geometries, especially thinner oxides and small spacings between parallel interconnects introduce proportionately larger coupling capacitance with other nets. The fast signal edges which can be expected, at least at the beginning of nets as a consequences of smaller transistors, then raise a concern for noise immunity. The ratio of coupling capacitance to load capacitance at the end of a long resistive interconnect run very quickly becomes an unwelcome additional complication to the design process.

4.4.2 The Layout Problem

Over the 70's Intel developed good statistics on the productivity of layout draftsman. [3]. Their figures show that for microprocessor design early results for less complex parts indicated a productivity of 10 transistors per day per draftsman. As the complexity of the components increased, this productivity rate decreased as one might expect. By encouraging repetitive use of designs, they were successful in bringing the productivity rate back up again.

Assuming one were able to bring the productivity rate back up to 10 transistors per day per draftsman (certainly possible with some use of repetitive structures), a draftsman would still have to start work at the time of the American Revolution (1776) to complete a design using .2μ geometries by the time they become available at the end of this decade! Alternatively one could assign over 200 draftsmen for a year to the same design, an equally unacceptable alternative. Obviously we will have to seek other approaches.

4.4.3 The Problem with Mistakes

Everyone knows that some things are too complex to ever be done. Balancing the federal budget or halting inflation seem to fall in this category. Unfortunately we are very unsuccessful in identifying some jobs that ultimately fall in this category ahead of time. Some software programmers in the early 70's, while very capable in their own right, and working with others who were equally capable, had spent a decade without ever working on a successful project. We are now beginning to see the same phenomenon in LSI design. The product simply refuses to work, or is abandoned before completion because it is over budget, or when it finally does work the market opportunity is missed.

While it is easy to speculate that a little more manpower might have solved the problem or perhaps a change in management might have helped, it is seldom as simple as that. Bright, talented, completely competent people make mistakes. For example, a 10% error rate on punched card data entry is not unusual and even verification (duplicate key stroking by another operator) did not bring the error rate down to zero. Further, the more complex the task, the more opportunity for errors exists.

We all known that as long as you're doing more things right than you're doing wrong, that you're making progress, right? Wrong! It depends on the interdependency of the things you're doing and how long it takes to find out that you've made a mistake.

As an example of this, the following is grossly oversimplified and completely fabricated, but illustrates the point:

Assume a reasonably complex design process consisting of many interrelated design activities. For simplicity assume each design activity is dependent on at least one previous activity and effects at most two future activities.

Finally assume an overall error rate of 15% where:

5% are caught before the next activity starts
4% are caught after one successive activity
3% are caught after two successive activities
2% are caught after three successive activities
1% is caught after four successive activities

An error in any activity implies that the activity must be repeated (to teach the designer a lesson perhaps). Of course any work based on erroneous work must also be done over. The first 5% are easily detected and corrected with only that small penalty in productivity. The next 4% inflict an additional 8% bad work before being detected resulting in a total of 12% of the work over two activities being rejected. (Of course, since we're only dealing with an academic example, we'll ignore the fact that some of this work would have been bad anyway). Likewise the next 3% becomes 21%, the 2% becomes 30%, and the final 1% becomes 31%.

Overall, a 99% error rate results! This is so close to zero productivity that the resulting poor morale would probably account for the last 1%.

The reason the final error rate is so high is the quadradic effect any error has in this example and the elapsed time before an error is detected. While the example takes very severe liberties in comparison with the real world, is it so farfetched? We know that the more complex things become, the more difficulty we have conceptualizing them and the more frequently mistakes are made. Likewise, the more complex the job, the less the people actually doing the work understand of the interdependencies of the parts. Consequently, the effects of mistakes propagate. The elapsed time to detect errors in the current design environment is frequently measured in days and may sometimes be months. Finally the most insidious mistakes, those which destroy previously good work, were not considered. If a design system is poorly managed, one mistake can destroy everything. For example, how often are "saves" performed on your turnkey graphics system, how long are they saved, and what assurance do you have that they are good?

4.5 Management of Complexity

Design style changed dramatically in the 70's. In the early part of the decade a component was designed largely by one person. Usually he had the assistance of one or two draftsmen and in rare cases, he might have the assistance of other junior engineers to do circuit analysis or to develop a test plan.

As more complex designs evolved toward the end of the decade, design teams were introduced. Design management became much more important as the number of people involved with a single design increased. Terms like "chip plan" and "silicon budget" were introduced as the need for a new discipline emerged. In microprocessor development the semiconductor industry demonstrated that it had no more luck in getting hardware and software personnel to communicate than the computer industry had shown before them. It is now common for a microprocessor to be released for sale with a list of known problems, hopefully accompanied with "work arounds" to circumvent design mistakes until they can be corrected. (This is strongly reminiscent of "white wires" in early model computers).

It is very tempting to draw parallels with software development in the 60's. There were many attempts to build large software systems at that time. There were also at least as many highly creative super egos among programmers at that time as we now have designing chips. Everyone did their job as they understood it and in their own unique style. The result was chaos. Mammoth software systems appeared that never worked, or if they worked, did so in such a poor fashion that their utility was severely curtailed. For a while, attempts were made to write very general purpose systems to distribute spiraling development costs over more applications. (A precursor of the "jelly bean" business?) The result was even larger systems that were less reliable, cost more to maintain, and performed worse than anyone had imagined. Another approach attempted to place the resources of large machines in the hands of relatively unsophisticated users through time-sharing. (A precursor of the semi-custom business?) The result was that service was unreliable, communication costs were high, and frequently these users would get themselves (and their companies) in more trouble than they know how to get out of.

The problem then was not one of bigness, but one of discipline. The intervening decade saw the painful evolution of Software Engineering, a more systematic approach to the management of large software systems development. Far from a mature technology, Software Engineering borrowed a great deal from another engineering discipline that put a man on the moon among other things -- The Systems Approach. For the first time software development was addressed from a much more global point of view. More time and effort was placed in design rather than in implementation and debug. In a more structured environment overall costs as well as development time were reduced.

Unfortunately, software engineering practices are far from universally understood, much less used. Even amongst the staunch advocates, few would go so far as originally proposed by DIJKSTRA [59] in imposing discipline on the programmer. Nevertheless, there is a great deal that can be drawn from the software experience and applied to VLSI design. Hopefully the software disasters of the 60's will not be repeated in silicon.

4.6 Structured Design

Structured Design begins with the concept of hierarchy. Any complex function may be divided into less complex subfunctions. These may, in turn, be subdivided into even less complex subfunctions. The process continues until a sufficiently simple function is defined to permit implementation. At each level the function to be performed and its interfaces with other functions are rigorously defined. The process just described is called top-down design.

Once a sufficiently detailed functional structure is defined, physical implementation may commence. Again a hierarchical structure may be defined. Beginning with the simplest physical entities, structures are tied together to form more complex physical entities. These, in turn, are assembled into yet more complex entities. The process continues until the total function is implemented. At each level the physical boundaries and interfaces with other physical entities are rigorously defined. This process is called bottom-up implementation.

Using these two processes, two similiar, but not necessarily identical hierarchical structures may exist. First, the functional or logical structure is associated with the functional portion of the design task. Second, the physical structure is associated with the physical implementation portion of the design task. Ideally, the tasks of specification, synthesis, simulation, and test development are performed in a top-down manner on the logical structure while partitioning, placement, wiring, and topological analysis are performed in a bottom-up manner on the physical structure.

If design were to proceed in a completely unconstrained manner through these two processes, the logical structure and the physical structure would probably appear quite different. A design, as first conceptualized, does not suffer from the same constraints imposed by the physical world. A variable stored in random access computer memory during simulation is equally as accessible as any other variable. In the physical implementation, however, a penalty is paid for the distance the data associated with that variable has to travel to be used. At the beginning of the design process, many of the physical constraints on the design are unknown. Consequently the two structures would tend to have different partitionings and probably even different hierarchical levels.

VANCLEEMPUT introduces yet a third structure, the behavioral structure [60]. Most high level design tools concentrate on the development of sequential processes. This might best be described as legacy of the field of Mathematics. We tend to think in terms of sequential processes and while this helps us to better understand the problem, it is a different mapping of the problem and its structure, while still hierarchical, is quite different from both the logical and physical structures.

Thus far in developing a simplified, less error prone, structured approach to design, we have introduced three different ways of thinking about the same design with the added disadvantage that mappings between them are computationally very difficult and proof of consistency practically impossible. One cannot help but question if this is the path to error free design.

In practice, one cannot ignore the behavioral structure in a design. To do so would be to ignore a great deal of very valuable work in digital design. Rather, it is useful to think of the behavioral structure as another dimension in design.

A simple example of a traffic light will help to clarify this. The concept of a traffic light begins with the desired behavior of the design shown below as a sequence of lights:

→ GREEN → YELLOW → RED ↩

The functional structure might show a timer connected through a decoder and drivers to the lights while the physical structure would identify specific components and locate them within assemblies.

Both the functional and physical representations of the design, when viewed in time space, perform according to the behavioral structure. One may then think of behavior as a sequential series of states which may be assumed by either the functional or physical design.

For this simple example, the behavioral model was trivial, but one might expect that, for more complex systems, structure and hierarchy are necessary. If one were to model the behavior of a traffic light system at a complex street intersection it would be necessary to consider the handling of interrupts (sensors) and their associated unique sequences. For example, an automobile approaching the intersection in a left turn lane might schedule a sequence of light changes ultimately allowing it to turn left while stopping conflicting traffic. The behavioral model of this system would contain several such sequences, probably under the control of a single "executive" in at least two levels of hierarchical structure.

It is at this point that the parallels drawn between structured software design and structured hardware design break down. The mapping of a behavioral model into software is relatively straightforward. Both are expressed as sequential processes. It is only when concurrency or parallelism is encountered that the process becomes difficult to conceptualize. Hardware structures tend to be inherently parallel, especially as the cost of logic circuitry declines.

The concept of state machines [61] has proved to be useful in mapping from a behavioral model to logic structures. The concept, developed at least two decades ago, has seen application in the design of some of the more complex LSI chips including calculators and microprocessors. The STATE of a machine being designed is the content of its static elements (flip-flops, memory, etc.) The NEXT-STATE FUNCTION is the combinatorial logic which defines the translation of one state of the machine to the next. The behavior of a design may thereby be modeled as a sequence of STATE CHANGES, each of which may contain many parallel operations.

The issue of how seriously the design process and the tools needed to support that design process should constrain the designer is currently an issue of heated debate. The vast majority of designs in the 70's were the product of a creative designer working with a minimum number of constraints placed on the design process. In particular, any data associated with behavior, function, or physical layout, if it existed on a computer at all, was rarely checked for consistency. Substantial changes might be made in the physical layout, for example, without making the corresponding changes in a simulation model. The designer, with his perception of the circuitry, could rationalize the change without feeling a need for additional analysis. It has only been the sheer volume of data that has finally discouraged the last minute editing of pattern generator commands rather than editing the graphic data at its source.

The imposition of reduced freedom in the design process, especially in the form of constraints on the behavioral, functional, and physical structures can no longer be avoided. The rapidly escalating design cost and time force a more global examination of the manner in which a design is performed and the tools which might be applied to that design process. Profitability is determined by overall cost savings and the timing of market entry. The costs associated with CAD figure heavily in that formula.

What we are witnessing is the natural evolution of an engineering discipline. A modern aircraft is not the product of a single designer, but the result of the coordinated efforts of a design team. Likewise, most engineering disciplines involving the design of complex systems have seen a metamorphosis from a highly individualistic art form to a structured, highly disciplined, profession.

A great deal of what is described in the remainder of this paper is based on the trade-offs that can be made between design freedom and the economics of automation. Obviously these trade-offs are highly dependent on the business environment. The semiconductor manufacturer designing a high volume part will certainly make different decisions than the computer manufacturer needing a small quantity of many high performance designs.

The most obvious trade-off in structure is between the functional and physical. If left to evolve independently, especially in a top-down design where the functional design is performed before the physical, the structures will differ greatly. If they are different, many computationally simple checks for consistency between the two structures become impossible. Consequently, errors in the design process remain undetected for a longer period of time (sometimes until after the part is fabricated) and design costs can be expected to escalate accordingly.

Current work at Cal Tech constrains the designer to a single structure [62]. As the design is expanded in a top-down manner, the designer anticipates the physical implementation by forcing communication between functions to be made vertically in the hierarchy to the maximum extent possible while minimizing horizontal communication between modules at the physical design by optimizing adjacency of connected functions. Throughout the design process minimizing communication, even at the expense of adding more logic, is emphasized.

The approach requires considerably different thought processes than accustomed to by most designers. The approach seems natural for problems which map easily into regular hierarchical structures. Some bus oriented structures appear well suited to the approach and form the basis for what is called a "silicon compiler." The work looks very promising, and will undoubtedly find application in many designs. A great deal more work needs to be done before it is a usable tool, however, and even then, most designers will find the constraints on his design to be unreasonable.

Ultimately, hierarchical structured design systems will be essential. Further, the benefits of these systems are most obvious when design is approached in a top-down manner. This implies that new skills and disciplines if not new thought processes must be adopted. One might reasonably expect that the acceptance of structured hardware design will be at least as difficult as that experienced in software.

4.7 Functional Design Tools

If we accept the requirement for a hierarchical structure in advanced computer aided design systems, we may then begin to address the tools which are required to operate within this structure. First, within the functional portion of the system, the tasks of specification, synthesis, simulation, and test development have been identified.

4.7.1 Specification

Graphical means of specification have become the most powerful and cost effective means of communication between the designer and the machine. Graphics equipment is now justified solely on the basis of reduced documentation costs, especially for the handling of engineering changes. A few systems now extract connectivity from a circuit schematic or a logic diagram, thereby providing the basic forms of data necessary to drive the remainder of an automation system. While currently only used for PC level systems, the extension of this approach to integrated circuits is straightforward. There is reasonable hope that at least this form of documentation and the resulting design will agree.

A designer will most certainly be constrained by these systems. If they are used for data entry to an automation system, the designer will be required to document his design. Further, any changes to the design would preferably be entered as changes to the documentation. The costs associated with the entry of new primitives in a system such as this might discourage the introduction of a new or slightly modified circuit. In the extreme, the designer might be restricted to primitives which simplify subsequent synthesis or test development.

4.7.2 Synthesis

Synthesis remains one of the most elusive functions in modern automation systems unless the primitives used by the designer are seriously constrained. PLA minimization techniques, macro expansion, and silicon compiler approaches are beneficial to enough designs to warrant considerable use in the near term despite their narrow applicability.

More general approaches are intended to insure functional correctness. In the past these approaches were largely unsuccessful because they required primitives other than those produced as standard components. Interest in this area is increasing as the cost of logic on a single monolithic die is reduced and the difficulty in producing correct designs increases.

Overall automatic synthesis remains uncompetitive with manual means coupled with simulation to build confidence in correctness. For the foreseeable future, manual techniques are likely to dominate.

4.7.3 Simulation

Simulation tools at all levels have become absolutely essential as circuit complexity increases. The emergence of multi-level simulators is the most encouraging sign that consistent notation and tools are being considered in hierarchical design systems. Low end tools such as SPLICE [63], DIANA [64], and MOTIS [65] address simulation from the transistor level to the gate level while high end tools such as ADLIB [66] and LOGIC V [31] address simulation from the gate level to the register transfer level.

If asked, many designers would still prefer to simulate their design in hardware. Breadboards or brassboards are still heavily used despite rapidly increasing costs. The attractiveness of this mode of simulation is two-fold: First, it permits real time simulation. The designer can usually approximate actual system performance to a high degree of accuracy. Second, it leaves control of this phase of design clearly in the hands of the designer. He is not dependent on another organization to keep his simulator operating. Nor is he dependent on the schedule of a central computer to get his job done. If he has a problem, he can work extra hours to solve it.

Ultimately breadboards will become too expensive and too error prone to be used. Software simulators are already gaining in popularity. If the architects of new automation systems ignore current experience with the reluctant acceptance of simulators, however, they will have missed a major signal in the reasons for acceptance or rejection of new automation tools. The utility of a new tool is directly related to its availability and ease of use.

4.7.4 Test Development

Testability has emerged as the most immediate problem of increased circuit complexity. Automatic test generation for arbitrary logic is not expected to be feasible for LSI levels of integration, although work is still continuing. The new tools for testability prediction are proving to be statistically accurate and will undoubtedly see extensive use. Fault simulation for test grading becomes compute bound at LSI levels of integration. Statistical sampling of the faults to be considered and the elimination of faults on redundant paths can be expected to extend application of this technique to VLSI complexities, but is still ultimately limited.

Testable design rule enforcement has generated a great deal of interest, but places the most serious constraints on the designer. This approach is the only known technique which is not ultimately compute bound as circuit complexity increases to VLSI levels.

Functional testing as described by Johnson [67] addresses both proof of correctness as well as testing for faults. While it places similar constraints on the designer as discussed earlier under synthesis, the approach is ultimately the most powerful. Current work in this area is very preliminary and a great deal of further development will be required before its utility is known.

There is no question that the demand for testability will constrain the designer. There remains only a question of degree. Most of the industry is already addressing testability as a design goal. At the other extreme, the use of testable design rule enforcement by IBM illustrates the severity of the problem for equipment developers. Testability is a field of computer aided design most seriously in need of a break-through.

4.7.5 Physical Constraints on the Functional Design

A principal weakness of the top-down design, bottom-up implementation process has been the physical constraints which ultimately conflict with and require change to the functional design. In practice, design is seldom performed in a purely linear fashion beginning with functional design and ending with physical implementation of that design. Any system which is intended to be used in this purely linear fashion can be expected to incur major recycles in the design task. Physical constraints force modification of the assumptions made during the functional design. It may not be possible to implement a functional algorithm within the constraints of performance, power, or area, for example.

While many designers maintain rough guidelines for estimating chip size and power dissipation as a function of logic complexity, automated tools are just beginning to become available. The MOSTRAN program, a transient analysis tool, automatically calculates average power dissipation as well as peak and average current requirements for each power source for any circuit being simulated. Similarly, for well behaved approaches to layout, statistical estimates of chip size for standard cell layouts are produced before layout is attempted. Likewise, wirability can be predicted for master slice arrays.

As one might expect, the more constraints that are placed on the entire design process, the more predictable the physical design becomes. The programming of a ROM, for example, is a highly useful, highly predictable, (and highly uninteresting) design process. At the opposite extreme, a completely custom, random logic, design in a new technology is completely unpredictable as painful experience will attest.

As the number of different components being designed increases, the requirements for predictability becomes increasingly important. While some segments of the semiconductor industry seem to thrive on a few products, most rely on a learning curve to reduce the cost of each subsequent new design. This is especially true for system designers were many custom chips are required. Those automation tools which improve the predictability of a design can be expected to significantly affect cost.

4.8 Physical Design Tools

Just as we have seen basic tools applied to the hierarchical structure in the functional phase of design, a set of tools also exist which are required to operate within the physical structure. The tasks of partitioning, layout, topological analysis, and timing analysis have already been identified.

4.8.1 Partitioning

In the past, physical constraints on partitioning, such as the increased drive necessary to communicate between levels of packaging, while severe, did not dramatically perturb the design. For high density circuits, the problems of partitioning are becoming more complex, and can be expected to become important in organizing logic on-chip as well as between chips.

The high penalty for connectivity inherent to high density circuitry described previously is most obvious in this area. Most designers are familiar with the tapering of geometries that is required to drive off-chip. Each successive level of logic is increased in geometric size until the desired drive capacity is achieved. The effects of partitioning may be seen to penetrate high density logic to a significant depth and have severe implications on the architecture of the design. It is not an accident that most memory devices are configured to have the maximum address space with the minimum number of bits output.

As geometries are reduced, the partitioning of logic on-chip begins to demonstrate the same charactertistic problems previously encountered in going off-chip. It is obvious that the practice of expanding a functional design down to primitive elements and then partitioning these into assemblies in the physical design is a less-than-optimum way to do design. The repercussions of the physical partitioning on the design architecture force major recycles in the design task. The desirability of addressing partitioning earlier in the design process by minimizing connectivity between functional elements increasingly indicates that similar if not identical functional and physical structures will evolve.

The role of automated aids to partitioning in this process is not clear. Certainly, the problem will be different than that addressed in the past. With the problem no better defined than it is at present, if any tools are likely to be developed, they will be interactive aids to the designer, possibly only accounting for and reporting the number of interconnection paths in a trial partition.

4.8.2 Physical Layout

Aids to physical layout, being the most pervasive tools in the industry now, can be expected to proliferate further and address a broader range of CAD functions. It will suffice here to project the major changes that can be expected.

Interactive Graphics

The most significant change expected in interactive graphics will be brought about by the combined need for increased resolution and higher performance, fortunately coupled with further reductions in the cost of equipment. Current 16 bit computers are limited (in single precision integer arithmetic) to a resolution of one part in ±32K or a limit of $\frac{1}{4}\mu$ resolution for current chip sizes. This limits practical use to 2μ geometries, the feature size of circuits now being designed. Further, the 16 bit word size severely restricts direct addressing, thereby limiting the performance of these machines.

The recent introduction of low cost 32 bit machines coupled with the availability of low cost, high performance semiconductor memory can be expected to dramatically change the utility of the turn-key graphic systems. Now compute power on a scale previously only available from a large scale central computer facility can be dedicated to the CAD function exclusively.

Further, the coupling of low cost semiconductor memory to color raster displays is making possible a much more dynamic and highly readable interface with the user. These displays can be expected to improve operator speed as well as reduce errors.

Symbolic layout is the most exciting development in interactive graphics and the recent announcement of a project to develop a symbolic system by one of the major commercial equipment producers is an encouraging indicator of a trend in that direction. The approach offers both the potential for interactive layout (where previous systems were used primarily for manual digitizing and editing) and for the capture of key data useful in reducing the work factor for topological analysis. Potentially, some degree of interactive net check and design rule check is also possible. Once again, the key to improved productivity is the detection and correction of errors as soon as possible.

Symbolic layout places one critical constraint on the designer. To achieve the full advantages of the approach, the designer must communicate with the system in terms of pre-defined symbols: interconnect, vias, or transistors. Detailed manipulation of the geometric shapes, while still possible, is undesirable because it negates the checks performed during layout.

Automated Layout

Automated approaches to layout will certainly gain in popularity in the semi-custom and custom business inherent to the computer, defense, and communication industries. Of major concern is whether these approaches can be extended to VLSI complexity circuits.

Known techniques for standard cell approaches to layout can extend the technique to 1000 cell complexities. Progress beyond that is questionable without significant changes in the approach. A major difficulty with this approach lies with the interconnect. Without a significant reduction in polysilicon interconnect resistance or additional layers of metal interconnect, the approach becomes interconnect limited. Hierarchical approaches to design which seek to minimize interconnect between functions are not easily mapped into the rigid physical architecture of this technique as long as all functions are expanded into primitive cells for placement.

The introduction of macro-cells into this layout approach is not obvious. Most programs use initial placement algorithms which assume unit cell size for efficiency. These programs perform poorly where large differences exist between actual cell sizes.

A significant market exists for these programs at lower complexity levels and proposed new algorithms, if successfully implemented, may extend these programs to VLSI complexity. They remain the only automated approach to layout which approaches 100% wirability.

Master slice approaches to automated layout suffer many of the same disadvantages of standard cell with even more serious constraints on the interconnect. The primary advantage of the approach, rapid turn-around for fabrication, remains valid, but it is doubtful that the technique will be used extensively beyond current complexities due to the rapidly declining efficiency of placement and wiring.

The approach may well find application for families of chips containing large, predefined functions which are personalized through the wiring of gate arrays coupled with the programming of ROMs or PROMs.

Work on custom cell approaches to automated layout will undoubtedly continue, but its utility will depend heavily on how well it competes with manual interactive layout where more arbitrary shapes can be accommodated. A combination of the two may ultimately hold the most promise.

All automated approaches to layout constrain the designer to the use of pre-designed structures. All are criticized for inefficient use of silicon. On the other hand, they provide the least costly means of performing design of components intended for low quantity production. As such, their increased use for custom and semi-custom design can be expected for the foreseeable future.

Other Layout Aids

With an anticipated dramatic increase in computer capacity local to the interactive graphics equipment, one might reasonably expect the natural evolution of many aids to the layout process. Many of these will be developed by an emerging merchant software industry, riding the ready-made market established by the more popular equipments. As an example of this, Digital Equipment Corporation has already established a user referral service and a third party software marketing service for the enormously popular VAX 11/780 computer. While this machine is not yet sold in a turn-key graphics system, at least one company has announced its intention to do so. Other manufacturers of this class of computer are likely to introduce similar services. Layout aids are a natural product in this environment due to the proliferation of turn-key graphics equipment.

4.8.3 Topological Analysis

Topological analysis tools in the form of design rule checks are already firmly established in the industry despite the relatively primitive status of current software. No program can yet accommodate all design rules and false violation messages are still a problem.

A great deal of work is evident in this area and it is only a matter of time until very powerful tools become available. Current approaches across all areas of topological analysis are ultimately computer bound and application to VLSI will depend on the capability of new software to accommodate hierarchy.

Current controversy surrounds the question of where in the design process this class of software is best applied. Interactive checks applied during layout conceptually reduce the delay to correct an error to the absolute minimum. It is unlikely that a full spectrum of rules checks, connectivity checks, and electrical rule checks can be performed interactively on detailed geometries. Symbolic layout offers promise for implementation of a limited number of checks, especially when coupled with compaction where design rule checks are avoided completely. Use of different classes of checks at the different levels of hierarchy offers advantages in efficiency as well as providing checks early in the design process [68]. There remains an enormous demand for a final level of comprehensive checks on completely unstructured layout just prior to artwork generation.

While the last approach offers the fewest constraints on the designs, it is ultimately doomed to failure for VLSI circuits. As geometries are reduced, the rules become more complex, and the computer work factor increases. Further, checks performed at this point are much too late in the design process. One could envision a complex design which is never completed when the rate of induced errors equals those corrected.

A second area of tradeoff involves the geometries permitted in a layout. Known software algorithms are much more efficient for orthogonal geometries than for non-orthogonal geometries. For structures other than memory arrays and other similar highly regular structures, it is questionable whether diagonal shapes save enough area to warrant their use. Performance of interactive rules checks is likely to be highly dependent on a restriction to orthogonal geometries.

4.8.4 Timing Analysis

As circuit performance becomes dominated by the interconnect, automated aids to timing analysis will become increasingly important. Unfortunately most current work predicts circuit delay as a function of load capacitance and sometimes input rise and fall time. Future application will need additional information and analysis of the interconnect path.

Since this class of software is seldom used at current levels of integration due to the amount of data required to characterize the circuit elements, the additional information required for interconnect will probably discourage more extensive use. A great deal of work needs be done here. Automated approaches to circuit characterization involving the use of transient circuit analysis programs as well as automatic parameter extraction from the physical layout are needed. This level of effort is probably only justified for standard cell or master slice approaches where it can be amortized across many designs.

4.9 Design Management

The previous two sections of this paper identified eight general classes of computer aided design tools. Many of these classes potentially involve several major software programs. It is easy to envision a system of twenty to thirty major tools and innumerable smaller "custom tools."

Further, there is an unmistakable trend toward a proliferation of computers in a typical design environment. Computer aided design is no longer the provenance of the large central computing facility. New, low cost hardware appears everywhere. The turn-key graphics system is most obvious, but turn-key simulators are now available, almost all testers are computer driven, all pattern generators are computer driven, and the personal computer has already placed basic circuit analysis tools in the very hands of the designer.

We are presently in the midst of a hardware and software explosion. Most of the software is of dubious origin and even more questionable reliability. Communication paths between the hardware (if they exist at all) are usually slow and also unreliable. Design procedures are handed down from engineer to engineer like old Indian legends.

Most companies are not so much concerned with the design of one component, or even with the design of many components in one technology. Rather, it is most desirable to build a design environment where each new design is "PREDICTABLY" more cost effective than the previous one.

If we accept the premise that VLSI designs will be the product of design teams, probably organized across organizational lines, and if we accept the premise that a component must be without mistakes to be usable, it is difficult to be optimistic about the prospects for successful VLSI design at any cost in the current environment. The probability of a second successful VLSI design is even more remote.

The single, most important work to be done in computer aided design is the management of the design system itself. It is encouraging to note that considerable work has been done already in this area, although not for semiconductor component design. The most recent development in this area is Designer's Workbench [69], a system that goes part of the way toward integrating the components of a PC level system.

The remainder of this section will concentrate on the author's perception of the components needed in a VLSI design management system. While each of these components have been described publicly, they have yet to be integrated into a system for the design of semiconductor components.

4.9.1 Data Base

The item of first concern in design management is the data associated with a design. For the duration of the design process, the design exists only as data stored on computers. The integrity of the design is the integrity of this data. A mistake in the data will result in a component which is, in some respect, faulty.

Successful VLSI design of the future will depend on the designer's ability to store and retrieve data reliably and efficiently. This implies that he will know where it is at when he needs it, that it will be where he expects it to be, that it will be there in the status that he expects, that he will be able to retrieve it in the form that he wants, that he will be able to get it when he needs it, and that he will be able to update it when he needs to. Further, he must be able to do all these things without interfering with the work of others in a job requiring a high degree of interaction.

In addressing the development of a data base the following questions must be answered:

1) Who needs data? What user and what application program?
2) What data is needed? What data items are required for each application?
3) When is it needed? What response time is required?
4) Where is it needed? What computer or what remote site?
5) How is decided later.

The gathering of this information will lead to an initial determination of the basic configuration of the data base from among the following three options:

Central Data Base

The simplest configuration to build and maintain is a central data base where all design data resides at a single site, usually on a single machine. Every attempt is made to store an item of data only once in the data base. The existence of redundant data seriously complicates update procedures and is avoided except where overriding performance requirements dominate. This form of data base is most applicable to systems where the majority of CAD tools run on one computer or where access to that computer can be made when needed from other computers without serious loss of performance.

Satellite Data Bases

When multiple computers are involved and performance requirements preclude remote access of data from a central data base, data may be copied and reside for an extended period of time on another computer. In this case the existence of redundant data in the system is tolerated in order to achieve essential performance. A turn-key interactive graphics system might be integrated into a CAD system in this manner. For this example, the real time performance requirement of the system, as well as the constraint to use an existing software system, prevent direct access of a central data base.

Communication with a satellite data base need not be on-line although that is usually desirable. Where no modification is made to data copied from the central data base, data translation and control need only be done in one direction. One could, for example. transfer data to a pattern generator for production of reticles. For this example, no data translation back to the central data base is required.

The system becomes more complex when two way communication is required, especially where data originating in the central data base is modified and it is desired to update the central data base. Seriously complex audit procedures are needed for this latter example, especially where that data is shared by multiple users.

Distributed Data Base

Distributed data bases attempt to avoid the problem of redundant data in a satellite data base system by physically moving data to the machine where it is needed or actually never bringing data to a central site if it is not needed there. Redundant data is only allowed to exist where coincident requirements for the same data exist at different machines. The system is greatly simplified if read-only access of the data is permitted in that case. One computer is always "in charge" and keeps track of the current location of all data.

The example of a turn-key interactive graphics system might alternatively be implemented in a distributed data base environment. A physical design in progress at the graphics system may never be transferred to a central site. In particular, if several graphics systems are being used on a given design, it is only necessary to know where each portion of the design resides until it is needed by another designer. This approach is much simpler and less dangerous than attempting to maintain a duplicate copy at a central site.

4.9.2 Data Management System

Once the tradeoffs involved in the data base configuration are understood, the basic software necessary to manage data in the desired configuration may be addressed. (Note that this software may play a key role in selecting the configuration. Commercial data base management systems are only available for central data bases, for example.)

The data base management system (DBMS) performs the fundamental tasks of:

a) Storage allocation and deallocation
b) Logical organization of data
c) Symbolic identification of data (schema)
d) Actual movement of data to and from storage
e) Access control of who is permitted to read or write data
f) Control of data sharing

Each of these tasks is exhaustively treated in any good text on DBMS and will not be discussed in detail here.

The author specifically excludes the following functions from the DBMS:

a) Archiving
b) Data manipulation packages such as SORT
c) Report generators
d) Any other utility function which depends heavily on the DBMS but is not an essential part of it.

These functions are excluded so that the DBMS may be kept as small, simple, and efficient as possible. This software, more than any other in a CAD system, will dictate the performance and reliability of the overall system. It must, therefore, be restricted to only those functions essential to the movement of data to and from the data base. Other functions may then be implemented as independently supported utilities. All access of data in the data base is constrained to be made through the DBMS.

For satellite or distributed data base systems, each computer may have its own DBMS software. Each CAD tool thereby interacts with stored data through a single DBMS on that machine. The author prefers to treat the management of data between machines at a higher level to be described later, again to maintain simplicity in the DBMS.

The architect of a multi-machine CAD system will be immediately confronted with the question of whether a single DBMS should be used or whether different systems should be used on different machines. In the case of turn-key systems, he has little choice but to use that supplied by the vendor. Further, no commercial systems are machine independent. If he has the option to write his own, every effort should be made to write a single machine independent system to simplify future maintenance. *In no case should he modify turn-key or commercial software unless he is prepared to divorce himself from the vendor entirely.*

4.9.3 Data Configuration Management

Most large electronic equipment manufacturers already have well-established configuration management systems which are usually invoked when a design is transferred to manufacturing. These systems are typically large, cumbersome, often manual paper systems, and almost universally avoided by design engineers. The purpose of these systems is to control change to a completed design where a mistake could have very serious financial repercussions.

In VLSI CAD systems a similar requirement exists to control change to a completed design, especially where other engineers' work is dependent on that design or where designs are cataloged for subsequent use on future chips. Here also, the consequences of uncontrolled changes to data have severe repercussions on the confidence that can be placed in the data representing a completed design. Control in the design environment must be much more responsive than in the manufacturing environment, however, and the CAD system must be configured accordingly. Two aspects of control are involved:

Initial Design Completion

The intent of this function is to provide the opportunity for an event or milestone in the design process for a designer to "sign off" on his portion of the design task. It is predicated on a CAD system where each design file contains all the data unique to a given step in the design process and where an individual designer has responsibility for its correctness. In "signing off" on his job, the designer is saying in effect, "I have done my job, this data is now available to others to use."

File permissions change at the time of control:

During Design: Read Permission--granted by the designer
(uncontrolled) Write Permission--only the designer

Design Complete: Read Permission--anyone
(controlled) Write Permission--no one

At the time of "sign off", a two way relationship is established between the designer's data and input data on which he is dependent. A reference is made to the data he on which is dependent while a "where used" pointer shows which new data is dependent.

Rules for the use of data are:

1) Only the originator of data may control it.
2) Use of uncontrolled data is at a designer's own risk.
3) New data may not be controlled until all input data not generated by the designer is controlled.
4) Controlled data may not be modified as long as other data is dependent on it.

The control procedures, when applied across the design process provide a great deal of information to the CAD system:

1) It shows that each individual has completed his portion of the design task and that no subsequent task was based on incomplete data.
2) When an error in the design process is encountered, all possible consequences of that error are identified.

Change Control

There are frequent reasons to change data in a CAD system where it is desirable to retain earlier versions of the same data in a systematic fashion.

A designer may wish to merely retain several trials of a new design until he selects the best. Alternatively, data may be perfectly good for one application, but require minor changes for another. In both cases, it is undesirable to create completely new names for the new data.

Instead revisions to the original data are cataloged. The original version is considered to be revision Ø. Subsequent revisions are numbered consecutively and are each based on the previous one. In its most complex form, this approach could assume a tree structure of dependencies. Usually a simple system where each revision modifies only the previous is adequate.

Revisions are controlled in the same manner as previously described with the constraint that each revision is dependent on the previous one. This provides the added benefit that the system may periodically be cleaned-up by compressing revisions that are unused except to support a subsequent revision. It is always possible in this system to make conscious decisions as to which data may safely be deleted due to the known dependency relationships.

4.9.4 Product Development Plan

While the Data Base, the Data Base Management System, and the associated utilities provide the basic tools to capture the data associated with a design, and the Data Configuration Management System provides the tools to control data, we do not yet have a system capable of predictably producing more than one design. In an environment of many software tools, some developed by the designer himself, and many computers, the design process is not reproducible. In fact, it may not even work the first time, unless the design process is well understood.

A designer may be very creative in getting a job accomplished. The next time he is asked to do the same job, he is likely not to remember how and will again have to improvise. Where these steps include the development and use of special purpose software, the problem is much more severe. The fourth component of design management is the set of tools necessary to develop, capture, use, and refine design procedures.

The following concepts are developed largely from the Bell Designer Workbench System [69] and the Denver Bell Configuration Management System [70-71]. Both systems are based on the definition of an interrelated set of individual design tasks necessary to complete the overall project. The Designer Workbench System provides a "friendly guide" through the maze of machines, job control languages, design description languages, software tools, and associated procedures. The creators of this system recognized that while each step of the design process might be simple and completely understood by the developers of the CAD tools, the system as a whole was impossibly complex to a designer who was, among other things, expected to be an expert in his own field.

The Denver Bell system was built to operate in an environment where many designers were contributing to the same project, each responsible for performing his own portion of the design task. While both systems were developed for PC level design, a great deal applies to VLSI design.

The goal of a Product Development Plan is to assemble, on the computer, a set of design procedures which will ultimately lead to a complete design. These procedures include the file naming conventions, allocation, and deallocation of storage, the preparation of task dependent job control language, the sequencing of automated, semi-automated, and manual tasks, and the configuration control of resulting data.

The intent is to create an environment of individual responsiveness to each task contributing to the overall project. The Product Development Plan is a network of interrelated design processes, each leading to a specific goal. Many of these processes will have been used before, are well understood, and can be expected to produce predictable results. Others will be new, a product of the innovation needed to produce this new product, or simply an improvement over an old procedure. It is key that the Product Development Plan develop along with the design in a top-down manner. Likewise, it is critical that the plan be implemented on the computer in a Designer Workbench "guide" manner. It is the responsibility of the CAD system to provide the necessary facilities for the designer to describe the manner in which the job will be done and then to follow that procedure each subsequent time that job is done.

The system just described does not exist in a manner acceptable to perform VLSI design. Designer's Workbench requires that the procedures be programmed into the controlling computer. While this is satisfactory for a stable, well understood, design process, what is needed here is the flexibility to easily capture new procedures in a highly dynamic environment. What is desired is the ability to provide a designer the option of following an existing procedure or defining a better procedure. In either case, the design task is repeatable.

4.9.5 Management Information System

The fifth and final component of design management is the collection, cataloging, and reporting of information relating to design progress and cost. The following concepts are developed from the Bell-Denver Configuration Management System [70-71] and the IBM Design Process Analysis System [72].

The components of design management previously described capture all the critical information needed by management to track project progress, identify costs, and estimate future projects. The network of design procedures bears strong resemblance to PERT [73] and provides the added advantage that data input can be made automatically, a principal advantage over previous manual data input to Management Information Systems.

Both the Bell and the IBM systems provide several significant advantages to management:

1) They facilitate communication on a real time basis between all organizations involved in a project.

2) They analyze and report the resources being used.
3) They predict the effect of any proposed change.
4) They track progress in real time.
5) They permit the automatic generation of statistics and predictive models.
6) They require minimal independent involvement by the designers.

It is no accident that Bell and IBM are the only companies to approach the five components of design management described here. Both companies have made substantial investments in CAD for at least two decades. While systems of this complexity are not essential to VLSI design, they are essential to a system where each design must be predictably less expensive then the previous one.

4.10 Conclusion

This paper has provided a brief review of the state-of-the-art in CAD and projected the requirements and probable developments needed for VLSI design. Those elements of CAD which were discussed were only the most obvious. In a field that is now a quarter of a century old, considerable progress might be expected. It should be obvious at this point that the development of a CAD system to support VLSI design is no small undertaking.

This paper did not address, in any large degree, where the CAD tools to support an all-pervasive electronics industry will come from. A reasonable estimate of the effort to develop such a system is in the neighborhood of 200 to 300 man-years! Fortunately some components are already commercially available, but even for the remaining effort, few companies have anticipated the cost nor have the qualified personnel. In commenting on "the problem with CAD/CAM", LLEWELYN states: "What is now certain is a critical shortage of scarce skills, nowhere more evident than in CAD/CAM, which will inhibit implementation and widespread or effective use for at least the next decade." [74]

In view of the current high demand for CAD, the limited availability of personnel, and the emergence of CAD in the merchant software market, there is a renewed danger of new CAD systems not performing "as advertised." Performance has been a chronic problem for CAD and the current situation is strongly reminiscent of the 60's when the effort to build new CAD systems was seriously underestimated and oversold. Hopefully the same mistakes will not be repeated.

The implementation of these new CAD systems is the most serious challenge we face in the next decade. It will likely take that period of time for these systems to evolve. The extent to which we are successful will largely determine the impact of VLSI technology.

References

1. MOORE, Gorden; "VLSI: Some Fundamental Challenges"; IEEE Spectrum, Apr 79, pp. 30-37.

2. WEIMANN, Warren; "VLSI Design at Motorola"; unpublished presentation at The 1979 Design Automation Workshop, East Lansing, MI, Oct 79.

3. LATTIN, William; "VLSI Design Methodology, The Problem of the 80's for Microprocessor Design"; 16th Design Automation Conference Proceedings, San Diego, June 79, pp. 548,9.

4. IEEE Computer Society Design Automation Technical Committee definition of Design Automation.

5. CRAY, S. R. and KISCH, R. N.; "A Progress Report on Computer Applications in Computer Design", Proc. Western Joint Computer Conference; Feb 56, p. 82.

6. SUTHERLAND, I. E.; "SKETCHPAD", Proc. AFIPS 1963 SJCC, Vol. 23.

7. BERNHARD, R.; "Technology '80: Solid State Looks at VLSI"; IEEE Spectrum, Jan 80, p. 44.

8. VANCLEEMPUT, W. M.; Computer Aided Design of Digital Systems, a Bibliography; Vol. I-IV, Computer Science Press, Potomac, MD.

9. BREUER, M. A. (ed.); Design Automation of Digital Systems: Theory and Techniques. Vol. 1: Hardware; Prentice Hall, Englewood Cliffs, NJ, 1972.

10. BREUER, M. A. (ed.); Digital System Design Automation: Languages, Simulation, and Data Base; Computer Science Press, Potomac, MD, 1975.

11. IEEE and ACM; 1st through 16th Proc. Design Automation Conference (Workshop prior to 1975), 1964-1979.

12. VANCLEEMPUT, W. M. (ed.); "TUTORIAL, Computer-Aided Design Tools for Digital Systems"; IEEE Computer Society, Long Beach, CA, 1979.

13. MEAD, C. and CONWAY, L.; Introduction to VLSI Systems; Addison-Wesley, 1980.

14. FOWLER, B. R.; "XYMASK"; Bell Laboratories Record, July 1969, pp. 204-209.

15. HIGHTOWER, D. W.; "Can CAD Meet the VLSI Design Problems of the 80's; Proc. 16th Design Automation Conference; June 1979, pp. 552,3.

16. KORENJAK, B. J.; "PLOTS: A User Oriented Language for CAD Artwork": RCA Engineer, Vol. 20 No. 4, Dec 1974, p. 20.

17. SHERWOOD, W.; "PLATO-PLA Translation/Optimizer"; Proc. Symp. on Design Automation and Microprocessors, Palo Alto, CA, Feb 77, pp. 28-35.

18. HONG, S. J., et al; "MINI: A Heuristic Approach for Logic Minimization"; IBM Journal of R&D, Vol. 18, No. 5, Sept 74, pp. 443-458.

19. BEKE, H., SANSEN, W.; "CALMOS: A Portable System for the Automatic and Interactive Layout of MOS/LSI"; Proc. 16th Design Automation Conference, San Diego, CA, June 1979, pp. 102-105.

20. EADES, J.; "GAELIC Auto-layout User's Manual"; Wolfson Institute, University of Edinburgh.

21. JOHANNSEN, D.; "Bristle Blocks: A Silicon Compiler"; Proc. 16th Design Automation Conference, San Diego, CA, 1979, pp. 310-313.

22. GORDEN et. al.; "General Purpose Simulation System/360 CS Version 2 User's Manual"; SH20-0694-0 IBM Corporation.

23. MARKOWITZ et. al.; "SIMSCRIPT, A Simulation Programming Language"; Prentice Hall, 1963.

24. ARNDT, R. L., DIETMEYER, D. L.; "DDLSIM-A Digital Design Language Simulator"; Proc. NEC (1970), pp. 116-118.

25. MESZTENYI, C. K.; "Computer Design Language Simulation and Boolean Translation"; Technical Report 68-72, Computer Science Center, University of Maryland, June 68.

26. MARSH, R. E.; "Logic Simulation Made Simple with LOGSIM"; IEEE Computer Group Repository, R70-63, 1970, 30 pp.

27. CASE, G. R.; "SALOGS- A CDC 6600 Program to Simulate Digital Logic Networks - Vol. 1 Users Manual;" Sandia Labs, Albuquerque, NM, Report 74-0441, Mar 75.

28. ROMBEEK, H., WILCOX, P.; "Interactive Logic Simulation and Test Pattern Development for Digital Circuitry;" Proc. Electro/76, Boston, MA, May 76, pp. 26.2.1-5.

29. <u>CC-TEGAS4 Test Generation and Simulation System--An Introduction</u>, Comprehensive Computing Systems and Services, Austin, TX, 1979.

30. CHAPPELL, S. G. et al "LAMP: Logic Analyzer for Maintenance Planning--Logic Circuit Simulators" Bell System Technical Journal Vol. 53 No. 8, Oct 74, pp. 1451-76.

31. STRUNGE, J. J.; "Fault Modeling in a Hierarchical Simulator"; Proc. 14th Design Automation Conference, New Orleans, LA, Jun 77, pp. 118-127.

32. COHEN, E.; "Program Reference for SPICE2"; Univ. of Calif., Berkeley, Memo ERL-M593, June 1976.

33. BRANIN, F. H., et al; "ECAP II--A New Circuit Analysis Program"; IEEE Journal of Solid State Circuits, 1971 Vol. SC-6, pp. 146-166.

34. McMILLAN, B. B.; "SCEPTRE Users Manual," Univ. of Calif., Berkeley, Lawrence Radiation Lab, Mar 71.

35. SCELL/MOSTRAN Cell Layout and Transient Analysis Programs Users Manual, Westinghouse Electric, Friendship, MD, Apr 77.

36. DUTTON, R. W.; Private communication on TANDOM-2D program; Sanford University, 1979.

37. ANTONIADIS, D. A. et al; "SUPREM II--A program for IC Process Modeling and Simulation;" Stanford Electronic Labs, Stanford University, Tech Rep. SEL 78-020, June 1978.

38. VAUGHN, G. D.; "CDALGO--A Test Pattern Generation Program"; Proc. 13th Design Automation Conf. San Francisco, CA, June 76, pp. 186-193.

39. RAMAMOORTHY, C. V. and CHANG, L. C.; "System Modeling and Testing Procedures for Microdiagnostics;" IEEE Trans-Computers, Vol. C-21 No. 11, Nov 72, pp. 1169-1188.

40. JOHNSON, W. A. et al; "Digital Circuit Test Analysis and Failure Diagnosis using the TITAN System"; Journal of Design Automation and Fault Tolerant Computing, Vol. 1 No. 3, May 1977, pp. 287-309.

41. GOLDSTEIN, L. H.; "Controllability/Observability Analysis of Digital Circuits"; Sandia Laboratories, Albuquerque, NM SAND 78-1895, Nov 78.

42. CORREIA, M.; "Introduction to an LSI Test System;" Proc. 14th Design Automation Conference, New Orleans, LA, June 77, pp. 460-1.

43. VANCLEEMPUT, W. M.; Private communication on partitioning--PSL/PSA, Stanford University, 1979.

44. LARSEN, R. P.; "Versatile Mask Generation Techniques for Custom Microelectronic Devices"; Proc. 15th Design Automation Conference; Los Vegas, NE, June 1978, pp. 193-8.

45. GIBSON, D.; NANCE, S.; "SLIC--Symbolic Layout of Integrated Circuits"; Proc. 13th Design Automation Conference, San Francisco, CA, June 76, pp. 434-440.

46. WILLIAMS, J. D.; "STICKS--A New Approach to LSI Design"; MIT Dept of EE, MS Thesis, May 77.

47. NEWTON, R.; Private correspondence on symbolic layout--CABBAGE, Univ. of Calif., Berkeley, 1979.

48. IC LAYOUT Users Manual. Vol. VII, Westinghouse Electric Corp., Friendship, MD, Apr 77.

49. EDGE, T. M.; The PR2D (Place, Route in 2-Dimensions) Automatic Layout Program Handbook; NASA Ref. Pub. 1029, Sept 78.

50. FELLER, A. "Automatic Layout of Low-Cost Quick-Turnaround Random-Logic Custom LSI Devices"; Proc. 13th Design Automation Conf., San Francisco, CA, June 1976, pp. 79-85.

51. SCHWEIKERT, D. G.; "Combined Interactive and Automatic Layout of IC's"; Proc. Int. Symp. on Circuits and Systems, Phoenix, AZ, Apr 77, pp. 432-5.

52. PREAS, B. T. and VANCLEEMPUT, W. M.; "Placement Algorithms for Arbitrarily Shaped Blocks" Proc. 16th Design Automation Conference, San Diego, CA, June 79, pp. 474-480.

53. LAUTHER, U.; "A Min-Cut Placement Algorithm for General Cell Assemblies Based on a Graph Representation", Proc. 16th Design Automation Conference, San Diego, CA, June 79, pp. 1-10.

54. AGRAWAL, P. "On the Probability of Success in a Routing Process"; Proc. IEEE, Vol. 64 No. 11, Nov 76, pp. 1624-5.

55. FOSTER, J. C. "Prerouting Analysis Programs" Proc. 12th Design Automation Conference, Boston, MA, June 1975, pp. 306-310.

56. CYO, Y. E. et al; "FLOSS: An approach to Automated Layout for High-Volume Designs"; Proc. 14th Design Automation Conference, New Orleans, LA, June 1977, pp. 138-141.

57. SEITZ, C. L.; "Self-timed VLSI Systems"; Proc. Very Large Scale Integration: Architecture, Design, Fabrication; Cal Tech, Pasadena, CA, Jan 79.

58. SUTHERLAND, I. E. and MEAD, C. A., "Microelectronics and Computer Science" Scientific American, Sept 77, pp. 210-228.

59. DIJKSTRA, E. W. "Programming Considered as a Human Activity" Proc. IFIP, 1965, Spartan Books, Wash., DC.

60. VANCLEEMPUT, N. M.; "Computer Hardware Description Languages and their Applications;" Proc. 16th Design Automation Conference, San Diego, CA, June 1979, pp. 554-560.

61. CLARE, C. R.; Designing Logic Systems Using State Machines, McGraw-Hill, 1973.

62. MEAD, C. A.; Private communication on VLSI design structure, 1979.

63. NEWTON, R. A., "The Simulation of LSI Circuits" Univ. of Calif., Berkeley, Memo No. UCB/ERL M78/52, July 1978.

64. DIANA v5 Principles of Operation; LISCO Palo Alto, CA, Report D04, Apr 79.

65. CHAWLA, B. R., et al; "An MOS Circuit Simulator for Integrated Circuit Design", Proc. Int. Symp. on Circuits and Systems, Newton, MA, Apr 75, pp. 228-230.

66. HILL, D. D.; "ADLIB-SABLE User's Guide" Computer Systems Lab Tech. Report, Stanford, 1979.

67. JOHNSON, W. A.; "Behavioral-Level Test Development"; Proc. 16th Design Automation Conference, San Diego, CA, June 1979, pp. 171-9.

68. LOSLEBEN, P. and THOMPSON, K. T. "Topological Analysis for VLSI Circuits"; Proc. 16th Design Automation Conference, San Diego, CA, June 1979, pp. 461-473.

69. O'NEIL, L. A. et al; "Designer's Workbench--Efficient and Economical Design Aids"; Proc. 16th Design Automation Conference, San Diego, CA, June 1979, pp. 185-199.

70. PISTILLI, P. O.; "Configuration Management (CMS)"; 1978 IEEE Design Automation Workshop on Engineering the Integrated Design Automation System; East Lansing, MI, Oct 1978.

71. PEGRAM, D. W. Jr.; "Configuration Management Disciplines for Engineering Design and Data Base Administration" IEEE/ACM Design Automation Data Base Workshop, Santa Barbara, CA, Feb 1980.

72. YATES, K. D.; "Design Process Analysis"; IEEE Design Automation as a Dependable Business Operation", East Lansing, MI, Oct 1977.

73. WIEST, J. D. and LEVY, F. K.; A Management Guide to PERT/CPM; Prentice-Hall, Englewood Cliffs, NJ; 1969.

74. LLEWELYN, A.; "So What is the Problem with CAD/CAM?"; CAD/CAM Digest, Nov/Dec 79, pp. 5-6.

5. GaAs Digital Integrated Circuits for Ultra High Speed LSI/VLSI

R. C. Eden and B. M. Welch

With 21 Figures

The principal requirements of a digital integrated circuit technology to make possible the development of ultra-high speed, very large scale integration (VLSI) circuits are: (1) very high density (low chip area per gate), (2) low gate power dissipation, (3) extremely low dynamic switching energy (speed-power product), (4) high speed (very low gate propagation delay) and (5) very high process yield (sufficient to achieve useable chip yields of such complex parts). These factors are not listed here according to priority; no priority is possible since all of these requirements must be met if such very high performance VLSI circuits are to be realized. The origins of most of these requirements are obvious. Clearly, large numbers of gates (10^4 to 10^5) cannot be placed on a reasonable sized ($\sim 1\ cm^2$) chip unless the gate areas are small ($<\sim 1000\ \mu m^2$/gate). The power per gate must be low ($<<1$ mW) if chip dissipations are to remain manageable. As illustrated in Table I, the requirement on dynamic switching energy for high speed VLSI is especially severe because this low power must be maintained at high clocking frequencies. Since the dynamic switching energy or speed-power product, $2P_D\tau_d$, is the minimum energy that a gate can dissipate during a cycle (two transitions), the power dissipation for a chip with N_g gates with an average gate clocking frequency of f_c will be

$$P_{chip} \geq 2\ N_g f_c\ (P_D \tau_d) \quad . \tag{1}$$

Table I shows the maximum speed-power products consistent with maintaining a two watt chip dissipation for chips with levels of complexity from $N_g = 1$ to 10^5 gates for average gate clocking frequencies from f_c = 100 kHz to 10 GHz. We can see in Table I, or from (1) that very high speed VLSI places extreme demands for low $P_D\tau_d$ products because the high speeds ($\tau_d << 1$ ns, $f_c \sim 1$ GHz) and the high gate counts ($N_g \sim 10^4$ to 10^5) combine in pushing up chip power. Even allowing for the fact that power dissipation for large chips could safely be much higher than the 2 watts assumed in Table I, dynamic switching energies of $P_D\tau_d < 0.1$ pJ appear essential for achieving practical very high speed VLSI.

The work on D-MESFET SDFL was supported in part by the Advanced Research Projects Agency of the Department of Defense and was monitored by the Air Force Office of Scientific Research under contract F49620-77-C-0087.

Table 1

MAXIMUM ALLOWABLE GATE DYNAMIC SWITCHING ENERGIES ($P_D\tau_D$) FOR VARIOUS AVERAGE GATE CLOCKING FREQUENCIES ASSUMING A 2 WATT CHIP DISSIPATION

		F_{CLOCK}					
GATES/CHIP		0.1 MHz	1.0 MHz	10 MHz	100 MHz	1 GHz	10 GHz
ULSI	10^5	100pJ	10pJ	1pJ	0.1pJ	0.01pJ	0.001pJ
VLSI	10^4	10^3pJ	100pJ	10pJ	1pJ	0.1pJ	0.01pJ
LSI	10^3	10^4pJ	10^3pJ	100pJ	10pJ	1pJ	0.1pJ
MSI	10^2	10^5pJ	10^4pJ	10^3pJ	100pJ	10pJ	1pJ
SSI	10	10^6pJ	10^5pJ	10^4pJ	10^3pJ	100pJ	10pJ
DEVICE	1	10^7pJ	10^6pJ	10^5pJ	10^4pJ	10^3pJ	100pJ

The progress made in silicon integrated circuits over the past two decades is extremely impressive, generating enormous confidence in the future potential of this technology. In some cases, this level of confidence has replaced careful examination of physical principles and limitations as a basis of projection of future performance. Historically, the two decades of monolithic integrated circuit technology development have been almost exclusively devoted to silicon. The achievements of improved performance levels in this silicon IC technology have been based principally on: (1) improvements in fine-line lithography, (2) remarkable cleverness in fabrication techniques, (3) selection of new types of active device approaches (NPN and PNP bipolar, PMOS, NMOS, CMOS, I^2L, Schottky diodes, MESFETs, etc.), (4) new logic circuit approaches and, (5) improvements in material quality (freedom from defects, etc.). In the areas of overall functional complexity per chip in medium speed logic, and in cost/gate on such chips, the improvements in this silicon IC technology have been almost astounding. In other areas, such as the maximum switching speeds in silicon ICs, the rate of progress has been quite limited. For example, 1 ns switching speed bipolar ECL has been commercially available for over a decade (the commercial introduction of MECL III was in 1968); today, only a few faster silicon ICs are commercially available (with propagation delays up to 50% lower than those of MECL III). The point here is that projections of progress need to carefully reference the means by which such progress can take place and the physical principles and limitations involved.

The key point to be conveyed in this paper is that the list of "tools" for improving IC performance cited in the previous paragraph is incomplete -- omitting a potentially extremely powerful method for improving IC performance. That method is to replace silicon with some other semiconductor having superior electronic properties. This approach, which may seem shocking to those who equate "integrated circuits" with "silicon integrated circuits" is not without precedent in IC history. Most early MOS work used PMOS because the fabrication technology for enhancement mode PMOS FETs and

ICs was relatively easy. However, in order to gain the advantages of the ~2.5 times higher mobility of electrons in silicon as compared to holes, the sizable effort to develop suitable NMOS IC processes were undertaken. The successful implementation of these N-channel procesess has led the higher performance NMOS to displace PMOS from much of the commercial market. An even more radical example, of course, is the development of CMOS ICs using heteroepitaxial silicon on sapphire (CMOS/SOS) in order to achieve lower parasitic capacitances for higher switching speeds and improved dynamic switching energies in logic circuits. Again, it appears that the successful development of this CMOS/SOS technology will also be rewarded with a share of the total integrated circuit market consistent with the applications areas it can serve and the costs involved. The question to ask is, then, if it was worth going from silicon PMOS to NMOS for a factor of 2.5 in mobility, and to CMOS/SOS for the benefits of an insulating substrate, is it worth going on to using GaAs as the semiconductor for high performance integrated circuits in order to attain the advantages of both a semi-insulating substrate and electron mobilities a factor of 5 or more higher than silicon? The answer, of course, involves a determination of the performance advantages to be realized in integrated circuits if GaAs, rather than silicon, were used, and a realistic assessment of the costs involved.

5.1 Performance Advantages Expected for GaAs ICs

For the purposes of this paper, we confine our focus to the application of GaAs for ultra high speed VLSI digital logic circuits. Let us first consider the performance advantages to be accrued in such high speed VLSI by going to GaAs, or in other words, how the use of GaAs rather than silicon can impact our chances of achieving those prerequisites for ultra high speed VLSI cited in the first paragraph. Of particular concern here are the issues of obtaining the subnanosecond switching speeds and extremely low ($P_D \tau_d$ < 100 fJ) dynamic switching energies that such high performance VLSI will require. We must start by specifying the types of active devices and circuits being considered. We will confine our attention here to FET logic, principally using GaAs metal-semiconductor (Schottky-gate) FETs (MESFETs) in comparison to silicon MESFETs or MOSFETs. Fig. 1 shows the structure for a planar, double-implanted GaAs MESFET fabricated by localized implantations into semi-insulating ($\sim 10^8$ Ωcm) GaAs [1]. The basic structure is very simple, comprising a thin (a ~ 1000 Å) n-type (typically $N_d \sim 2\times10^{17}$ cm^{-3}) active region joining two ohmic contacts (separated typically by a distance of L_{ch} = 3-5 µm) with a narrow (L_g ~ 1 µm) metal Schottky barrier gate separating the source and drain. The GaAs FET operation is similar to that of a normal JFET, with the conducting n-channel confined between the gate depletion region and the semi-insulating GaAs substrate. By varying the thickness and doping level of the active region, the pinchoff voltage, V_p, and saturated drain current (at V_{gs} = 0), I_{dss}, for the GaAs MESFET can be varied, even down to the point where the normal built-in voltage of the Schottky barrier gate can pinch off the FET (enhancement-mode MESFET operation).

The GaAs MESFET strucure of Fig. 1 is attractive not only for its exceptionally high performance, but also for its simplicity and ease of fabrication. The source-drain ohmic contact spacing in a depletion-mode MESFET is typically of the order of 4 µm, so that the alignment accuracy required for the placement of the L_g = 1 µm gate in this S-D gap is of the order of ± 1 µm. Note that present 2.5 to 3 µ Si NMOS IC processes require more stringent (typically ± 0.75 µm) layer to layer alignment accuracies than the L_g = 1 µm GaAs MESFET. Inasmuch as it is usually easier in photolithography to obtain a fine (~ 1 µm) linewidth than it is to be able to repeatably obtain precise layer to layer alignments, it appears that attaining L_g = 1 µm device performance levels in GaAs ICs should be a great deal easier than in Si NMOS. The fact that L_g = 1 µm MESFETs fabricated with projection optical lithography were demonstrated a decade ago [2], and GaAs microwave

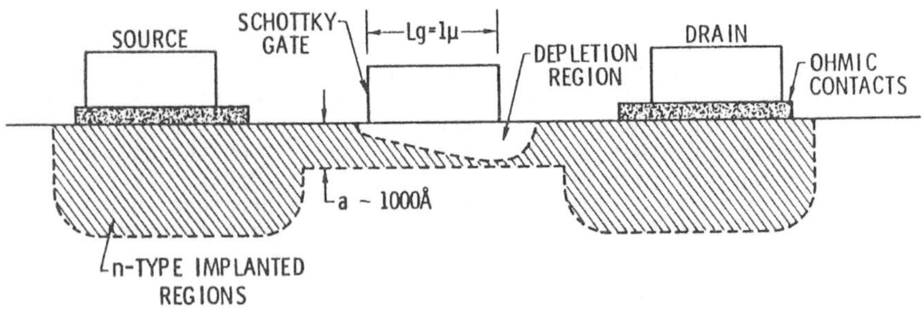

Fig. 1 Cross-section of a planar GaAs Schottky-Gate Field Effect Transistor (MESFET). The separate localized ion implantation for source and drain regions (as used in [1]) are not essential.

FETs with L_g = 1 µm (and even 0.5 µm) have been commercially available for many years supports this conclusion.

In a field effect transistor (FET), the gate voltage simultaneously controls the charge density in the conducting channel, Q, as well as the electric field, E_c, acting on the charge. The drain current of the FET, then, is given from the transit time, τ_t, and total controlled charge in the channel, Q_cL_gW, as

$$I_{ds} = \frac{Q_cL_gW}{\tau_t} = \frac{Q_cL_gW}{(L_g/v_d)} = Q_cW\,v_d \tag{2}$$

where Q_c is the charge per unit area, L_g the channel length in the direction of conduction, W the channel width, and v_d the average carrier velocity. If we assume the notion of carriers in the control region of the channel (under the gate) to be ohmic, then their velocity will be given by

$$v_d = \mu_n E_c \tag{3}$$

where μ_n is the channel mobility for carriers. The average control field E_c is given by

$$E_c = \frac{V_{gs} - V_p}{L_g} \tag{4}$$

and, assuming a constant or average thickness, a, of dielectric with permitivity ε separating the gate electrode from the channel, we have for the control charge density,

$$Q_c = \frac{\varepsilon}{2a}(V_{gs} - V_p) \tag{5}$$

(where $1/2(V_{gs} - V_p)$ is the average potential difference between the gate and the channel). Substituting (3) - (5) to (2), we have

$$I_{ds} = K(V_{gs} - V_p)^2 \tag{6}$$

where

$$K = \frac{\varepsilon\, \mu_n W}{2aL_g} \tag{7}$$

This is simply the Shockley model for the FET (which has, of course, somewhat limited applicability for short-gate length FETs). The maximum logic switching speed from an active device loaded (principally) by the inputs of N similar devices is limited by the ability of the active device to furnish the current gain required to drive the N inputs. Consequently, it is the current gain-bandwidth product, f_τ, which provides an upper limit to the switching speed (Appendix, Eq. A1). In the Shockley model, from (6) and (7), we have

$$f_\tau = \frac{g_m}{2\pi C_{gs}} = \frac{\mu_n}{2\pi L_g^{\;2}}(V_{gs} - V_p) \tag{8}$$

This shows that the f_τ, and hence, switching speeds, for FET logic circuits will be dependent on the gate voltage above threshold, $V_{gs} - V_p$ (which would usually be comparable to the logic voltage swing, V_m), the gate length, L_g, and the carrier mobility, μ_n, in the channel. Increasing the logic swing to improve switching speeds is, of course, at the expense of increased dynamic switching energy, since, for a voltage swing, V_m, on a capacitance, C_L,

$$P_D \tau_d > 1/2\; C_L\, V_m^{\;2} \quad . \tag{9}$$

On the other hand, reducing the gate length, L_g, and using a higher carrier mobility, μ_n, material allow achieving higher switching speeds without paying the price in dynamic switching energy.

While the simple Shockley FET model would project switching speeds of the order of 5 to 10 times higher for GaAs FETs than silicon FETs with similar dimensions, it is often argued that carrier velocity saturation effects will make them similar. In the case of velocity saturation, (3) is in gross error and, in fact, for very high fields sustained for long times we have $v_d = v_{sat}$. To get the FET transconductance we must differentiate (2):

$$g_m = \frac{d}{dV_{gs}}(Q_c W v_d) = W \quad Q_c \frac{dv_d}{dV_{gs}} + v_d \frac{dQ_c}{dV_{gs}} \tag{10}$$

In the fully velocity-saturated channel, $dv_d/dV_{gs} = 0$ (since $v_d = v_{sat}$ is constant), so that, using $C_{gs} = WL_g\,(dQ_c/dV_{gs})$, we have

$$f_\tau = \frac{g_m}{2\pi C_{gs}} = \frac{W\ v_s \frac{dQ_c}{dV_{gs}}}{2\pi\ WL_g \frac{dQ_c}{dV_{gs}}} = \frac{v_s}{2\pi L_g} \quad . \tag{11}$$

On this basis, it is argued that since the saturated electron velocity in bulk silicon is about 10^7 cm/sec and the saturated electron velocity in bulk GaAs is only slightly higher, there will be no difference in performance between short gate silicon and GaAs FETs operated far above threshold (E_c in (4) large; $E_c >> v_s/\mu_n$). The counter-argument to this is that for VLSI, the severe $P_D\tau_d$ requirements necessitates using low logic voltage swings ($V_m \sim 1$ V or less). Assuming logic swing from threshold ($V_{gs} = V_p$) to $V_{gs} = V_p + V_m$, the maximum value of the control electric field strength for a $L_g = 1$ µm short gate FET will be, from (4), $E_c = 1$ V/1 µm $= 10^4$ V/cm, which is insufficient to achieve carrier velocity saturation in silicon.

At this point we have two different simple FET models; the Shockley model which gives a transconductance

$$g_m = \frac{\varepsilon\mu_n W}{a\ L_g}(V_{gs} - V_p) \qquad \text{[Shockley]} \tag{12}$$

which predicts over 5 times higher g_m for GaAs than silicon for similar geometries and biases, and the saturated drift velocity model for which the transconductance, from (10) is

$$g_m = \frac{v_s}{L_g} C_{gs} = \frac{\varepsilon\ v_s W}{a} \qquad \text{[Velocity Saturated]} \tag{13}$$

implying similar transconductance for GaAs and silicon devices if the published bulk saturation velocities apply ($v_s \sim 10^7$ cm/sec). While one might argue at considerable length the relative limitations or merits of either of these two viewpoints, the more fruitful approach to resolving this issue is to simply compare the experimental data for short ($L_g = 1$ µm) gate silicon and GaAs FETs. Fig. 2 shows experimental curves of saturated drain current vs gate voltage above pinchoff, I_{ds} vs ($V_{gs} - V_p$) for $L_g = 1$ µm, W = 10 µm, depletion-mode MESFETs of GaAs and silicon. The GaAs MESFET data is typical for planar, double-implanted devices of the type shown in Fig. 1 with low pinchoff voltages ($V_p \sim -0.5$ to -1.5 volts) as used in low power GaAs ICs [1]. The 1 µm gate N-channel silicon MESFET data is for the microwave MESFETs described in [2] (representative characteristic curves from [3]). Silicon N-channel MOSFETs with $L_g = 1$ µm [4] show, as would be expected, very similar I_{ds} vs $V_{gs} - V_p$ characteristics to the silicon MESFET curve in Fig. 2. The drain currents for both the silicon and GaAs MESFETs exhibit approximate square-law behavior as predicted from [6], but with the K value (at 25% maximum I_{ds}) for GaAs (0.7 mA/V^2) much higher than that for silicon (0.09 mA/V^2), as predicted by the higher GaAs mobility [7]. Comparing the GaAs and silicon MESFETs at equal gate biases, $V_{gs} - V_p = 1$ V, the GaAs FET has about 6 times the drain current (0.65 mA vs 0.11 mA) and 6 times the transconductance (1.07 mS vs 0.17 mS) of the silicon MESFET. At higher drain current levels, the I_{ds} vs $V_{gs} - V_p$ curves linearize noticeably, presumably a reflection of the influence of velocity saturation effects, but still the GaAs FET is much better than the silicon MESFET (e.g., at $V_{gs} - V_p = 1.5$ V, the GaAs I_{ds} is 6.5 times that of silicon and g_m is 7.25 times higher). The superior dc performance of the GaAs MESFET is matched by its superior RF performance. While silicon MESFETs [2], with $L_g = 1$ µm achieve maximum frequencies of oscillation (unity power gain frequencies) in the $f_{max} \approx 10$ GHz range, even the first 1 µm gate length GaAs FETs had $f_{max} \approx 30$ GHz values and more recent $L_g = 0.9$ µm GaAs FETs give extrapolated $f_{max} \approx 75$ GHz. It was, of course, the greatly superior performance of the GaAs FETs, in terms of noise figure and power gain, that led to a rapid decline in interest in the silicon MESFET for microwave use.

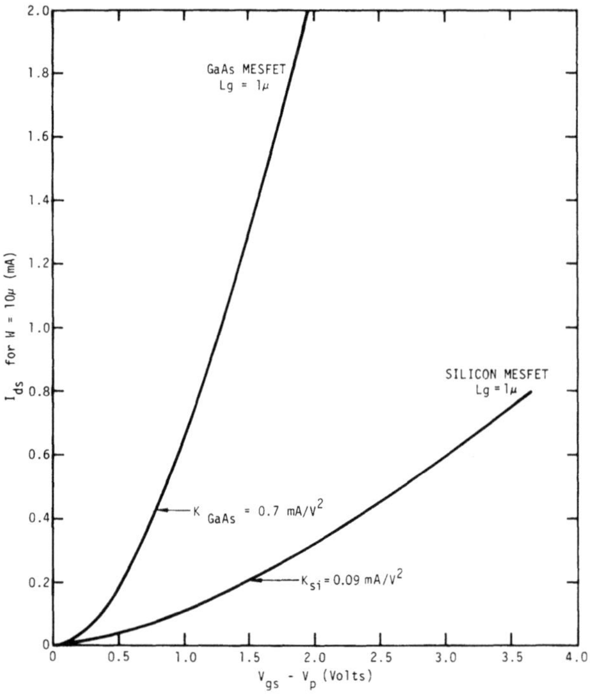

Fig. 2 Comparison of the saturated drain current vs gate voltage (above pinchoff) characteristics for silicon [3] and GaAs [1] depletion-mode MESFETs with the same $L_g = 1$ µm geometries. While both devices give approximately square-law, $I_{ds} \sim K(V_{gs} - V_p)^2$ curves near threshold, the higher GaAs mobility gives much higher K-values for the GaAs MESFET than the silicon MESFET.

The superior performance of these low power (low pinchoff) GaAs MESFETs as compared to silicon MESFETs or MOSFETs is a consequence of the superior electron dynamics in GaAs. This is reflected, for example, in the steady state velocity-field characteristics for electrons in GaAs and silicon FET channels shown in Fig. 3. Inasmuch as the silicon NMOS, as opposed to MESFET, logic technology would appear the closer competitor to GaAs IC's, the silicon data shown is for n-channel MOS, <100> surfaces in heavy inversion taken from the pulsed velocity-field measurements of Fang and Fowler [5].

The silicon MESFETs of [33] have similar channel mobilities to the NMOS data in Fig. 3, but the bulk saturation velocity ($v_d \sim 7.7 \times 10^6$ cm/s @ 3×10^4 V/cm) is higher than that in the n-channel ($v_s \sim 6.5 \times 10^6$ cm/s). This value is in good agreement with the short-channel NMOS data of [4], for which $g_m = 3.9$ mS (from $V_{gs} = +3$ to $+4$ V) was obtained for $L_g = 1$ µm, W = 100 µm MOSFETs with an a ≅ 500 Å (SiO_2) gate oxide thickness, giving from (13), $v_s = 5.8 \times 10^6$ cm/s. The GaAs steady state velocity-field data in Fig. 3 were measured on high resistivity bulk GaAs by Ruch and Kino [6]. The mobility, $\mu_n \sim 8000$ cm^2/V • s, for that material is considerably higher than expected for typical MESFET channel dopings ($\mu_n \sim 4000$ cm^2/V • s at $N_D = 2 \times 10^{17}$ cm^{-3}), so an additional GaAs curve extrapolated from this low-field mobility value is also shown in Fig. 3. (There is no published experimental v_d vs E data for $N_d \sim 10^{17}$ cm^{-3} doped GaAs.) The striking feature in Fig. 3 is that at the $E_c \sim 10^4$ V/cm fields in low power short channel FETs (for $V_{gs} - V_p \sim 1$ V, $L_g = 1$ µm), the steady state electron velocities in GaAs are much higher than those in silicon FET channels.

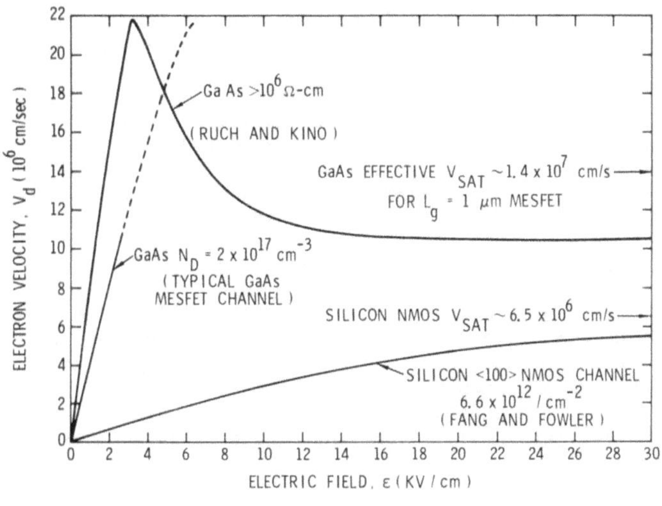

Fig. 3 Comparison of the velocity-field characteristics for electrons in silicon MOS channels and GaAs. The high resistivity GaAs data (dashed curve) exhibits larger mobilities than typical GaAs MESFET channels (the curve for which is simply extrapolated from the low-field mobility).

In fact, the electron dynamics in GaAs FETs is even better than suggested by Fig. 3. Whereas the steady-state velocity-field curve for GaAs shows a saturation velocity of $v_s \sim 1.05 \times 10^7$ cm/s, analysis (based on [10]) of the actual GaAs MESFET data of Fig. 2 gives an effective saturation velocity [applicable to (13)] of $v_s = 1.4 \times 10^7$ cm/s. (Ref. [17] quotes a value of 1.7×10^7 cm/s; other workers give values as high as 2×10^7 cm/s.) There are at least two fairly simple reasons why experimental short-channel GaAs MESFETs perform better than expected from the steady-state velocity-field characteristic, both of which derive from the fact that the low power short gate FET does not represent a "steady state" case. The steady state velocity-field curves assume long times, distances and total voltage drops very large in comparison to internal energy exchange mechanisms in the semiconductor. In GaAs, the dropoff in v_d for $E > 3.1 \times 10^3$ V/cm is a reflection of the transfer of electrons from the lowest, $\Gamma(000)$, conduction band minimum to higher minima, 0.3 eV above the Γ minimum [7]. In a low power MESFET where the voltage drop across the undepleted channel, $V_{gs} - V_p$, is in many cases not large in comparison to 0.3 eV, the static velocity-field curve of Fig. 3 is meaningless and the low-field mobility would be expected to determine the conduction characteristics. Hence in low voltage, short-channel FETs, electron velocities much higher than those predicted from Fig. 3 could be obtained (particularly with L_g = 0.5 μm devices). Further, the transit times of electrons through the high field regions in short gate FETs is not sufficiently long in comparison to the times involved in the energy equilibration processes in the semiconductor to guarantee that the steady state velocity-field curves will apply even when $V_{gs} - V_p$ and V_{ds} are >> 0.3 V. These "velocity overshoot" effects can give transient electron velocities many times those predicted from the steady state velocity field characteristics [8], [9] with consequent influence on FET characteristics.

Given that short-gate GaAs FETs do give much higher transconductance and K-values (in (3) and (4)) than silicon FETs of the same dimensions, what influence will this have on the speed and speed-power performance of logic circuits? More specifically, to what degree would using GaAs rather than silicon enhance our ability to meet the requirements for ultra-high speed VLSI, particularly the requirements for $P_D\tau_d$ < 100 fJ dynamic switching energies with very low, τ_d << 1 nS, propagation delays? An analysis of the dependence of the switching speed and speed-power product for FET logic circuits as a function of logic voltage swing, V_m, the load capacitance, C_L, and the K-value of the switching FET is presented in the Appendix. This analysis assumes a more or less square-law I_{ds} vs $V_{gs} - V_p$ relationship (6), with the facility to handle experimental I_{ds} vs $V_{gs} - V_p$ curves which may depart from square-law (by using (A4)-(A7)). The fastest, lowest $P_D\tau_d$ switching case is treated: that for hard-limited logic swing from right at the FET threshold ($V_{gs} = V_p$) to $V_{gs} = V_p + V_m$, with the active load (current source) current always adjusted to give half of the maximum FET drain current (for equal positive and negative slew rates). The principle results for this analysis are that the propagation delay will be given by (A9):

$$\tau_d = \frac{4C_L}{3KV_m} \tag{14}$$

and the dynamic switching energy ($P_D\tau_d$ > 1/2 C_L $V_m{}^2$) expressed in terms of its explicit τ_d dependence (assuming $V_{dd} \cong 1.7 V_m$ and choosing V_m to give the desired τ_d from (14)),is given by (A11):

$$P_D\tau_d = \frac{2C_L{}^3}{K^2\tau_d{}^2} \tag{15}$$

This analytical relationship between dynamic switching energy, $P_D\tau_d$, and the propagation delay, τ_d, with load capacitance, C_L, and FET K-value as parameters, is derived assuming a square-law, $I_{ds} = K(V_{gs} - V_p)^2$, drain current characteristic. In real FETs (Fig. 2), only limited ranges of $V_{gs} - V_p$ are practical, so that (from (14)), only limited τ_d ranges are available for a given FET and load, and hence only a finite range over which we can vary the logic swing, V_m, to execute the speed-power tradeoff given by (15) is achievable. We can learn a considerable amount from (15), however. It shows, for example, the "price" in terms of switching energy of increasing speed (reducing τ_d); everything else being constant, going to higher speeds by increasing V_m costs speed-power product proportional to the square of speed, $P_D\tau_d \propto (1/\tau_d)^2$. It also shows that going to smaller device geometries, thereby increasing FET K-values (7) and reducing C_L, can be very advantageous in reducing $P_D\tau_d$ products for any given τ_d. Further, improving the K-values of the FETs by going to a higher performance semiconductor such as GaAs can greatly reduce the dynamic switching energy for a given propagation delay and geometry, since $P_D\tau_d$ varies as the square of (1/K). Hence, in going from a silicon L_g = 1 µm FET to a GaAs L_g = 1 µm MESFET, from the data of Fig. 2, (15) would predict that one might achieve up to a factor of 50 to 100 reduction in speed-power product for the equal switching speed operation (because that operation would be at much lower voltage swings with GaAs).

As an example of the improvement in performance expected with GaAs FET logic compared to silicon FET logic, Fig. 4 compares the calculated dynamic switching energy ($P_D\tau_d$) vs propagation delay (τ_d) relationships for the W = 10 µm, L_g = 1 µm MESFETs of silicon and GaAs of Fig. 2, assuming a load capacitance of C_L = 30 ff (reasonable for circuits using an insulating substrate such as GaAs or silicon-on-sapphire). Here, the actual I_{ds} vs $V_{gs} - V_p$ relationships of Fig. 2 were used. The logic swing, $V_m = V_{gs} - V_p$ (scale at the right in Fig. 4) is taken as the independent parameter, obtaining I_{dm} from Fig. 2 and then calculating τ_d from (A4) and $P_D\tau_d$ from (A7) for each V_m. We see in Fig. 4 that for low values of logic voltage swing, the $P_D\tau_d \propto (1/\tau_d)^2$ behavior of (15) is observed; for logic swings above V_m = 1V or so, where velocity saturation effects in the FETs become noticeable, the relation between speed power product and propagation delay becomes even stronger [$P_D\tau_d \propto (1/\tau_d)^3$]. (This, of course, reflects the fact that the transconductance does not increase as rapidly with V_m as carrier velocity saturation is approached.)

Figure 4 illustrates the dramatic improvements in logic switching speeds and in speed-power products in FET logic which can be obtained by going to a higher performance semiconductor. For the same logic voltage swing or dynamic switching energy values, the L_g = 1 µm GaAs MESFET would give about 6 times higher switching speeds than the L_g = 1 µm silicon MESFET (or L_g = 1 µm N-channel silicon MOSFETs, since they have I_{ds} vs $V_{gs} - V_p$ characteristics similar to that of the silicon MESFET). For a logic swing of V_m = 3.5 V (from threshold to the onset of Schottky gate conduction), the silicon MESFET should achieve a switching speed with the C_L = 30ff load of τ_d = 183 pS ($P_D\tau_d$ = 0.45 pJ). With the same gate length MESFET in GaAs and the same load, this same switching speed should be achievable with only a V_m = 300 mV logic swing, for a calculated reduction in dynamic switching energy of over two orders of magnitude (~3fJ for GaAs, x 150 less). Actually, for the very low logic swings, the assumption of a supply voltage of V_{dd} = 1.7 V_m would probably be impractically low so the $P_D\tau_d$ products would be somewhat higher than given by (15) or in Fig. 4. Comparison of N-channel GaAs to p-channel silicon (as used in CMOS or CMOS/SOS) would be even more dramatic because of the lower hole mobility and consequent lower PMOS FET K-values.

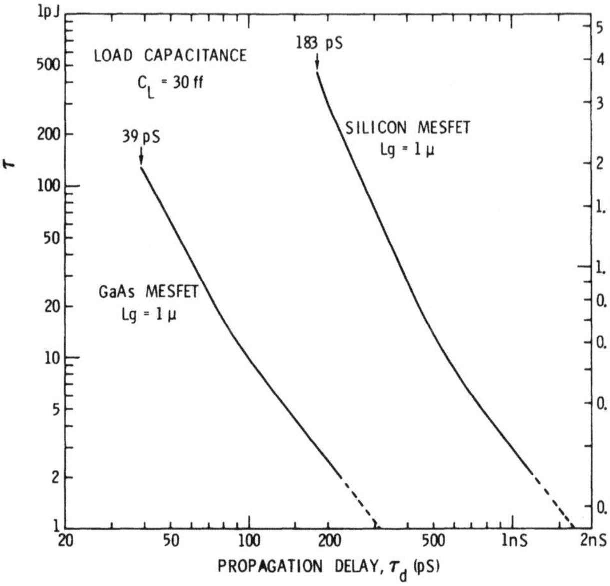

Fig. 4 Optimized switching performances for the silicon and GaAs MESFETs of Fig. 2 with a C_L = 30 ff load capacitance for varying logic voltage swing, V_m. Note the strong dependence of the dynamic switching energy, $P_D\tau_d$, on the desired propagation delay, τ_d (as selected by varying V_m), and the marked superiority of GaAs in achieving lower energies at any given switching speed and in attaining higher speeds.

The foregoing analysis, as illustrated in Fig. 4, shows that the most promising approach to meeting the very demanding speed and speed-power product performance requirements for ultra-high speed VLSI is to combine small (L_g = 1 µm or less) FET geometries with a very high performance semiconductor, such as GaAs, preferably on an insulating substrate for low parasitic capacitances (readily available with GaAs). From a performance standpoint, GaAs appears to be an almost ideal choice for meeting these requirements for ultra-high speed VLSI. The next question to be asked is what is the state of this GaAs digital integrated circuit technology? Are there practical circuit approaches to GaAs digital ICs capable of meeting these high density, ultra high speed, low power and ultra-low speed-power product requirements? What is the state of fabrication technology in GaAs? Are there practical GaAs IC fabrication approaches that appear capable of achieving the kinds of yields and gate densities required for VLSI? These questions will be dealt with in the following sections.

5.2 Circuit Approaches for GaAs Digital Logic ICs

Inasmuch as the superiority of GaAs over silicon for high-speed low-power logic circuits is principally a consequence of the very high electron mobility in GaAs, it is not surprising that most GaAs IC circuit approaches utilize majority carrier devices with n-type GaAs active regions. The principal choice has been the field effect transistor (FET), but with the

Schottky barrier diode valuable either as a nonlinear logic element or for voltage shifting. The very low hole mobility in GaAs (μ_p ~ 250 cm^2/Vs typically) discourages the use of p-channel or bipolar devices in GaAs, so that complementary logic types analogous to CMOS in silicon have not been extensively considered. While the bulk of the effort in high speed GaAs digital ICs has been based on n-channel FETs, the negative differential mobility of GaAs (Fig. 3) offers another alternative not available in silicon: the transferred electron logic device (TELD). While these TELD devices do not appear at present compatible with VLSI requirements, we will briefly discuss them at the end of this section.

The largest amount of work on GaAs digital logic ICs has been done using metal-semiconductor (Schottky barrier) FETs (MESFETs) as the active devices, although circuits have also been made using junction FETs (JFETs), and the use of heterojunction FETs (HJFETs) has been proposed. Historically, this is hardly surprising, considering the great development efforts that have gone into GaAs MESFETs as microwave transistors either for low-noise or high-power applications [11].

The direction of GaAs IC efforts toward the use of MESFET (or, to a lesser extent, JFET or HJFET) active devices has also been influenced by the difficulty of making useable insulated-gate FETs (MISFET or MOSFET) in GaAs. In spite of considerable efforts to obtain a satisfactory oxide or insulator on GaAs, it has proven extremely difficult to achieve stable insulator systems of MIS device quality on GaAs. Some recent promising results demonstrating microwave MOSFETs have been published [12] and the attempts at the utilization of GaAs MOSFETs for digital ICs have been published [13], but application of this technology for practical GaAs ICs would still appear some time away. While the lack of an insulated gate FET (MOSFET) technology does limit the logic swings available in enhancement-mode GaAs FET logic to something less than the built-in voltage of the MESFET, JFET or HJFET gate junction, and correspondingly somewhat limits the device speeds (as per (14) or Fig. 4), the low logic swings of the GaAs enhancement-mode FET circuits achieve good speeds with extremely low $P_D \tau_D$ products. Larger logic swings, and hence the highest speed ranges, can be achieved with the depletion-mode GaAs MESFET circuit approaches.

The choice of a particular type of FET device is associated with a choice of circuit and fabrication approach. The depletion mode metal-semiconductor (Schottky barrier) FET (D-MESFET) is the most widely used device, and also the one that has given the highest performance to date. Circuits employing depletion mode MESFETs pose the least fabrication problems (see Section III) because Schottky barriers on GaAs are easier to fabricate than p-n junctions, and the larger (typically > 1 V) logic swings associated with D-MESFET circuits avoid excessively stringent requirements for FET pinchoff voltage uniformity. Because any regions of the source-drain channel not under the gate are automatically strongly conducting in D-MESFETs (Fig. 1), precise gate alignments are not required, nor are special gate recess etch processes or other means to avoid parasitic source and drain resistances necessary. The MESFET fabrication simplicity makes it considerably easier to achieve high yields than with more complex device structures. On the other hand, however, logic gates employing depletion mode active devices necessarily require some form of voltage level shifting between FET drains and gates to meet turnoff requirements, and usually require two power supplies, imposing some penalty in terms of wafer area utilization. An exception to the two power supply requirements for

D-MESFET circuits is the enhancement-depletion logic approach [35], which uses $-0.4\ V < V_p < 0.1\ V$ MESFETs with diode level shifting in single power supply logic circuits. (They utilize device structures similar to E-MESFETs, however.)

Enhancement-mode MESFETs (E-MESFETs) offer circuit simplicity because the logic gates require only one power supply, but the permissible voltage swing is rather low because Schottky barrier gates on GaAs cannot be forward biased above 0.6 to 0.8 V without drawing excessive currents. A $V_m \sim 0.5$ V swing is, after all, a quite desirable goal for the operating range of ultra low power circuits, but very tight control is required in order to fabricate uniform, very thin active layers, so that they are totally depleted at zero gate bias voltage and yet give good device transconductance when the device is turned on. For reasonable noise margins and good dynamic performance, standard deviations of FET pinchoff voltage of the order of 25 mV would be required - a very difficult goal for GaAs FETs.

This situation may be improved somewhat by the use of a p-n junction gate FET (JFET), since the larger built-in voltage of the p-n junction allows the GaAs JFET to be biased up to $V_{gs} \sim 1$ V without excessive conduction [15], [16]. As discussed in Section 3 , GaAs JFET IC fabrication represents a more difficult processing problem and is at a somewhat earlier stage of development than the MESFET work [15]. The allowable positive gate voltage can be made even larger than that of a regular JFET by making the gate from a p-type semiconductor with a band gap larger than GaAs, forming a heterojunction gate FET (HJFET). An HJFET using a p-type $Ga_{0.5}Al_{0.5}As$ alloy can be biased up to $V_{gs} = 1.4$ V without drawing significant current [17]. However, the potential for high process yield would appear limited due to the need for an epitaxial growth and subsequent etch fabrication steps in this approach [57]. Finally, implementation of a MOSFET or MISFET technology in GaAs would eliminate the logic swing limitation completely, but attaining such devices has proven difficult. Some simple ring oscillators have been fabricated with directly coupled FET logic implemented with buried channel GaAs MOSFETs with resistor loads [13]. However, at this point, stable oxides have not been achieved in such circuits so that gate thresholds shift according to the prior input signal history. This allows the demonstration of ring oscillators and other simple circuits in which the input waveform has a precisely symmetrical (50% duty cycle) nature, but would not support their use in general digital circuits. Efforts to improve this oxide technology are being pursued [55,56].

5.2.1 Enhancement Mode FET Logic Approaches

The choice between E-MESFET, JFET, or HJFET approaches implies a choice between substantially different fabrication processes. The main differences have been outlined above, and they will be discussed with greater detail in Section 3 . From the standpoint of circuit implementation, the enhancement-mode approaches are essentially identical. (An exception is for MOSFETs and HJFETs where the larger allowable logic voltage swings will make practical transfer-gate and other circuit approches used commonly in silicon MOS.) The simplest circuit approach, direct-coupled FET logic, is illustrated for a 3-input (positive) NOR gate in Fig. 5a. This circuit configuration is called directly coupled FET logic. The normally-off FETs start conducting when their gate voltage becomes positive. A logic "0" corresponds to a voltage near zero. A logic "1" corresponds to a positive voltage capable of fully turning on the normally off FETs, a value usually limited by the onset of gate conduction in the FET; typically on the order of 0.6 V to 1.4 V depending on what technology is used (MESFET, JFET, or HJFET). It has been proposed to place input FETs in series, generating the NAND function [15]. However, the implementation of such design would appear impractical because the on resistance of the conducting FETs would cause larger threshold shifts than could be tolerated with the very low logic swings of E-MESFET or E-JFET logic. (It would probably work with HJFETs, however, and should represent no problem with MOSFETs.)

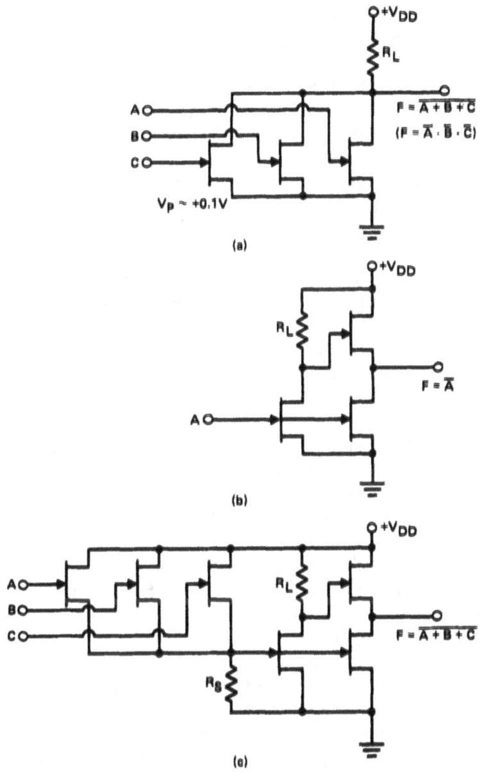

Fig. 5 Enhancement-mode JFET or MESFET circuits. a) Simple direct-coupled FET logic (DCFL) NOR gate with resistor load. b) Psuedo-complementary buffered inverter gate. c) Combination of source-follower logic with the circuit of b) to give a buffered NOR gate. This type of approach has been extended to two-level gates as well [36].

A significant improvement to the directly coupled logic gate shown in Fig. 5a would be to substitute for load resistor R_L, an active load current source made with a normally on (depletion mode) FET, with its gate tied to the source. Such a nonlinear load would sharpen the transfer characteristic and significantly improve the speed and speed-power products of the circuits (by perhaps a factor of 2). The fabrication of the depletion-mode active load requires a carrier concentration profile different from that of the enhancement mode devices. Although all the demonstration circuits using enhancement mode devices have been fabricated using only one type of active layer, a multiple localized implantation fabrication technique used for the Schottky-diode FET logic (SDFL) discussed later in this section, which relies on implanting different doping profiles for different devices, could probably be applied to such enhancement-mode circuits.

From a static point of view, the fanout capability of the directly coupled FET logic is excellent since it is determined by the very low gate leakage currents. However, from a dynamic point of view, the switching speeds are reduced by the gate capacitance loadings by a factor of approximately 1/N where N is the number of loading gates, as in silicon MOS. In general, the current through the resistor, R_L, or active load is kept fairly low in DCFL in order to reduce static power and improve noise margin by reducing the "on"- voltage drop of the FET (output "low" voltage). Consequently the output risetime under heavy fanout loading conditions is very poor. This can be greatly improved with the psuedo-complementary output buffer configuration of Fig. 5b, at very litte increase in static power dissipation, but this circuit performs only logic inversion [15]. By combining this inverting buffer with a source-follower positive-OR input structure as shown in Fig. 5c, a general multiple-input NOR gate can be achieved which has excellent fan-in and fanout drive capabilities at very modest static power levels [36]. Unfortunately, this source-OR/psuedo-complementary inverter gate configuration is also quite complex, requiring 7 FETs and 2 resistors for a 4-input NOR gate, which can be expected to consume considerable chip area and have significant self capacitance.

Demonstration circuits have been fabricated both for the E-MESFET and E-JFET approach. With the JFET circuit of Fig. 5b, measurements on 9 stage ring oscillators (see Section IV) with L_g = 1 µm show propagation delays per gate of τ_d = 150 ps with power dissipation of P_D = 400 µW, resulting in power delay products of $P_D \tau_d$ = 60 fJ [42].

E-MESFET demonstration circuits have, so far, shown the lowest power dissipation for the enhancement-mode FET technology. For a 13-stage ring oscillator, the power dissipation was ~ 0.1 mW per gate. Propagation delays are ~ 300 ps, larger than the values achieved with D-MESFETs, but yet yielding a lower power speed product (~ 30 fJ) [18]. With larger propagation delay (1.1 ns), a speed power product of only 1.6 fJ has been demonstrated on a 15-stage ring oscillator [19]. Logic circuit development has been carried further with this E-MESFET technology. A complementary clocked binary divider has been implemented with 8 NOR gates. As described in Section V, this device shows speed and power dissipations comparable to those of the 13-stage ring oscillator [20].

5.2.2 Single Supply Enhancement-Depletion Mode MESFET Logic

Because of the nonlinear, approximately square-law nature of the FET I_{ds} vs V_{gs} relationship (6), it is not always necessary to completely turn off the FET (i.e., make V_{gs} more negative than V_p) in order to obtain switching behavior.

Drain dotting of many FETs as in Fig. 5a necessitates turning all of the FETs nearly off so that the sum of all of their drain currents is substantially less than the load current (through R_L in Fig. 5a), I_L. However, if only a single FET switches the load, it is only necessary to reduce its drain current in the off state to a value significantly smaller than I_L, while its "on" current is well above I_L. This can be achieved in depletion-mode MESFETs with reasonably small pinchoff voltages ($V_P \approx -0.4$ V) with zero or slightly positive gate-voltages, so that only a single power supply is required. For example, using (6) with $V_P = -0.4$ and V_{gs} (on) = +0.7 V, V_{gs} (off) = +0.1 V, we have I_{ds} (on) = 4.84 x I_{ds} (off), an ample margin for switching.

A number of circuit approaches for single supply E-D MESFET logic have been proposed and analyzed [35]. Fig. 6a shows the circuit diagram for an elemental 3-input NOR gate in the most promising of these published approaches. This uses source follower logic to obtain the positive OR function, with single diode level shifting and resistor pulldown, R_S, to drive the output inverter FET. The analysis in [35] indicates proper gate operation for MESFET pinchoff voltages in the $-0.4 < V_P < +0.1$ V range; several times the allowable range width for E-MESFET logic and much more reasonable in terms of practical fabrication control. The supply voltage ($V_{dd} \sim 3$ V) and logic voltage swing ($V_{out} \sim 0.2$ V to 2.4 V) values used are even larger than those used in the SDFL D-MESFET approach, so that very low $P_D \tau_d$ products would not be expected. The gate output of Fig. 6a has the same drive problems as that of Fig. 5a, but this should be improved for heavily loaded gates with the buffer structure of Fig. 6b. This structure is, of course, very similar (except for the two voltage shifting diodes) to the enhancement circuit of Fig. 5c.

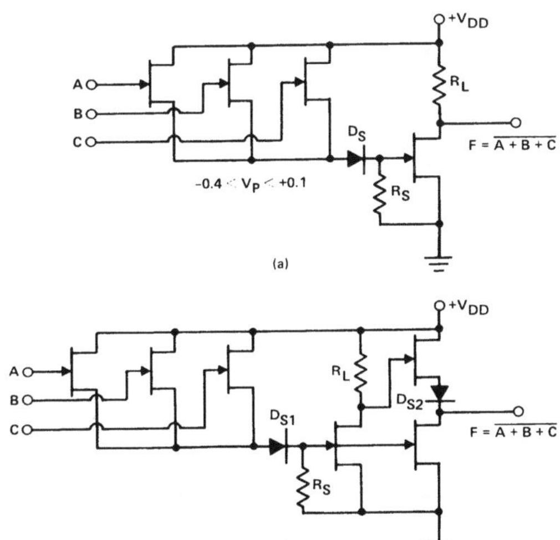

Fig. 6 Single-supply enhancement-depletion MESFET NOR gate circuits. a) Quasi-Normally-Off 3-input NOR gate [35]. b) 3-input NOR gate with Psuedo-complementary buffer; note similarity to Fig. 5c.

5.2.3 D-MESFET Approach/Buffered FET Logic

The first GaAs MESFET IC work reported utilized a D-MESFET circuit design best termed buffered FET logic [21, 22]. The circuits for an inverter and several basic logic functions (NAND and NOR) are shown in Fig. 7 [22]. This circuit typically uses -3 V < V_p < -1 V depletion mode MESFETs and hence requires two power supplies. Since it requires a negative gate voltage to turn off an n-channel D-MESFET, while its drain voltage is positive, level shifting must be introduced at some point so that the output logic levels match the input levels. In the buffered FET logic approach, the choice was made to operate with negative logic swings by level-shifting the positive drain voltages at the gate output. This is accomplished by the three-level-shifting diodes in the source-follower output stage of the gate (Fig. 7a). The diodes are always forward biased, each one giving a voltage drop of ~ 0.8 V. In the example shown in Fig. 7a, the three diodes in series accomplished the required ~ 2.5 V level shift.

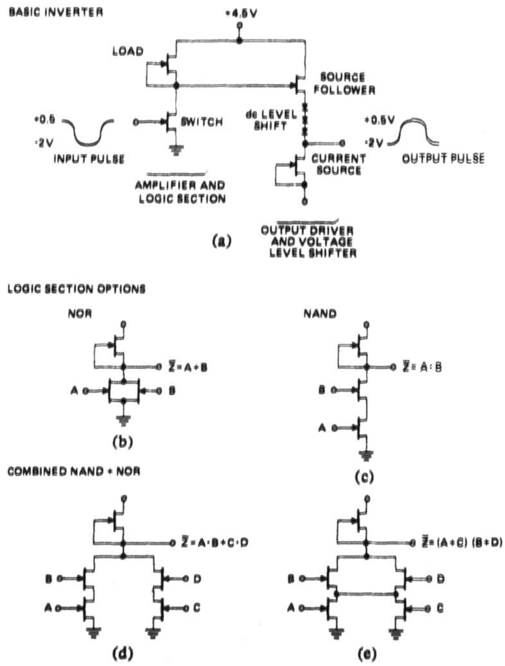

Fig. 7 Basic circuit configurations for buffered FET logic. (a) Basic inverter circuit. (b)-(e) Options for the input section for NOR, NAND, and combined NAND-NOR functions (reprint from [22]).

Placing the level shifting diodes at the output has a drawback. Since the current through the output driver is relatively high, the power dissipation caused by the level shifting diodes is also high. Furthermore, the level shifting diodes must be rather large in order to be able to handle the large current (particularly in a single active layer process), resulting in the consumption of a significant wafer area. On the other hand, an attraction of this circuit approach is that the circuit shown in Fig. 6 can be easily implemented with a fabrication process paralleling that of a conventional microwave GaAs MESFET (see Section III). Using large diodes that operate only in forward bias does not require diode doping profiles different from those of the MESFETs, so that originally it was possible to fabricate the early circuits on wafers which had a uniform epitaxially grown layer on the semi-insulating substrate [21] (later changed to a uniformly implanted layer [22]), on which the device structures were isolated from one another by mesa etching.

In the buffered FET logic approach described here, the nonlinearity of the FET itself is used to implement logic functions. As shown in Figs. 7a and 7b, the positive OR function is performed by FETs in parallel, while the AND function is performed by FETs in series (or equivalently by multiple-gate FETs) (Figs. 7a and 7c). Higher level logic functions (see Figs. 7d and 7e) are also possible by combining series and parallel arrangements of FETs. There is a practical limitation to the number of FETs that can be placed in series to accomplish a NAND function, since more than two or three FETs in series would give excessive voltage drop due to "on" resistance. (This problem is less severe than in the enhancement FET circuits, however, because the voltage swings can be much larger.) Also, since the full gate-source capacitance of higher gates appears as a Miller capacitance in the switching of gates near ground, the dynamic performance practically limits NAND fan-in to two. Similarly, the sizeable parasitics also limit practical drain dotting fan-ins to 2 or 3 (two in published circuits).

A number of types of demonstration circuits have been made using this buffered FET logic circuit approach, starting from simple ring oscillators and proceeding into more complex sequential logic circuits, including complementary-clocked master-slave flip flop stages used in ÷2 and ÷8 binary ripple counter configurations. Propagation delays of $\tau_d \sim 100$ ps were obtained in ring oscillator measurements on two-level NAND/NOR logic gates fabricated with 1 µm photolithography, performance which translates to 4.5 GHz maximum clocking speeds for frequency dividers (Section 5) [21], [22]. Much simpler chains of $L_g = 0.5$ µm inverters fabricated using E-beam lithography give, as expected, faster ring oscillator speeds, as low as $\tau_d = 34$ ps [23]. This is much faster than silicon ECL, although the gate dissipation ($P_D \sim 41$ mW/gate) is similar, and certainly far too high for LSI or VLSI applications. A radically different circuit approach to depletion-mode GaAs MESFET logic called Schottky diode FET logic (SDFL) has been developed to greatly reduce these power dissipation levels while retaining the $\tau_d \sim 100$ ps logic speeds, as discussed next.

5.2.4 D-MESFET Schottky Diode-FET Logic (SDFL)

The SDFL approach retains the high speed of the GaAs logic demonstrated with the D-MESFET buffered FET logic approach, but leads to reduction of the power dissipation by almost two orders of magnitude [1], [24], [37]. As the name Schottky diode-FET logic implies, SDFL utilizes clusters of small, high performance Schottky diodes to perform the logical positive-OR function on groups of inputs which may then be further processed with the normal FET logic functions (series-NANDing, drain dotting, etc.). Fig. 8 shows SDFL gate circuit diagrams for single, two- and three-level logic gate configurations [37]. Note that the SDFL gate structure allows virtually unlimited fan-in at the first (positive-OR) logic level (SDFL circuits

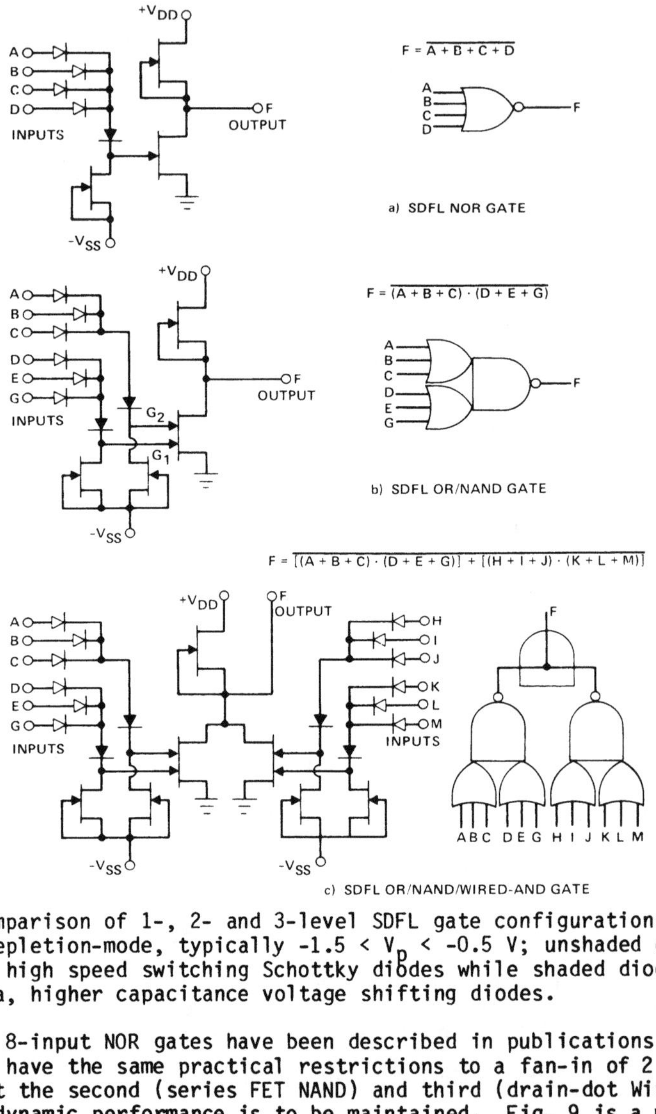

Fig. 8 Comparison of 1-, 2- and 3-level SDFL gate configurations. All FETs are depletion-mode, typically $-1.5 < V_p < -0.5$ V; unshaded diodes are very small high speed switching Schottky diodes while shaded diodes are larger area, higher capacitance voltage shifting diodes.

with up to 8-input NOR gates have been described in publications [38], [39]), but have the same practical restrictions to a fan-in of 2 (or possibly 3) at the second (series FET NAND) and third (drain-dot Wired-AND) levels if dynamic performance is to be maintained. Fig. 9 is a microscope picture of a portion of an SDFL circuit (that of Fig. 4 of [37]) showing both 2-level OR/NAND and 3-level OR/NAND/WIRED-AND gates (Figs. 8b and 8c, respectively), along with some NOR gate inverters. In this fast ripple adder circuit, the SDFL OR/NAND/WIRED-AND gate is used to generate the full adder carry output in only one gate propagation delay time (1 τ_d).

The SDFL circuit approach offers large savings, not only in power, but also in circuit area, over previous D-MESFET approches. The circuit area savings comes about because of the simplicity of the gate design and replacement of (large) FETs with very small (typically 1 µm x 2 µm) Schottky diodes for most logic functions. The fact that the diodes are 2-terminal devices also significantly reduces the number of vias and overcrossings

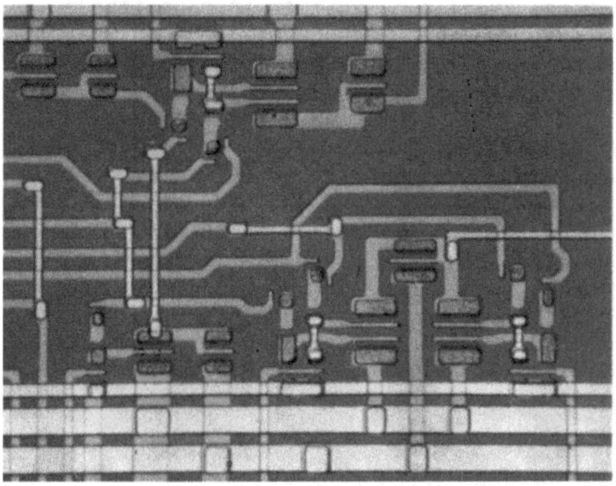

Fig. 9 Microscope photograph of section of planar SDFL circuit (Fig. 4 of [37]) showing 1-, 2- and 3-level SDFL gates. At the lower left is an inverter (1-input NOR gate), at the top center is a 2-level OR/NAND gate, while at the lower right is a 3-level OR/NAND/WIRED-AND gate.

required in most circuits as compared to that required with 3-terminal devices such as FETs. SDFL NOR gate areas are typically in the 600 μm^2 to 1000 μm^2 range. The small gate areas required for the SDFL gates also implies that the parasitic capacitances associated with these gates tends to be very small. This allows for the attainment of excellent speeds even in gates implemented with low power ($V_p \sim -1$ V) FETs with small widths. For example, the best speeds obtained with $L_g = 1$ µm, W = 10 µm, SDFL fan in = 2, fan out = 1 NOR gate ring oscillators is $\tau_d = 62$ ps at P_D = 1.1 mW/gate ($P_D\tau_d$ = 68 fJ). This is even faster than E-beam fabricated L_g = 0.5 µm, W = 10 µm buffered FET logic inverters operating at over 5 times higher power levels (τ_d = 82 ps at P_D = 5.7 mW/gate or $P_D\tau_d$ = 470 fJ) [23], [40].

It should be noted that although the fan-out is limited to a maximum of three or four (without buffering) for typical SDFL gate designs, the propagation delays are not expected to be degraded very much by fan-out loadings (as would be expected for direct coupled FET logic approaches like NMOS or CMOS). The reason for this is that the gate turnoff current for the switching FET is not provided by the preceding FET drain current, but rather by the current of the much smaller pulldown (PD) to $-V_{SS}$, since the previous FET drain is isolated by the switching diode. Hence, in effect, a fan-out of 2 to 3 is built into the SDFL logic gate itself, so that the propagation delay should be relatively weakly dependent on fan-out loading. Further, since the switching Schottky diode capacitance (~ 2fF for a 1 µm x 2 µm diode) is so small, the propagation delay should vary very little with the fan-in of the SDFL gates (number of gate inputs connected). Heavily loaded gates are buffered with source follower or quasi-complementary buffers.

The SFDL circuit could not have been implemented with a standard microwave MESFET fabrication process because the small logic diodes require doping characteristics different than those of the MESFET devices. The diodes require moderately low carrier concentration so that their capacitance in reverse bias is low, but their doping profile must be deeper than that of the MESFET, so that the diode series resistance is low. The SDFL approach has been made possible by the development of a process that makes use of localized implantation into selected areas of the semi-insulating substrate, therefore allowing the optimization of the doping profile of each device area [25]. A further benefit derived from the fact that the fabrication process does not require compromises on device doping, is that FETs with very low pinchoff voltage (below 1 V) can be fabricated. Such D-MESFETs operating at low bias voltages can have power dissipations nearly as low as enhancement-mode devices. SDFL circuits have been implemented into 7-stage and 9-stage ring oscillators for simple measurements of speed and power dissipation, as well as into more complex logic circuits with up to 100 gate MSI circuit complexities [34]. Circuits of 260 and 550 gates are under test at the time of this writing and demonstration of 1000 gate complexities is expected in 1980.

5.2.5 Transferred Electron Logic Device (TELD) Approach

Fundamentally, the exceedingly high gain-bandwidth products attainable in negative-resistance, transferred electron device operation should offer extremely high logic speeds. At present, however, transferred-electron logic devices (TELDs) appear to be unsuitable for VLSI because of their high operating power levels and the great difficulty in getting large numbers of critical threshold devices like TELDs on one chip to have operating voltages close enough to work from the same power supply. Over and above this practical yield problem, the power dissipation problem seems fundamental in that the devices require high fields over many microns device lengths (as opposed to a FET in which the high field region can be very short). The unique properties of TELDs can offer exceptional performance in specific unique applications [26], but do not appear generally usable in VLSI.

5.3 GaAs Integrated Circuits: Fabrication Technology

While the superior electronic properties of GaAs made it of interest for integrated circuits even as Silicon IC technology was emerging, the lack of a process-useful native oxide and a controllable n-type diffusion technology doomed the early GaAs IC efforts to failure. In that perspective, the state of fabrication technology is probably the most critical factor in determining if GaAs IC's can achieve a useful LSI or VLSI status.

The objective of this fabrication overview is to answer two key questions; what is the state of fabrication technology in GaAs and are there practical fabrication approaches capable of achieving LSI/VLSI compatible densities, power levels and yields? In order to focus on these issues, this review of GaAs IC fabrication processes will be limited to GaAs technologies which have already demonstrated, through published results, workable GaAs integrated circuits. Very new technologies such as HJFET and MOSFET whose propsects appear premature for VLSI considerations will not be presented.

Discussion on the merits and trade-offs relative to achieving VLSI will be presented for fabrication technologies based on both depletion and enhancement GaAs FETs. However, since the major thrust of GaAs technology development over the last decade has been on depletion-mode MESFETs an overview of the evolution of MESFET technology is presented before continuing with a discussion on various competing process technologies.

5.3.1 Evolution of GaAs Process Technology

This discussion will describe how GaAs devices have slowly evolved from very crude mesa structures to the present day sophisticated planar implanted "silicon-like" integrated circuits. As already indicated, nature did not reward GaAs with the inherent advantages common to Si, namely a stable, passivating native oxide and convenient diffusion technology. These basic limitations of GaAs in comparison to Si severly restricted the widespread usage of GaAs during the 1960s. Pressures for GaAs development came mainly from areas where Si could not fill the void such as in high frequency microwave devices. Therefore, during the early 1970s initial GaAs microwave MESFET process developments moved in considerably different directions than silicon process technology. Since an n-type GaAs diffusion technology was not available and ion implantation in GaAs was not well developed, GaAs MESFETs were fabricated on epitaxial layers with mesa-defined isolated regions. The common MESFET structure which was produced using this crude fabrication approach during the early 1970s is shown in the left portion of Fig. 10(a). More complicated double epitaxial structures, which were developed later for power MESFETs, are shown in the right side of Fig. 10(a). These early fabrication approaches provided significant momentum for successful development of liquid phase epitaxy (LPE) and vapor phase epitaxy (VPE) methods, which, ironically, inhibited the early development of ion implantation technology in GaAs. In order to take full advantage of the high electron mobility of GaAs for high speed operation, MESFETs were fabricated by necessity with 1 μm gate lengths. This factor severely inhibited GaAs progress during this early development since photolithography was limited to contact mask techniques resulting in marginal 1 μm resolution capability. Since 1 μm GaAs devices were totally out of place with Si requirements (5 μm geometry) at the time, little benefit was gained from the developing Si IC industry, leaving GaAs to progress slowly on its own. These initial GaAs efforts which employed epitaxial techniques for ultrathin (~ 2000A) layers also encountered great difficulty in achieving the required active layer uniformity and control necessary for ICs.

Parallel to microwave device research, GaAs ion implantation material research [45] fostered the early application of ion implantation for replacing MESFET epitaxial active layers [46]. Subsequently, implanted layers began to be substituted for epitaxial layers, with the implant made into a high resistivity epitaxial buffer layer or directly into the semi-insulating GaAs. These fabrication approaches are shown in Fig. 10(b). Recessed gate structures shown in the right side of Fig. 10(b) are often used. Today, both techniques are being used in the mesa-implanted D-MESFET fabrication of GaAs buffered FET logic (BFL) integrated circuits [22].

Simultaneous with these ion implanted material advances, more sophisticated pattern delineation techniques were becoming available. Two important advances, projection photolithography and electron beam lithography have allowed GaAs devices to be routinely fabricated at the 1 μm level with EBL providing MESFET gates down to the .5 μm level [47].

The strength of the mesa fabrication approach has been its simplicity, but in its present form, it is restricted to circuit applications requiring only a single active layer (implanted or epitaxial). This, along with the ultimate density and yield limitations associated with mesa structures, inhibits extension of these approaches to complex VLSI. This conclusion is

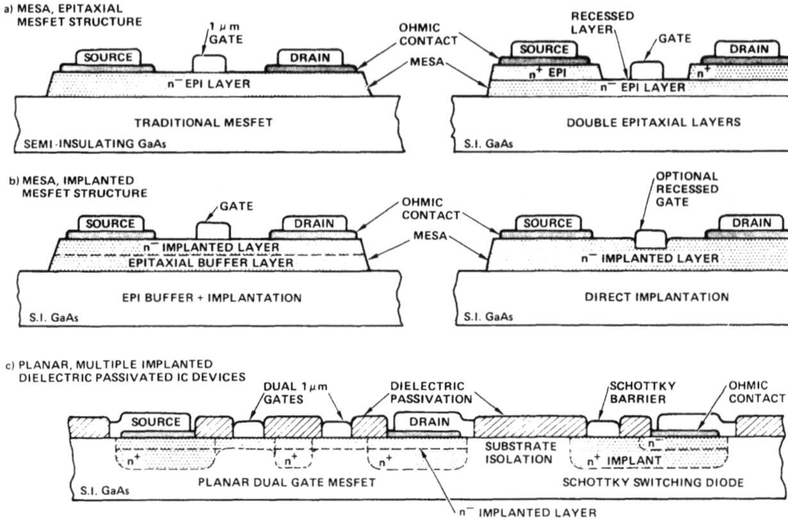

Fig. 10 Schematics showing the evolution of GaAs technology from the simple mesa-epitaxial D-MESFET to the present sophisticated planar multiple implanted GaAs IC.

supported by noting that the main application of these mesa approaches has only been directed toward ultra high-speed MSI digital logic [22,47].

Observing that the key developments in Si IC technology which led to rapid progress toward LSI were characterized by planar, ion implanted, dielectric-passivated fabrication approaches suggests that similar developments in GaAs are crucial issues for progressing to LSI/VLSI. With these basic concepts in mind, workers at Rockwell specifically set out to develop a planar GaAs fabrication technology capable of producing the integrated circuit structures shown in Fig. 10(c) [25,48,49]. These planar circuits are fabricated by using multiple localized ion implants directly into semi-insulating GaAs substrates. Hence, individual devices can be optimized by using different implants with the unimplanted GaAs substrate directly providing isolation between adjacent devices. The fabrication method conveniently compliments the Schottky diode-FET logic (SDFL) circuit approach, since the optimization of both the MESFET and Schottky diodes requires at least two different implanted active layers. An example of the sophistication of this fabrication technology is illustrated in Fig. 10(c) by the use of a 1 μm wide n^+ implant region placed between dual 1 μm MESFET gates in order to minimize the channel resistance between the dual gates. Another important aspect of this fabrication technology is the use of dielectrics. The dielectric regions shown in Fig. 10(c) are used for post implantation annealing, passivation and protecting the GaAs surfaces during processing. A thorough discussion of the features and capabilities of this planar implanted D-MESFET IC technology will be presented later.

Along with planar D-MESFET technology several other IC process technologies at various stages of development will now be presented. Five such processes are schematically depicted in Fig. 11, including process variations of depletion and enhancement MESFET and enhancement JFET technologies. Fabrication processes using GaAs HJFET [17], MOSFET [12], [13], or TELD devices are not represented since these technologies are either in initial stages of development (HJFET, MOSFET) or not appropriate (TELD) for VLSI applications.

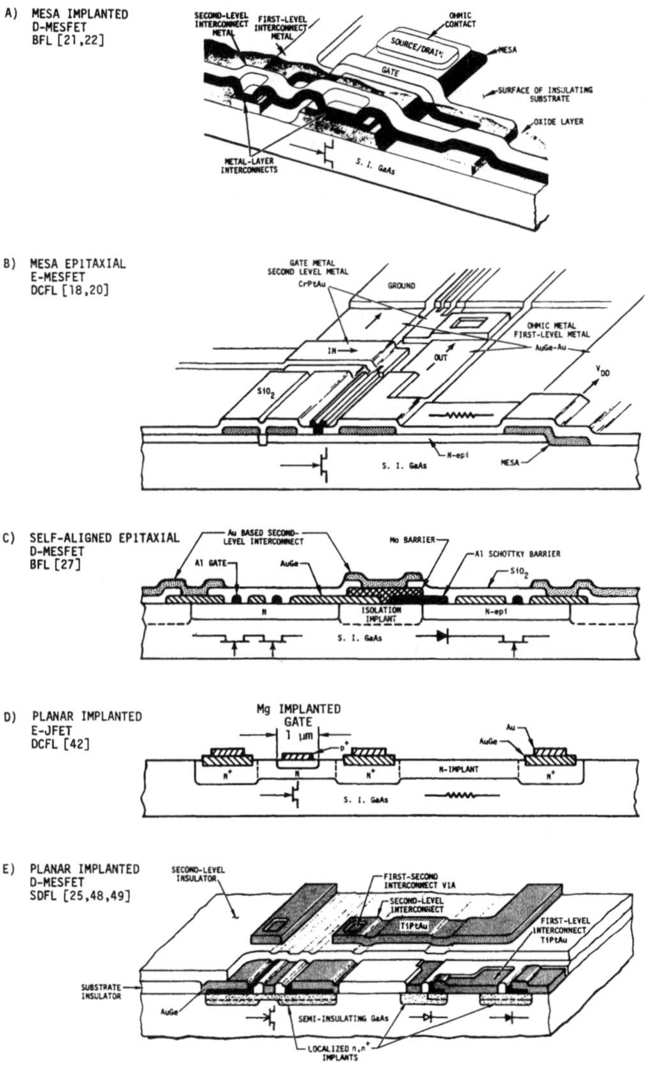

Fig. 11 Schematic representation of various GaAs IC fabrication technologies.

5.3.2 Mesa Implanted D-MESFET

Initial GaAs IC efforts were based on the well-developed depletion-mode GaAs MESFET mesa fabrication technology used in the production of low noise microwave FET's [22]. Basically, as shown in Fig. 11(a), a discrete MESFET fabrication process is used with the addition of multilayer interconnect process steps for interconnecting the various circuit elements. Initial work employing epitaxial techniques for the ultrathin (~ 2000Å) active layers encountered difficulties in achieving the uniformity and reproducibility required for IC's. Subsequently, an implanted layer was substituted for the epitaxial layer, with the implantation made into a high resistivity epitaxial buffer layer or directly into the semi-insulating substrate. Both techniques are currently being used in the mesa-implanted D-MESFET fabrication of buffered FET logic (BFL) circuits shown in Fig. 11(a).

While this mesa fabrication approach has the important advantage of process simplicity, it is, at least in its present form, restricted to applications in which only a single active layer (implanted or epitaxial) is required, i.e., circuits in which only a single type of device needs to be optimized. The density and yield limitations associated with mesa structures may inhibit the extension of this technology to VLSI. Most of the work using this approach has been directed toward high-speed MSI logic, with excellent results achieved with 1 μm contact photolithograph fabricated circuits [21] with gate areas ~7000 μm^2.

Recently, direct writing electron-beam lithography (EBL) has been utilized in conjunction with this basic mesa approach to fabricate high speed GaAs FET integrated circuits with .5 μm long single and dual-gate FETs [23,47]. The motivation for this EBL approach is to achieve higher operating speeds through shorter (.5 μm) gate lengths normally considered beyond the practical range of optical photolithography techniques. However, when one examines (see Table II) the resulting propagation delays using .5 μm long gates in BFL circuits there is not any significant speed performance advantages gained over nominal 1 μm gate SDFL IC results. It appears that the real advantage from using e-beam lithography results from the inherent flexibility built into the software pattern definition systems used in e-beam writers. Circuit pattern flexability allows rapid exploration of the impact of device design, circuit layout and logic design on circuit speed and power consumption.

Table II

TABLE II
RING OSCILLATOR PERFORMANCE FOR SEVERAL GaAs IC APPROACHES

GaAs IC APPROACH	GATE WIDTH	GATE LENGTH	POWER DISSIP.	PROPAGATION DELAY	SPEED POWER PRODUCT	FAN IN	FAN OUT
DEPLETION-MODE BUFFERED FET LOGIC (HP) [22]	20 μm	1 μm	40 mW	86 ps	3.9 pJ	2	2
DEPLETION-MODE BUFFERED FET LOGIC (HUGHES) [23]	10 50	0.5 0.5	5.6 41	83 34	0.46 1.4	1 1	1 1
DEPLETION-MODE SCHOTTKY DIODE FET LOGIC	5	1.0	0.17 0.62	156 87	0.027 0.054	2	1
(ROCKWELL) [37]	10	1.0	0.34 1.10	120 62	0.040 0.068	2	1
ENHANCEMENT-MODE FET, DIRECT COUPLED (FIJITSU) [20]	20 20	1.2 1.2	0.10 0.10	300 430	0.03 0.10	1 2	1 2

5.3.3 Mesa Epitaxial/Implanted E-MESFET

Enhancement-mode GaAs MESFET devices fabricated on epitaxial layers and isolated by mesa etching have been used in Directed Coupled FET Logic (DCFL) circuits as shown in Fig. 11b [18,20]. Other similar processes not illustrated [19, 49, 50, 51] have used both epitaxial and ion implanted layers.

The enhancement-mode devices offer single power supply circuits capable of very low power operation, though with some sacrifice in speed compared to the depletion-mode approaches. The principal difficulty with this approach is that the small (~500 mV) logic swings involved, necessitate precise control of the FET pinchoff voltages, to within ±50 mV to ±100 mV of nominal design threshold. To achieve reasonable LSI/VLSI yields, this would require standard deviations of threshold voltage of the order or σ_{V_p} = 25 mV. The best published uniformity data for GaAs MESFET's, even for very uniform implanted devices, have a standard deviation of σ_{V_p} = 70 mV [34], implying that significant improvements in threshold control (such as through improved ion implantation, or perhaps molecular beam epitaxy techniques) will be required to make this otherwise attractive approach practical for VLSI. Future improvement is anticipated since unreported results from this laboratory have demonstrated standard deviations of σ_{V_p} = 37 mV with devices fabricated on ion implanted layers.

Many of the GaAs E-MESFET IC efforts reported [19, 20] have used vapor phase epitaxial layers, for which precise thickness control has proven very difficult. This has necessitated going to controlled thinning of the layers using self limiting anodization and stripping techniques which do not appear practical for VLSI. Workers employing ion implanted layers [50] appear to omit this step. However, all of these E-MESFET approaches require the use of recessed gate structures. Recessed gate structures circumvent some of the difficulties associated with the high series resistance layers required of normally-off E-MESFETs. The process approach used is to initially provide a deeper implant or thicker epitaxial layer (lower sheet resistance) and recess the Schottky gate into the GaAs surface by using a chemical etchent or chemical anodization method. This process approach provides improved FET characteristics by minimizing source/drain to gate series resistance but the uniformity, control and yield of the resulting devices for VLSI applications is in serious question. For example, the difficulties encountered in obtaining adequate device uniformity using implanted layers and recessed gate structures for E-MESFET devices has led workers to explore innovative, less demanding (in terms of device uniformity) circuit concepts such as quasi-normally-off MESFET logic [35].

Both contact photolithography and EBL [19] have been employed in the fabrication of L_g = 1 µm circuits of this type. The fabrication of these E-MESFET circuits [19, 40, 50] is quite similar to that of the mesa D-MESFET approach except for the variation shown in Fig. 11(b) which uses the ohmic contact metallization for the first-level interconnects, with the Schottky gate metallization also serving for the second-level interconnects [20]. While the use of alloyed ohmic contact metal on the semi-insulating substrate might be anticipated to lead to current leakage problems, successful operation of SSI circuits of this type [18, 20] have been demonstrated.

In summary, the advantages of the E-MESFET fabrication and circuit approach are extremely low power dissipations, and the need for only a single power supply, therefore simplifying complex logic layouts and minimizing the number of interconnect crossovers. Densities of ~ 400 gates/mm^2 are claimed to be achievable with this circuit and fabrication approach. (Packing densities > 1000 gates/mm^2 have not been demonstrated to date). From a fabrication point of view, the main objection to the E-MESFET approach is whether the precise control of active layers required for LSI/VLSI can be realized with practical high yield fabrication methods.

5.3.4 Self-Aligned Epitaxial D-MESFET

The development of a planar technology in Si IC's marked the turning point which led to rapid progress toward LSI, suggesting that planar developments in GaAs will also have a similar impact. Recent work toward the development of planar structures has led to the fabrication of D-MESFET GaAs IC's shown in Fig. 11(c) [27]. In this fabrication process, the structure can be made planar by replacing the mesa isolation step with a selective proton, oxygen or boron bombardment which renders the underlying epitaxial material semi-insulating. This fabrication method may use epitaxial or implanted layers, however, it is still limited to a single active layer. The quality of the electrical isolation and the long-term reliability in high temperature operation of this high resistivity ion bombarded layer needs to be investigated in greater detail for VLSI applications.

Unique to this fabrication approach is the use of a self-aligned FET gate scheme [27] which does not require an FET Schottky gate masking step. This process requires the use of Al gates since ohmic contacts are fabricated after the formation of the gates and Al provides a suitable Schottky barrier material capable of withstanding the subsequent 450°C alloying cycle. Since the ohmic contacts are composed of AuGe and the gates are made of Al, a Mo barrier layer is required to separate the Al and Au based metallization systems, adding some complexity to the process and raising some concern regarding the long-term reliability of mixing Al and Au metallizations. Circuits of this type have been fabricated using contact photolithography and have yielded promising results in the lower MSI level of complexity, with reasonable promise of extension to larger circuits. However, the use of self-aligned schemes often proves restictive in terms of circuit flexibility. For instance, multiple-gate MESFET's used for NAND logic functions cannot be easily realized because of the self-aligned gate structure.

5.3.5 Planar Implanted E-JFET

A planar enhancement-mode E-JFET fabrication technology [15,42] currently under development is illustrated in Fig. 11(d). This technology uses a junction FET (JFET) in direct coupled FET logic (DCFL) with resistor load configuration similar to that of the E-MESFET circuits. The FET channel and heavily doped regions under the ohmic contacts are produced by using multiple selective n-type implantation steps and the gate region is fabricated using a p-type implant resulting in a planar structure.

The principal interest in the E-JFET is for its larger allowable voltage swing before the onset of gate conduction, although the structure also offers potentially lower parasitic source and drain channel resistances than the E-MESFET. The E-JFET might also offer better control over gate threshold voltage than the E-MESFET in that V_p can be controlled both by the n-implant and by controlling the location of the p^+/-n junction (in a manner similar to controlling the emitter depth of a Si bipolar transistor). This fabrication technology has been used to produce E-JFET devices with gate lengths of 1 µm, using a Mg implant for the p^+ gate. This work is at an earlier stage of development than the D-MESFET or E-MESFET efforts, but some inverter ring oscillator results have been published with gate densities of 200 gates/mm^2 with projected [15,42] gate densities of 800 gates/mm^2. In principle this appears to be an attractive fabrication and circuit approach for enhancement logic. The main draw back for practical VLSI results from the yield limiting additional processing required for JFETs in comparison to D-MESFETs. In particular the additional p^+ implant and subsequent self-aligned gate electrode pose difficult fabrication problems at the 1 µm level and below. Also, it would appear that a depletion-mode active load would be an attractive alternative to the currently used resistor load configuration, leading to additional processing is the form of implants and Schottky barriers. Optimized development of planar JFET logic will possibly require 2 (or more) more mask levels than D-MESFET fabrication approaches.

5.3.6 Planar Implanted D-MESFET

The planar implanted D-MESFET GaAs IC fabrication approach is illustrated in Fig. 11(e). Planar circuits are fabricated by using multiple localized ion implants directly into semi-insulating GaAs substrates. Hence, individual devices can be optimized by using different implants and the unimplanted GaAs substrate directly provides isolation between devices. This fabrication method conveniently complements the Schottky diode-FET logic (SDFL) circuit approach[1] used in this work, which requires the use of at least two different implantations for optimizing both D-MESFET and high speed Schottky barrier switching diodes.

Various aspects of this fabrication method parallel Si planar IC processes very closely. For instance, the GaAs substrate is totally protected by dielectic layers throughout the fabrication process; windows are opened in the dielectric only where ohmic contacts, Schottky barriers, or interconnect metallizations are required. Resolving the micron features is accomplished by using reduction projection photolithography in conjunction with lift-off, plasma etching, and ion milling techniques. The metallization used in these IC's are AuGe for ohmic contacts and TiPtAu for gates and first- and second-level interconnects. At the present level of development this fabrication technology has demonstrated gate areas as low as 600 µm^2/gate or circuit densities of ~800 gates/mm^2 (including interconnects) in test ICs. MSI/LSI circuits with up to 250 gate complexities have been successfully demonstrated with this fabrication technology. Further discussion will support the conclusion that this combination of fabrication (planar multiple implanted D-MESFET) and circuit approach (SDFL) is clearly the most advanced GaAs IC technology for LSI/VLSI.

5.3.7 Fabrication Technology Review: Conclusions

After evaluating the various GaAs IC fabrication approaches and reviewing their present status we have reached the following general conclusions as to what we feel are the important issues relating to a practical near term GaAs VSLI fabrication approach:

1. D-MESFET devices have had the most work, are the easiest to fabricate and, therefore, should provide the highest yield ICs. VLSI compatible power levels can be achieved by using a SDFL circuit approach as opposed to a BFL approach. SDFL D-MESFET ICs have been demonstrated at high speed and low power with excellent yield. LSI demonstration is expected in 1980.
2. E-JFET devices are more difficult to fabricate and continued process development will be required in order to precisely control the p^+ gate length and junction depth. Advantages of this approach are low power and a single power supply. Disadvantages include lower process yield do to increased mask levels and difficult active layer control. Current level SSI;MSI expected 1980.
3. E-MESFET devices with very low logic swings should yield the lowest-power technology. However, noise margins will be small and achieving adequate FET threshold control will pose very difficult material and process control problems. Again the advantages of single power supply applies. Current level SSI/MSI.
4. E-MOSFET and HJFET fabrication processes are in early development. The prospect for developing a stable GaAs dielectric interface still appears poor. HJFET approach has physics on its side but may never be attractive from a yield and cost point of view. Status, very premature for any broad conclusions. VLSI issue not relevant at this time.
5. IC device active layers require precise control. Ion implantation appears the only technology capable of the necessary uniformity and reproducibility, combined with good cost effectiveness. Direct implantation into semi-insulating substrates is simpler and less expensive than implantation into epitaxial buffer layers.
6. Device flexibility and optimization favor the use of multiple localized implantations with the semi-insulating substrate providing device isolation.
7. The use of planar structures rather than mesas, with semi-insulating GaAs for device isolation should result in higher circuit densities and higher yields.
8. Fabrication approaches such as self-aligned techniques which limit device or circuit designs may not have the flexibility to satisfy VLSI requirements.
9. The enhanced reliability and high current capacity of an all-gold-based metallization system should be important for VLSI.
10. Contact photolithography will not meet the requirements of VLSI and in general has already given way to projection photolithography at the LSI level. Reduction projection photolithography (4X, 5X, and 10X systems are available) have advantages over 1:1 projection systems and will produce higher yields for VLSI.
11. EBL may best be utilized by integrating this technology (in a hybrid approach) with photolithography. EBL would be used when a particular mask level requires finer geometry or more precise alignment than consistent with photolithography techniques.
12. Circuit lithography processes should utilize dry etching techniques which appear necessary for achieving high yield circuits with 1 μm or finer features.

It should be apparent from these conclusions that there is a demonstrated GaAs fabrication technology which is addressing and is well on the way to meeting the goals of future GaAs VLSI. The remainder of this section will present a detailed description of the Rockwell planar GaAs IC fabrication technology which satisfies the majority of the foregoing conclusions.

5.3.8 Planar VLSI Compatible Fabrication Technology

Ion implantation is now an established material approach for GaAs ICs and for many discrete GaAs devices as well. However, many workers still appear to be using mesa, single implanted fabrication techniques which will limit device optimizaiton, circuit flexibility, circuit density and ultimate yields required of future VLSI. The Rockwell fabrication approach addresses the ultimate goals of GaAs VLSI development, using Si VLSI technology as an example, by utilizing extremely high density, planar structures for attaining the necessary high yield VLSI compatible processes. The planar GaAs process steps used to fabricate D-MESFET SDFL ICs are shown in Fig. 12.

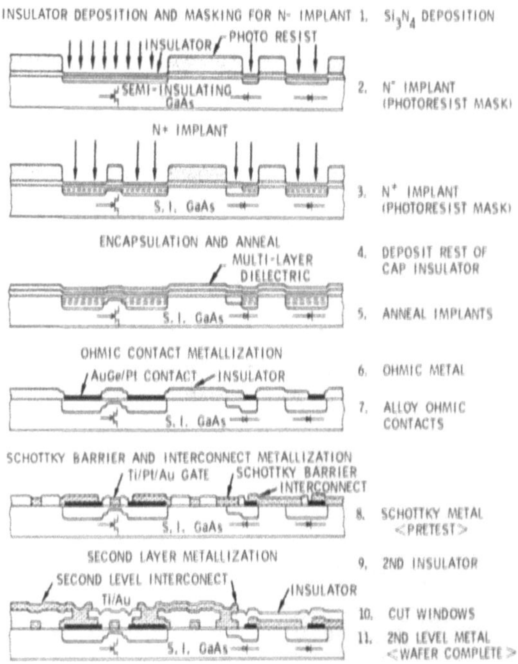

Fig. 12 Planar GaAs IC process steps used to fabricate SDFL circuits.

Initially, the GaAs is coated with a thin layer of Si_3N_4 which remains on the wafer throughout all of the subsequent processing steps. The first process steps are the two localized implantations (steps 1-5) carried out through the thin Si_3N_4 layer using thick photoresist as the ion beam mask. After each implant, a shallow step is plasma etched into the Si_3N_4 for registration of the implanted regions. Following the implants, additional dielectric is added prior to the post implantation annealing step. The same dielectric used in annealing also provides us with an intermediate layer lift-off medium used in the processing of first level metal interconnects.

Normally, a sequence of two implants is used, however, this process is capable of any number of implantation steps that may be necessary in order to optimize individual devices. An example would be an additional high dose implant step to heavily enhance the doping under ohmic contacts.

Typical implantation parameters and the resulting electron concentration doping profiles for both IC implants are shown in Fig. 13. The 400 keV Se implantation profile is peaked near the GaAs surface due to the energy absorbed in penetrating the thin Si_3N_4 layer. Little influence, except the shift of the peak of the implanted profile, has been observed as a result of implanting through thin dielectric layers. The theoretical predictions of enhanced doping from "knock-ons" from the Si_3N_4 cap during ion implantation has not been observed experimentally nor has any shift been observed in the peak of the Se profile as a result of small variations in dielectric

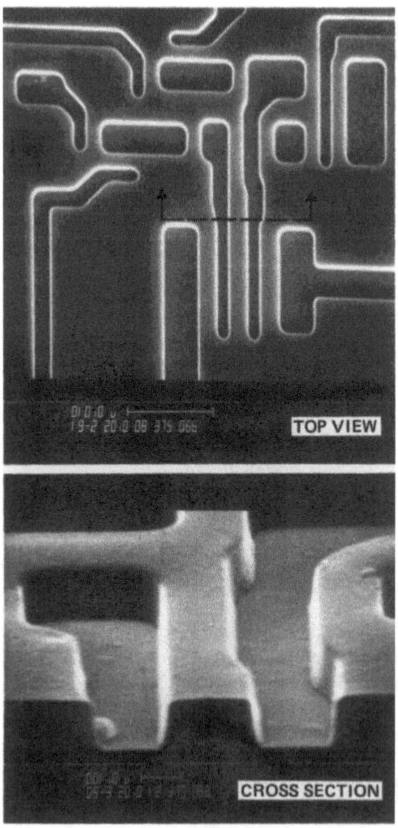

Fig. 13 Implantation profiles used for GaAs MESFET channels and high speed Schottky barrier switching diodes.

thickness. (Si_3N_4 thickness is held to ± 20Å.) The shallow (~1500Å) Se D-MESFET channel profiles shown in Fig. 13 were obtained from 16 different GaAs slices distributed across one ingot. Pinch-off voltages of devices made from these layers correlate well with the depletion voltage required to deplete these layers to the 10^{16} cm^{-3} electron concentration level. Acceptable control and reproducibility of the material and implantation process is evident in the histogram of depletion voltages (V_d) for these different slices yielding an averge 1.03 volts with a standard deviation of 0.11 volts. The spread of MESFET pinch-off voltages from wafer to wafer (reproducibility) would be expected to be of the same magnitude as these C-V profile results. The spread of pinch-off voltage on a single IC wafer (uniformity) is much better (typically ~80 mV).

Referring to Fig. 13 observe that the n^+ S implant provides a much deeper (~4000Å) profile, ideally suited for the high conductance low capacitance high speed switching diode required in SDFL. Both implants are used for level shifting diodes and enhancing the doping under all ohmic contact regions. The S profile is much deeper than the predicted, LSS calculated profile, as a result of a cap dependent enhanced diffusion mechanism [52]. The average sheet resistivities measured on these implanted layers are typically ~2500 Ω/ for the Se n^- layers, ~450 Ω/ for the S n^+ layer and ~ 350 Ω/ for the combination of both implants. As is suggested from the implanted profiles and sheet resistance values, the most difficult layer to control is the shallow lightly doped high resistance n^- MESFET channel layer. This implant step is critical since this layer directly controls the MESFET pinch-off voltages. The n^+ implanted layer used predominantly for the high speed switching diodes has a somewhat more relaxed specification since the main concern for this device is that it must be fabricated on a relatively high conductance low capacitance layer.

The circuit lithography process techniques used in conjuction with the planar implantation approach are designed with 1 µm features requiring a totally dry VLSI compatible process. There are no wet chemical etching steps and the only wet process steps are associated with photolithography and photoresist lift-off processes.

Common to all of the GaAs IC process steps is the delineation of fine line resist patterns required for the fabrication of these circuits. Over the last three years fabrication of GaAs ICs has been accomplished by using a Canon 4x projection mask aligner. The Si industry, due to the recent emphasis on VLSI, is also gradually adopting this state-of-the-art photolithography technique. Reduction projection photolithography has several advantages: there is no mask wear as in contact printing, any mask step-and-repeat error or small defect on the mask is minimized by a factor of 4 at the wafer level, and alignments are precise, typically within 0.5 µm. An example of the resolution capability of this photolithography technique is illustrated in the SEM photo of dual 1 µm wide lines shown in Fig. 14. In principle, as shown in this edge view, relatively vertical photoresist side walls can result from this technique.

Historically, replication of 1 µm geometries on GaAs has been accomplished by using direct photoresist lift-off methods. Direct photoresist metal lift-off techniques have basic limitations which are incompatible with the yield requirements of VLSI. The process usually is not 100% reliable, often resulting in coarse metal edge topography. A cross section of poor metal topography resulting from this technique is illustrated in

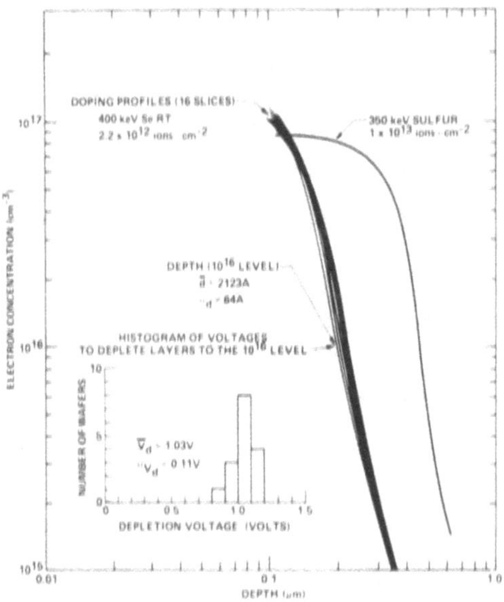

Fig. 14 SEM photograph highlighting dual 1 μ gate photoresist patterns defined by projection photolithography.

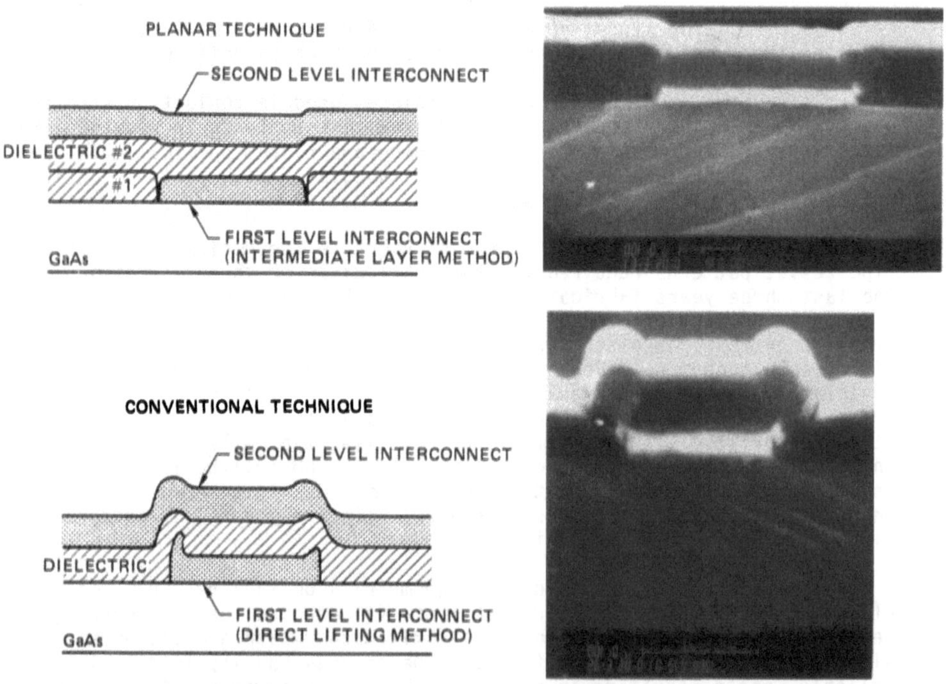

Fig. 15 Schematics and SEM photographs of GaAs planar vs conventional multi-level interconnect structures fabricated by intermediate layer and direct lift-off techniques.

the left lower portion of Fig. 15 with a scanning electron micrograph of an actual sample shown at the lower right.

Therefore, in order to insure high process yields, an alternative intermediate layer lift-off method [48,49] has been developed (shown in the upper portion of Fig. 15). This process can be described by referring back to Fig. 12 and observing that after the implantation steps the entire surface of the GaAs wafer is covered with dielectric. Following the post implantation anneal, in the subsequent processing steps for defining ohmic contacts, MESFET gates, Schottky barriers, and first level interconnects, the dielectric serves as the intermediate layer for the Rockwell high yield lift-off technique. The process of defining the first level metallization is accomplished by delineating photoresist patterns, plasma etching the underlying dielectic, evaporating the appropriate metal, followed by the photoresist lift-off step. High yields are achieved using this process with an added advantage that metal contacts and interconnects are automatically precisely registered within the plasma etched dielectric windows. The resulting smooth metal structures are demonstrated in the upper right scanning electron micrograph of Fig. 15. The sophistication of this powerful lithographic technique for fabricating planar, high yield, dense 1 µm structures precisely aligned within dielectric windows is highlighted in Fig. 9. This photograph of a portion of a planar GaAs IC shows an SDFL logic gate containing a dual 1 µm gate MESFET. The ability to fabricate such dual gate SDFL circuits has provided GaAs ICs with new multi-level logic approaches [37] for optimizing increasingly more complex circuits.

The multi-level interconnect process shown in the last step of Fig. 12 also illustrates other advances made in GaAs fabrication concepts. As shown in Fig. 15, fabricating the first-level metal within windows in the first-level dielectric and maintaining the first-level metalization thickness close to the dielectric thickness results in a planar structure which greatly facilitiates the fabrication of complex multi-level interconnects.

The process used to fabricate the planar multi-layer structure starts by depositing a 5000Å thick plasma enhanced CVD silicon nitride (plasma nitride) layer on the ICs. Via windows are then reactively ion etched through the nitride in order to interconnect the first level metal to the second level metal. A second level metal composed of Ti/Au is deposited over the second level dielectric and is subsequently defined by ion milling, completing the process.

The upper portion of Fig. 15, shows the smooth planar crossovers resulting from this planar multi-layer fabrication approach. The planar crossovers eliminate any potential for crossover problems such as shorts between first and second level interconnects, and high resistances or open interconnects resulting from poor step coverage. The lower portion of Fig. 15 illustrates how interconnect dificulties could easily exist when poorly defined first level metal is used in conjunction with conventional multilevel crossover techniques.

Figure 16 (left side) shows a very compact GaAs IC that utilizes, for the sake of density, parallel as well as conventional orthogonal first-to-second-level crossovers. The ability to accommodate optimized circuit layouts using parallel crossovers without any process yield or circuit performance problems demonstrates the power of this planar crossover approach. Furthermore, this planar multi-level inteconnect structure

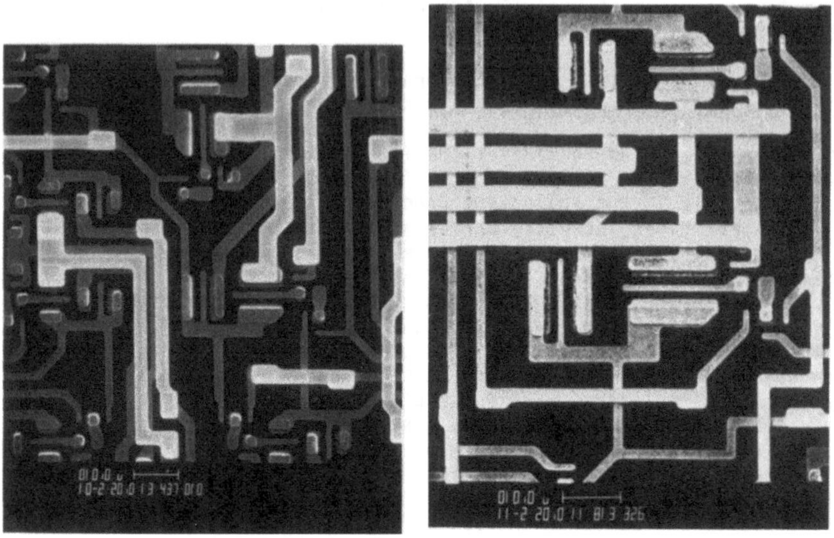

Fig. 16 a) SEM photograph of a high density planar GaAs IC designed with parallel first to second level interconnect crossovers. b) SEM photograph of a portion of a D flip-flop showing two 3-input NOR gates.

provides, in principle, the capability to successfully fabricate increased levels of interconnects. This is desired in VLSI to allow the simplification of complex circuit layouts and reduced chip areas.

The SEM photograph (Fig. 16, right side) of a portion of a GaAs IC, also highlights the planar, featureless GaAs surface and resolution and alignment capabilities of the overall process. These capabilities are demonstrated where 10 µm and 15 µm wide FET's have 1 µm long gates precisely aligned within the 4 µm source to drain gaps. Three, extremely small (1 µm x 2 µm) logic diodes are utilized here for a 3-input logic "OR" function. Diodes and FET's requiring different implantations are separated by as small as 3 µm isolation gaps provided by the semi-insulating substrate.

In summary, we have presented a GaAs fabrication process which is well on the way to meeting the goals of future GaAs VLSI. The excellent prospects of GaAs VLSI are enhanced by the fact the basic process steps used in fabricating GaAs IC's are generally fewer than those used in Si IC's. This is possible mainly because of the availability of semi-insulating substrates providing minimal parasitic capacitances and direct device isolation without any additional complicated processes as found in Si NMOS, COMOS, or CMOS/SOS . These factors should have a significant impact on the ultimate yield attainable for GaAs VLSI applications. Despite the differences in the materials, the fabrication techniques required for ultra-high density and extremely fine-line Si VLSI and GaAs VLSI are quite similar. Continued progress in the GaAs material technologies of substrate growth and ion implantation can be anticipated. Therefore, the prospects for GaAs VLSI from a fabrication point of view appear as realistic as for Si VLSI.

5.4 Performance Results for GaAs Digital ICs

In this section, the speed-power performance of various GaAs FET logic approaches will be presented and discussed. The comparisons of performance results should be viewed in the broader context of the complexity of circuits currently fabricable with these technologies. Many of these results were obtained from intensively-processed "proof of concept" samples, as opposed to circuits fabricated with production-compatible processes. Circuits fabricated with direct-writing E-beam lithography might be considered to fall close to this category. Most definitely of a "proof of concept" nature is some of the E-MESFET work in which chips (as opposed to wafers) have been individually processed in order to obtain the required pinchoff voltage control. The GaAs MOSFET results are for samples which do not have stable gate oxides and hence are not useable in general digital circuits in which the duty cycle of the input signal is not 50%. On the other hand, the L_g = 1 µm D-MESFET BFL and SDFL results were fabricated with optical lithography using production-compatible processes, and both have demonstrated good yields of circuits of MSI complexities. In this section we will put primary focus on the results for the GaAs IC approaches having demonstrated reasonable levels of circuit complexity.

The ring oscillator circuit provides a convenient and widely used method for evaluation of propagation delay and power dissipation since the ring oscillator provides its own signal source. A chain of an odd number, N, of inverters or gates is connected in a ring, so that the measured frequency of oscillation is related to the intrinsic gate propagation delay by $f = 1/(2N\ \tau_d)$. Oscillators with gate fanouts of 1 or 2 are generally used. Fig. 17 compares GaAs ring oscillator results with those of the more conventional silicon IC technologies. Note that, in general, GaAs devices provide much higher speed for the same power dissipation per gate as expected from the superior electron dynamics and lower substrate parasitics for GaAs devices. Some more recent high performance Si data points [29], [30], [41] with reduced geometries have been included on the plot. Very low propagation delay has been obtained with the Si NMOS device (65 ps/gate) [30], [32]; however extremely small channel geometries (0.25 µm) require use of very sophisticated, EBL-based processing techniques. Even the L_g = 1.3 µ NMOS work [41] uses a direct-writing EBL process which is much more complicated than that used for L_g = 0.5 µm GaAs work [23].

General discussion of the relative design tradeoffs resulting from each GaAs logic approach can be found in Section II along with projections of circuit performance under various loading conditions. However, a further comment should be made relating the comparison of these ring oscillator results with the propagation delays expected in actual complex logic circuits. In typical silicon ICs (NMOS, for example), there is at least a factor of five difference between small inverter ring oscillator speeds and the speeds in the real circuits fabricated from the same technology. About half of this speed loss results from the fanout loading in the real circuits (as expected for NMOS) and the rest comes from the parasitic capacitances incurred in the larger layout configuration. With semi-insulating GaAs substrates, this latter speed degradation source is greatly reduced, so that there is a much closer correspondence between ring oscillator speeds and the propagation delays measured in complex circuits. For example, the HP buffered FET logic ring oscillator measurements gave ~ 100 ps propagation delays for their 4-input NAND/NOR gates [22]. Table II shows that a propagation

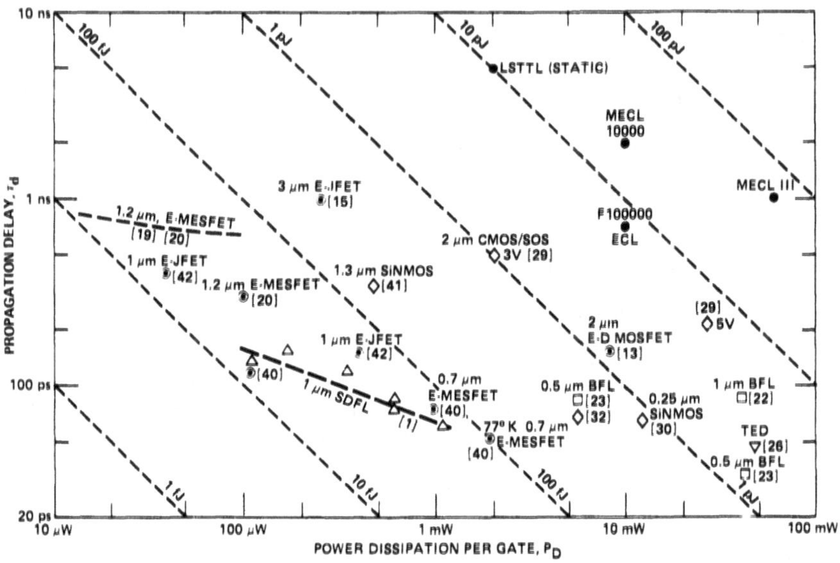

Fig. 17 Map of propagation delay vs power dissipation per gate comparing published results for GaAs and Si IC technologies.

Table II Comparison of ring oscillator performance from several GaAs IC approaches

GaAs IC Approach	Gate Width	Gate Length	Power Dissip.	Propagation Delay	Speed Power Product	Fan In	Fan Out
Depletion-Mode Buffered FET Logic (HP) [22]	20 μm	1 μm	40 mW	86 ps	3.9 pJ	2	2
Depletion-Mode	10	0.5	5.6	83	0.46	1	1
Buffered FET Logic	50	0.5	41	34	1.4	1	1
(Hughes) [23]							
Depletion-Mode	5	1.0	0.17	156	0.027	2	1
Schottky Diode			0.62	87	0.054	2	1
FET Logic							
(Rockwell) [37]	10	1.0	0.34	120	0.040	2	1
			1.10	62	0.068	2	1
Enhancement-Mode	20	1.2	0.10	300	0.03	1	1
FET, Direct Coupled	20	1.2	0.10	430	0.10	2	2
(Fijitsu) [20]							

delay of 111 ps was inferred from frequency divider performance, in good agreement with ring oscillator results.

The general features and dependencies of this ring oscillator data can be understood in terms of the simple switching model described in the

Appendix, as discussed in [39]. It is of interest to examine in more detail the ring oscillator results obtained with those GaAs IC approches for which experimental data on MSI sequential logic circuits (such as frequency dividers) has been published. Table II compares measured performance of ring oscillators fabricated by enhancement mode direct coupled FET logic [18], [20], depletion-mode Schottky diode FET logic [24], and depletion-mode buffered FET logic [22]. It can be seen from this data that enhancement-mode logic, while providing very low speed-power products, also suffers from a significantly slower propagation delay than the depletion mode approaches, particularly with reasonable gate loading levels. The low pinch-off voltage, SDFL ring oscillators have provided both low $P_D\tau_d$ products and propagation delays as low as 62 ps. The range of performance shown for each gate width reflects results from wafers with very low and moderate (-0.5 V and -1.2 V) pinchoff voltage. The depletion-mode buffered FET logic also provides excellent propagation delay, but speed-power product is approximately 40 times greater than the low pinch-off SDFL approach for equivalent FET widths. This is a consequence not only of the larger logic swing (V_m in (A7)) but also because the complex buffered-FET logic gates require much larger chip areas so that the parasitic load capacitances are much larger. Note that L_g = 0.5 µm, EBL-fabricated, depletion-mode, buffered FET logic [23], W = 50 µm inverters give the highest speeds, as would be expected. These devices have somewhat lower pinchoff voltages than the 1 µm BFL work [22], for somewhat reduced power dissipation. Note, however, that at W = 10 µm gate size, L_g = 1 µm SDFL NOR gates are faster (62 vs 83 ps) than the L_g = 0.5 µm BFL circuits at less than one fifth the power. This is a clear measure of the effect of the low gate pareasitic capacitances in SDFL. Fast OR/NAND dividers have been fabricated in this 0.5 µm BFL, but have yet to achieve as high toggle frequencies as the 1 µm BFL of [27].

While ring oscillators are useful devices for measurement of the basic speed power properties of logic gates, it is also important to demonstrate performance of gates in real sequential or combinatorial logic circuits. This avoids the possibility of being deceived by misleading ring oscillator results due to unrealistically low loading capacitances or misleading $P_D\tau_d$ figures due to operation in linear (Class A amplifier - as opposed to non-linear logic) modes at V_{dd} levels far below those required for reasonable noise margins. (This is particularly a problem in enhancement mode data.) Such circuits generally require use of gates with multiple inputs and fan-outs of 2 or more. Therefore, demonstration of GaAs circuits in sequential logic applications such as frequency dividers (e.g., binary ripple counters) is a more demanding and more realistic index of performance.

Table III contrasts the performance of three of the GaAs IC approaches that have been able to demonstrate at least divide-by-two functions. In the Schottky diode FET logic work [39] and the French BFL work [44] a standard NOR-implemented D flip-flop circuit was used T-connected ($\bar{Q}$ output connected to the data input) in order to achieve frequency division, while the other BFL work [22], [28] and the E-MESFET work [20] used complementary-clocked frequency divider circuits. The theoretical maximum toggle or clock frequency for the D-FF and NOR-implemented dividers was estimated by modeling on an analog-response logic analysis program, which results in an estimate relative to the intrinsic propagation delay of the logic gates. An equivalent propagation delay per gate can be inferred from the model when the maximum experimental clock frequency is determined.

Table III

TABLE III
GaAs L_g = 1 μm MESFET FREQUENCY DIVIDER IC RESULTS

GaAs IC TECHNOLOGY	CIRCUIT APPROACH	THEOR MAX TOGGLE f	MEAS'D MAX TOGGLE f	POWER DISSIPATION P_D	EQUIV $P_D\tau_d$	EQUIV τ_d
DEPLETION-MODE SDFL (ROCKWELL) [39]	D-FLIP FLOP T-CONN'D FOR ÷ 2 (NOR GATES)	$f_{tm} = \frac{1}{4.85\tau_d}$	1.87 GHz	15 mW/6 GATES P_D = 2.5 mW/GATE	0.28 pJ	110 ps
DEPLETION-MODE BFL (THOMSON-CSF) [44]	D-FLIP FLOP T-CONN'D FOR ÷ 2 (NOR GATES)	$f_{tm} = \frac{1}{4.85\tau_d}$	1.5 GHz	250 mW/6 GATES P_D = 40 mW/GATE	5.5 pJ	137 ps
DEPLETION-MODE BFL (HP) [28]	COMPLEMENTARY-CLOCKED ÷ 2 (NAND/NOR GATES)	$f_{tm} = \frac{1}{2\tau_d}$	4.5 GHz	160 mW/4 GATES P_D = 40 mW/GATE	4.44 pJ	111 ps
ENHANCEMENT MODE DCFL (FUJITSU) [20]	COMPLEMENTARY-CLOCKED ÷ 2 (NOR GATES)	$f_{tm} = \frac{1}{4\tau_d}$	0.610 GHz	1.66 mW/ 8 GATES P_D = 0.21 mW/GATE	0.085 pJ	410 ps

Table III indicates that the gate propagation delays determined from the divider frequencies of all three approaches are fairly close to those obtained in the ring oscillator evaluations and that, for the depletion mode approaches, the speed of the logic gates is not greatly reduced by fanouts of 2 or 3. Finally, it is obvious that the circuit design (complementary vs. single clocked) has a strong influence on the maximum clock frequency achievable for a given type of GaAs FET logic. Note that higher toggle frequencies were attained with the SDFL D-FFs, than with BFL D-FFs operating at over 15 times higher power levels. The higher toggle frequencies demonstrated by the HP buffered FET logic dividers are primarily a consequence of the use of the complementary-clocked, NAND/NOR-implemented, master-slave divider stage circuit design [22], [28], rather than shorter gate propagation delays. This circuit (actually its complement) could also be implemented in SDFL, by using the OR/NAND gate configuration of Fig. 8b.

The SDFL and BFL edge-triggered D-flip flops were implemented with six NOR gates as shown in Figure 18. The divide-by-two characteristic was accomplished by connecting the $\bar{Q}$ output to the data (D) input, thereby providing the necessary feedback for a state change to occur on only every clock input trailing edge. Logic simulation of this circuit indicates that correct operation should be possible up to a clock frequency of 1/4.85 τ_d. The maximum clock frequency at which correct frequency division occurred was 1875 MHz for the SDFL circuit. This implies an equivalent τ_D of approximately 110 ps, comparable to the buffered FET logic approach. If complementary-clocked multi-level logic versions of the SDFL divider had been employed, fewer gates would be required, and a maximum clock frequency roughly twice that observed would be expected for the same τ_d. The power dissipation of 2.5 mW/gate is somewhat higher than enhancement mode logic, but considerably lower than the buffered FET logic approach. It results in an equivalent $P_D\tau_d$ product of 0.28 pJ/gate for the SDFL divider. These T-connected D-FFs have also been cascaded (by connecting the Q output to the C input of the succeeding stage) into binary ripple dividers with ÷2, ÷4, and ÷8 outputs [34]. Fig. 19a illustrates the low frequency (100 MHz) operation of one of these dividers with the sinewave clock input and all 3 outputs shown. Fig. 19b shows the ÷8 output for this SDFL ripple divider, operating at its maximum toggle frequency of 1.9 GHz.

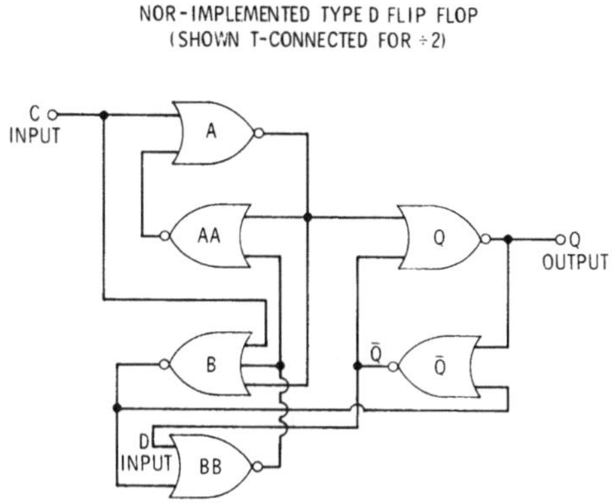

◄ Fig.18 Logic diagram for a NOR-implement edge-triggered type D flip-flop (shown T-connected for ÷2).

▼ Fig.19 SDFL D-FF binary ripple divider results [34]. a) Low frequency operation showing 100 MHz sinewave clock and ÷2, ÷4, and ÷8 circuit outputs. b) ÷8 output of this ripple divider operating at a clock frequency of 1.9 GHz.

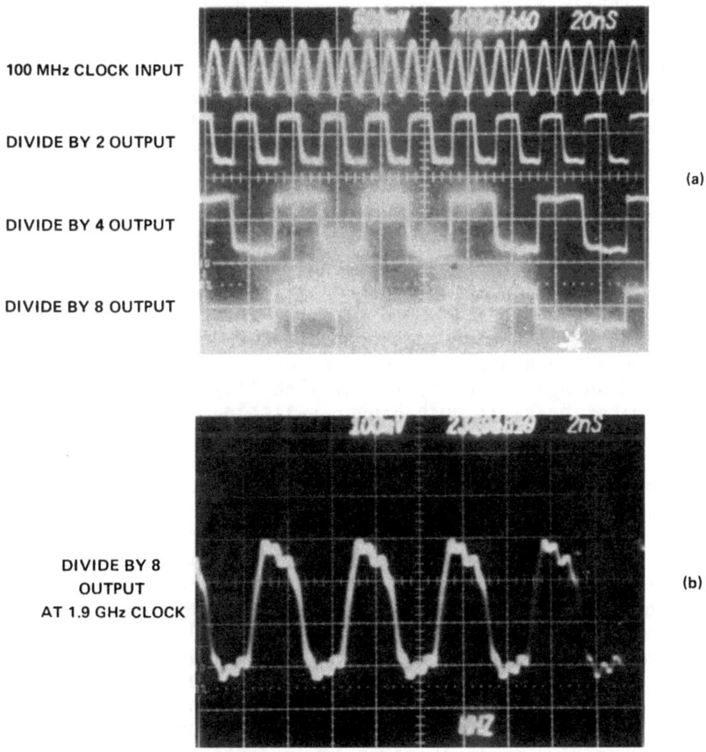

It is interesting at this point to return to the issue we began this paper with; the comparison of the performance of GaAs and Si ICs. Fortunately, there is now beginning to emerge a reasonable volume of data on micron and submicron gate Si NMOS ICs to join the profusion of L_g = 1 µm or less GaAs IC data published over the last few years. In addition to the NMOS and CMOS/SOS ring oscillator data points noted in Fig. 17, there has recently been published data on L_g = 0.7 µm Si NMOS frequency dividers [43], presumably of the same OR/NAND configuration as used in the HP BFL work [22]. Inverter ring oscillators with these 0.7 µm gates gave τ_d = 175 ps at V_{dd} = 5 V or $P_D\tau_d$ =0.24 pJ. (At this bias, these devices would appear to be above the threshold for long term MOSFET degradation by hot electron injection into the oxide.) The maximum toggle frequency for the 0.7 μm, ÷16 ripple dividers was f_c = 600 MHz, for $P_D\tau_d$ = 1.3 pJ (or somewhat better if we allow for buffer dissipation). Note that the τ_d = 883 ps in the divider is nearly 5 times higher than the τ_d = 175 ps unloaded ring oscillator result. This factor, typical for NMOS circuits, is much worse than for GaAs ICs, as seen from comparing Table III with Table II.

The bottom line is to compare the τ_d = 833 ps NMOS delays and $P_D\tau_d$ = 1.3 pJ dynamic switching energies in the actual L_g = 0.7 μm NMOS divider circuits with the GaAs IC results. Here the GaAs superiority becomes very clear. The HP L_g = 1 μm BFL results [22] on the same divider architecture, τ_d = 111 ps, are 7.5 times faster than the 0.7 μm NMOS, though at about 3.4 times the $P_D\tau_d$ product. Unpublished GaAs BFL work [44] using 0.7 μm EBL fabricated BFL NOR gates in D-FFs have given equivalent delays under 70 ps, which is twelve times faster than the Si NMOS results with the same gate length. Even the very low power L_g = 1.2 μm E-MESFET dividers [20] gave equivalent delays (Table III) of less than half that of the L_g = 0.7 μm Si NMOS at fifteen times better speed-power products. The L_g = 1 μm SDFL D-FFs gave equivalent NOR gate delays 7.5 times faster than the 0.7 μm Si NMOS and with a 4.6 times better speed-power product. These numbers speak for themselves regarding the reason for going to GaAs for ultrahigh speed VLSI.

5.5 Summary, Conclusions and Projections

The analysis and experimental data shown here indicate clearly that gallium arsenide offers major performance advantage over silicon for integrated circuit fabrication. The high mobility and electron velocities in short channel (L_g ~ 1 μm) GaAs MESFETs give transconductances or drain currents about 6 times those for silicon n-channel FETs at equivalent gate biases. This FET performance advantage, aided by the availability of a semi-insulating GaAs substrate, makes possible much higher switching speeds for GaAs ICs or greatly reduced $P_d\tau_d$ products in comparison to silicon ICs. The experimental data for GaAs ICs both for simple ring oscillators and for more complex sequential logic circuits, live up to the projections based on FET performance. Depletion-mode MESFET GaAs ICs have shown propagation delays as low as τ_d = 62 ps for L_g = 1 μm NOR gate ring oscillators (34 ps for L_g = 0.5 μm inverters), while enchancement-mode MESFET circuits have given speed-power products as low as 30 fJ at τ_d = 300 ps switching speeds. Equivalent propagation delays as low as τ_d = 70 ps have been obtained in L_g = 0.7 μm depletion-mode MESFET frequency dividers or τ_d = 410 ps in the lower power enchancement mode GaAs MESFET ICs. This divider performance is far beyond that achieved in L_g = 0.7 μm silicon NMOS dividers (τ_d ~ 833 ps at $P_D\tau_d$ ~ 1.3 pJ). D-MESFET GaAs dividers give gate delays up to 12 times faster than the Si circuits for the same FET gate lengths and even L_g = 1.2 μm GaAs E-MESFET dividers give up to 15 times lower speed-power products at twice the speeds as the 0.7 μm Si NMOS.

It has been the significant improvements in GaAs IC processing techniques which have made possible the fabrication of complex, high performance GaAs ICs after earlier attempts failed. While many of the GaAs IC fabrication approaches are specifically adapted to a particular circuit approach (or the circuit approach adapted to the limitations of a fabrication process), more recent developments in planar, ion-implanted GaAs ICs offer silicon-like process versatility. Using localized ion implantations for the active devices and using the unimplanted semi-insulating GaAs substrate material for isolation, high circuit densities (as required for VLSI) can be achieved. The multiple-implantation aspect of this approach allows more than one active device to be optimized in a circuit, making possible, for example, use of the high density Schottky-diode FET logic (SDFL) circuit approach.

Of the several circuit approaches investigated for GaAs ICs, both direct-coupled FET logic (DCFL) using enchancement-mode MESFETs and SDFL (using low power depletion mode MESFETs) have demonstrated LSI and VLSI-compatible gate dissipations ($P_D \tau_d$ < 100 fJ). The depletion mode MESFET buffered-FET logic (BFL) work published thus far has exhibited gate powers and speed-power products too high for LSI, along with sizeable gate areas (7000 μm^2 as compared with < 1000 μm^2 for SDFL), so that this approach appears marginal for VLSI. Work on other GaAs IC approaches using JFET, HJFET, MOSFET or TELD active devices is in earlier development stage than the MESFET work although publication of L_g = 1 μm E-JFET MSI circuit results is anticipated in the near future.

In the area of demonstrated GaAs integrated circuit complexity, the lead technology is the planar, D-MESFET SDFL work [38]. Many different circuits, with up to 100 gate complexities have been successfully demonstrated in SDFL. A major goal of the ARPA program at Rockwell is the fabrication of a ~ 1000-gate 8 x 8 bit multiplier in 1980. Already successfully tested is the first step toward this chip, a 75-gate 3 x 3 bit multiplier [38]. An intermediate size 5 x 5 bit multiplier chip containing 260 SDFL NOR gates, shown in Fig. 20, has also been fabricated and is beginning evaluation as of this writing, while the mask set for the full 8 x 8 bit multiplier is in digitization. (As discussed in the addendum - see p. 296 - up to 20% yields were obtained on these 5 × 5 multiplier chips and fully working 8 × 8 bit parallel multipliers were in fact achieved in 1980 in this program.)

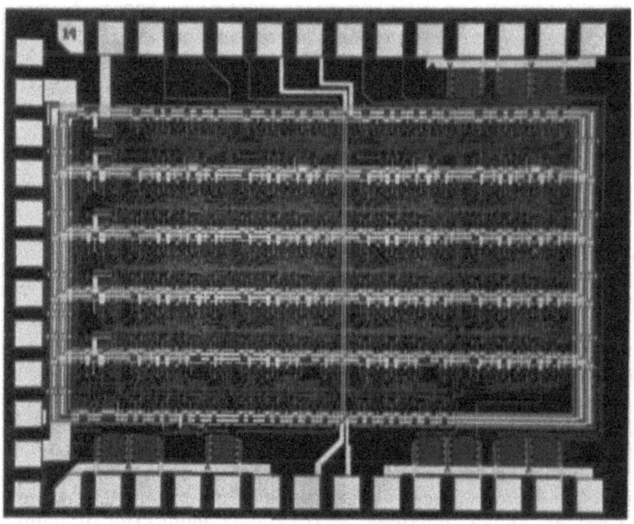

Fig. 20 Microscope photograph of 5 x 5 bit parallel multiplier chip implemented with 260 SDFL NOR gates. Chip power is 45 mW and area (including pads) is 1.125 mm x 1.35 mm (0.044" x 0.053").

Since the bulk of a large multiplier is full adder cells, the speed of these cells dictates the multiply time. In this GaAs multiplier, an SDFL NOR gate-implemented full adder cell was used, using a full minterm expansion for both sum and carry, so that the propagation delays are kept to 2 τ_d for the carry and 3 τ_d for the sum (where τ_d is the NOR gate delay). A total of 12 gates (9 NORs and 3 inverters) are required per full adder cell. For the 3 x 3 parallel multiplier chip, three half adders and three full adders are used, for a total gate count of 75. The initial lot of 3 x 3 multipliers fabricated gave fully operational circuits with power levels as low as 31.5 mW, or 420 μW/gate. The multiply times were evaluated using an on-chip test circuit to be as low as 1.55 ns in a 5 τ_d path for a worst-case (12 τ_d) rate of 0.485 giga-multiplies/sec or a NOR gate delay of 172 ps (at P_D = 750 μW/gate or $P_D\tau_d$ = 128 fJ). If we extrapolate even these initial 3 x 3 GaAs SDFL IC results to the 8 x 8 parallel multiplier chip (scheduled for fabrication in 1980), it would achieve a worst case multiply time of 6 ns, or a power level of 294 mW. Compare this to the best available silicon bipolar multipliers (TRW 1008J) having 70 to 100 ns multiply time and consuming 1.2 to 1.6 watts. It is anticipated that with expected improvements, multiply times for an 8 x 8 GaAs multiplier using the present (1 μm) lithography of the order of 2 ns should be achievable at the present (few hundred mW) power levels. The prospect of 500 million multiplies/s parts operating at reasonable power levels should open up exciting signal processing opportunities for system designers.

The characteristics of the GaAs integrated circuits demonstrated to date are in the range required for ultra high speed VLSI. This can perhaps best be illustrated by considering, not as any proposed design but rather as an exercise in arithmetic, the fabrication of an entire central processing unit from a mainframe computer on a single GaAs chip. For example, the CPU for a state-of-the-art Amdahl V470 computer contains about 10^5 gates in 1680 silicon bipolar "LSI" packages, dissipating ~ 3 kW (+ cooling power). If we take as a practical chip area 1 cm^2 and assume only 30% of the area will be useable for gates, the required area/gate would be < 300 μm^2; GaAs SDFL gates of 600 μm^2 have been fabricated, and areas as low as 150 μm^2 appear achievable with the present 1 μm optical lithography. Taking a conservative 20 W power dissipation limit for the 1 cm^2 chip, the gate dissipation limit would be 200 μW/gate; power levels as low as 150 μW/gate have already been demonstrated with GaAs SDFL. Assuming an average gate clocking frequency of f_c = 1 GHz, from (1) the dynamic switching energy would have to be less than 100 fJ; GaAs SDFL NOR gates with values as low as $P_D\tau_d$ = 27 fJ (16 fJ in inverters) have already been reported. While it is not the purpose of this example to propose it, if such a chip were in fact fabricable and practical for this application, it would be over one order of magnitude faster, and consume well over two orders of magnitude less power than the current silicon bipolar implementation.

While the performance results already achieved in GaAs ICs are exciting, it is interesting to speculate about what future improvements may bring. On the basis of (A1), speed-optimized depletion-mode GaAs MESFET ICs with L_g = 1 μm might be expected to achieve τ_d ~ 40 to 50 ps propagation delays. Using direct-writing E-beam lithography rather than optical lithography, L_g = 0.5 μm or smaller circuits might achieve propagation delays below 20 ps. These speeds are far beyond those anticipated for silicon ICs, and are in fact about the same as those achieved with superconducting Josephson junction technology. (The $P_d\tau_d$ products for Josephson logic is either better than GaAs ICs or worse, depending on whether or not one corrects for the He refrigerator power.) It should be noted that modest cooling (T ~ 77°K) of GaAs MESFETs, improves the mobility and drift velocity of the carriers for improved g_m and f_τ, so that considerably higher logic switching speeds should be obtained by cooling GaAs ICs. Of course, if operation at reduced temperatures were anticipated, even higher performance might be obtained by going to a smaller-bandgap III-V semiconductor such as InAs or InSb in which carrier mobilities in the $\mu_n \sim 10^5$ to 10^6 $cm^2/V \cdot s$ range are obtained at 77°K. Logic circuits using InAs or InSb MOSFETs should not only achieve excellent switching speeds, but should do so with very low operating voltages for ultra-low dynamic switching energies.

If the achievement of very complex ultra high speed GaAs VLSI circuits proves possible, it would define new dimensions in on-chip computational power. A 10^5 gate GaAs chip with τ_d = 100 ps speeds operating in some type of pipeline processor would have, assuming an $f_c = 1/5\ \tau_d$ clock frequency, a gate count-clock rate product of 2×10^{14} per second - a prodigious computation rate. The achievement of this level of GHz rate computational power in a low power integrated circuit could make possible systems approaches in high-speed signal processing, communications, etc., that one would hardly dare visualize with today's technology.

Acknowledgement

The authors are indebted to the many hard working individuals contributing to the GaAs integrated circuit programs at Rockwell. The analysis and results presented here are a tribute to them. In particular we wish to thank P. Asbeck, C. Kirkpatrick, and Y. D. Shen for their contributions in material studies, ion implantation and processing, and S. Long, R. Zucca, F. Lee, and C. P. Lee for circuit design, testing, and device analysis. F. Eisen has made invaluable contributions in the development of GaAs ion implantation technology and along with F. Blum has provided, valuable guidance, and support to the development of GaAs integrated circuit technology.

Appendix: Nonlinear Switching Analysis for $P_D\tau_d$ Dependence on τ_d and FET K-Value

Simple linear current gain considerations show that the switching speed of a transistor or FET with fanout of N similar devices, even in the absence of parasitic capacitances, must be limited by the current gain-bandwidth product, f_τ, to a value of the order of

$$\tau_d > \frac{N}{\pi f_\tau} \quad . \tag{A1}$$

As discussed in Section I, $f_\tau \sim g_m/2\pi C_{gs}$ so that both materials (ε, μ_n, v_s, etc.) and device geometry (L_g, etc.) parameters influence the maximum attainable switching speeds. Analysis at this linear level, however, does not consider the performance in the nonlinear switching (i.e., switching the FET from "off," where $g_m = 0$, to "on") mode usually used in FET logic. Further, the analysis gives little insight into the power or speed-power product (dynamic switching energy) trade-offs involved in attaining these switching speeds. In particular, we would like to know the "price" in terms of speed-power product, $P_D\tau_d$, which we must pay in order to achieve a given switching speed or propagation delay, τ_d, and how the FET material or geometrical parameters can influence the $P_D \tau_d$ "price." The following simple nonlinear FET switching analysis addresses these questions.

Consider the simple FET inverter circuit shown in Fig. 21a, consisting of a switching FET and a depletion-mode (current-source) load. In general, FET logic requires, for example, paralleling many FETs to obtain multi-input (positive) NOR functions. This means that each individual FET must be switched virtually "off" ($V_{gs} < V_p$) for the sum of the many drain "off" currents to be small in comparison to the active load current, I_o. We consider then, the lowest possible $P_D \tau_d$ product switching case meeting this "full off" criterion; switching between "hard" limits at $V_{gs} = V_p$ (FET conduction threshold) and $V_{gs} = V_p + V_m$, where V_m is the (peak to peak) logic swing. In the example of Fig. 15a, these "hard" switching limits are provided by ideal diodes; in most FET logic circuits, of course, the $I_{ds}(V_{gs})$ characteristics of the FET and active loads typically define the switching limits (near ground and V_{dd}). For simplicity, we assume both the switching FET and active loads are operated in the saturated drain current region at all times. This provides the fastest switching posible for the circuit. We have assumed here that the gate input range, like the output range is positive, typical for MOSFETs for example. For depletion-mode switching FETs such as depletion mode GaAs MESFETs, this requires that some constant shifting voltage be inserted in series with the actual FET gate to give it the apparent positive threshold. However, in principle, no static power need be taken from such a shifter, since the dc gate current is assumed zero, so the level shift circuitry need not be considered in the analysis, but rather just assumed in the FET characteristics.

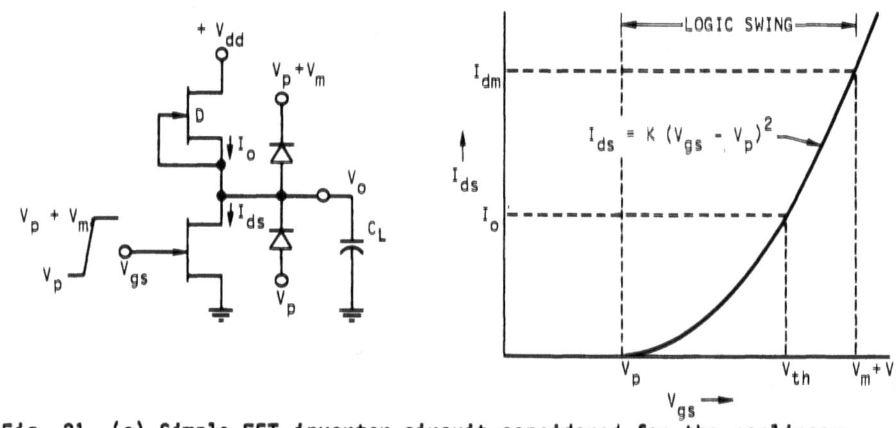

Fig. 21 (a) Simple FET inverter circuit considered for the nonlinear switching analysis. The ideal diodes used to define the output switching range to $V_p < V_o < V_m$ are conceptual; in most real FET logic circuits the FET and depletion mode load characteristics serve this function. (b) Drain current as a function of V_{gs} characteristics assumed for the switching FET, with switching limits and logic threshold ($V_o = V_{gs}$) conditions shown.

The starting point for the performance analysis for this idealized FET switching circuit of Fig. 21 is to determine the output transition time for a step input voltage to the gate. If we choose a value of load current, I_o, equal to half the maximum FET current I_{dm} (the value of saturated drain current at $V_{gs} - V_p = V_m$, such as might be obtained from Fig. 3, for example), $I_o = I_{dm}/2$, then the negative and positive transition times (output slew times from $V_o = V_p + V_m$ to $V_o = V_p$ and visa versa) for the step input will be equal and (from $I_L = I_o - I_{ds} = \pm 1/2\ I_{dm}$ for $V_{gs} = V_p$ and $V_{gs} = V_p + V_m$) given by

$$\Delta T = \frac{2C_L V_m}{I_{dm}} \tag{A2}$$

where C_L is the capacitance loading the gate. The logic propagation delay for chains of these inverters is related to this output slew time, ΔT. Computer analysis of gate chains with $I_{dm} = K(V_{gs} - V_p)^2$ gives the exact logic propagation delay, τ_d, as

$$\tau_d = \frac{2}{3}\,\Delta T \quad . \tag{A3}$$

This result should be fairly accurate for other $I_{ds}(V_{gs})$ relationships as long as they are reasonably "square law" in nature. Substituting (A3) into (A2) gives the explicit dependence of the propagation delay on the logic voltage swing, V_m, as

$$\tau_d = \frac{4C_L V_m}{3I_{dm}} \qquad \text{[Propagation Delay]} \quad . \tag{A4}$$

The peak power dissipation for this circuit is given by

$$P_D = V_{dd} I_o = V_{dd} I_{dm}/2 \quad . \tag{A5}$$

(Note that this is the maximum power; in normal 50% duty cycle switching when the limit diodes are actually the FET and active loads, the average power could be as low as half this, so the following expressions could be lowered by up to a factor of two.) The speed power product, from (A4) and (A5) is given by

$$P_D \tau_d = \frac{2}{3}\, C_L V_{dd} V_m \tag{A6}$$

where it should be noted that is the <u>dynamic</u> speed power product; that is, with power measured at maximum clocking rate. We can further simplify this $P_D \tau_d$ expression by noting that the power supply voltage, V_{dd}, must significantly exceed the logic swing, V_m. If, for arithmetic convenience, we choose $V_{dd} = (27/16)\ V_m \approx 1.7\ V_m$, a reasonable value, we have

$$P_D \tau_d = \frac{9}{8}\, C_L V_m^2 \qquad \left[\text{for } V_{dd} = \frac{27}{16} V_m\right] \quad . \tag{A7}$$

While (A2) - (A7) are valid for any I_{ds} vs $V_{gs} - V_p$ characteristics as long as it is a reasonable approximation of a square law curve, they can be considerably simplified if we treat the exact square law, $I_{ds} = K(V_{gs} - V_p)^2$ case [(3) and (4) in Section 1]. This gives

$$I_{dm} = KV_m^2 \qquad \text{(A8)}$$

which, substituted in (A4) gives for the propagation delay

$$\tau_d = \frac{4C_L}{3KV_m} \quad \text{[Propagation Delay]} \quad . \qquad \text{(A9)}$$

This gives the reasonable result that the propagation delay is proportional to the load capacitance, C_L, and inversely proportional to the product of the FET K factor [(4) or Fig. 2] and the logic voltage swing, V_m, since the available load current is $\pm 1/2\ KV_m{}^2$. Increasing V_m improves the switching speed, but at the expense of markedly increasing the power required. We can solve (A8) for the value of V_m required to achieve any given speed, τ_d, as

$$V_m = \frac{4C_L}{3K\tau_d} \quad . \qquad \text{(A10)}$$

Substituting this expression for V_m into (A7) gives us the explicit dependence of $P_D\tau_d$ product on the required switching speed, τ_d, as

$$P_D\tau_d = \frac{2C_L{}^3}{K^2\tau_d^2} \quad \text{[Dynamic Switching Energy]} \quad . \qquad \text{(A11)}$$

This equation shows the "price" in terms of switching energy,that must be paid in order to obtain high switching speeds (low values of propagation delay, τ_d). Note that this "price" for going to lower τ_d's can be offset by increasing the K-value of the FET. From (4), one way to increase K is to go to a higher mobility semiconductor, since K is proportional to the carrier mobility, μ_n, in the FET channel. Reducing the FET gate length, L_g, will also improve K, and at least to some degree, reduce C_L as well, if we assume part of the load to be input capacitances of other logic gates. In real circuits, a major contribution to C_L can be the stray substrate capacitances on interconnect lines, etc. These can be greatly reduced, improving the logic speed and speed-power products, by going to an insulating or semi-insulating substrate, as done with GaAs or CMOS/SOS. Of course, in LSI or VLSI there will in general be some fraction of interconnects which are relatively long and for which the fringing and overcrossing capacitances will be significant even with a semi-insulating substrate. From $P_d\tau_d > 1/2\ C_LV_m{}^2$, the only way to minimize the switching energy for driving such lines is to reduce the logic swing, V_m, increasing the width of the driving FET to maintain the required τ_d. Equation (A11) still applies, of course, so that the lower logic swings allowable in high speed switching with high-mobility semiconductor, high K-value FETs are of great advantage for VLSI.

Addendum (see p. 296)

References

1. R. C. Eden, B. M. Welch, and R.Zucca, "Planar GaAs IC Technology: Applications for Digital LSI," IEEE J. Solid-State Circuits, vol. SC-13, pp. 419-426, August 1978.

2. P. Wolf, "Microwave Properties of Schottky-barrier Field-effect Transistors," IBM J. Res. Develop., vol 14, pp. 125-141, Mar 1970, and other papers in that issue.
3. S. A. Roosild, private communication of silicon MESFET characteristic curve data from Air Force Contract F19628-70-C-0094 USAF, RADC, Hanscom Field, MA. (K. E. Drangeid, IBM, principal investigator).
4. J. H. Yuan and E. Harari, "Short-Channel CMOS/SOS Technology," IEEE Trans. Electron Devices, vol. ED-25, p. 989, Aug. 1978.
5. F. F. Fang and A. B.Fowler, "Hot Electron Effects and Saturation Velocities in Silicon Inversion Layers," J. of Appl. Phys., vol. 41, pp. 1825-1831, Mar. 15, 1970.
6. J. G. Ruch and G. S. Kino, "Measurements of the Velocity-Field Characteristics of Gallium Arsenide," Appl. Phys. Lett., vol. 10, p. 40, 1967.
7. R.C. Eden, J. L. Moll, and W. E. Spicer, "Experimental Evidence for Optical Population of the X Minima in GaAs," Phys. Rev. Lett., vol. 18, pp. 597-599, Apr. 1967.
8. R. S. Huang and P. H. Ladbrooks, "The Physics of Excess Electron Velocity in Submicron-channel FETs," J. Appl. Phys., vol. 48, pp. 4791-4798, Nov. 1977.
9. T. J. Maloney and J. Frey, "Transient and Steady State Electron Transport Properties of GaAs and InP," J. Appl. Phys., vol. 48, pp. 781-787, Feb. 1977.
10. M. S. Shur, "Analytical Model of GaAs MESFET's," IEEE Trans. Electron Devices, vol. ED-25, p. 612, June 1978.
11. J. A. Higgins, R. L. Kuvas, F. H. Eisen, and D. R. Chen, "Low-Noise GaAs FET's Prepared by Ion Implantation," IEEE Trans. Electron Dev., vol. ED-25, pp. 587-596, June 1978, and other papers in that issue
12. T. Mimura, K, Odani, N. Yokoyama, Y. Nakayama, and M. Fukuta, "GaAs Microwave MOSFET's," IEEE Trans. Electron. Dev., vol. ED-25, pp. 573-579, June 1978.
13. N. Yokoyama, T. Mimura, H. Kusakawa, K. Suyama and M. Fukata, "Low-Power High-Speed Integrated Logic with GaAs MOSFETs," Digest of Tech. Papers, 1979 Int. Conf.on Solid State Devices, Tokyo, Aug. 27, 1979, pp. 31-32; to be published in Japanese J. Appl. Phys., vol 19, Feb. 1980.
14. R. Zucca, "Electrical Compensation in Semi-Insulating GaAs," J. Appl. Phys., vol 48, pp. 1987-1994, May 1977.
15. R. Zuleeg, J. K. Notthoff, and K. Lehovec, "Femtojoule High-Speed Planar GaAs E-JFET Logic," IEEE Trans. Electron Devices, vol. ED-25, pp. 628-639, June 1978.
16. A. A. Immorlica, Jr., and F. H. Eisen, "Planar Passivated GaAs Hyperabrupt Varactor Diodes," Proc. Sixth Biennial Cornell Electrical Eng. Conf., August 1977, pp 151-159.
17. H. Morkoc, S. Bandy, R. Sankaran, G. Antypas, and R. Bell, "A Study of High-Speed Normally Off and Normally On $Al_{0.5}Ga_{0.5}As$ Heterojunction Gate GaAs FET's (HJFET)," IEEE Trans. Electron Devices, vol. ED-25, pp. 619-627, June 1978.
18. H.Ishikawa, H. Kusakawa, K. Suyama, and M. Fukuta, "Normally-Off Type GaAs MESFET for Low-Power High-Speed Logic Circuits," 1977 Int. Solid State Circuits Conf., Digest of Tech. Papers, Feb. 1977, pp. 200-201.
19. G. Bert, G. Nuzillat, and C. Arnodo, "Femtojoule Logic Circuit Using Normally-Off GaAs MESFETs," Electron. Lett., vol. 13, pp. 644-645, Oct. 1977.
20. M. Fukuta, K. Suyama, and H. Kusakawa, "Low Power GaAs Digital Integrated Circuits with Normally Off MESFETs," IEEE Trans. Electron Devices, vol ED-25. p. 1340, Nov. 1978.

21. C. A.Liechti, "GaAs FET Logic," 1976 Intern. GaAs Symp., Inst. Phys. Conf., Ser. No. 33a, 1977, ch. 5, pp 227-236.
22. R. L. VanTuyl, C. A. Liechti, R. E. Lee, and E. Gowen, "GaAs MESFET Logic with 4-GHz Clock Rate," IEEE J. Solid-State Circuits, vol. SC-12, pp. 485-496, Oct. 1977.
23. P. T. Greiling, C. F. Krumm, F. S. Ozdemir, L. H. Hackett, and R. F. Lohr, Jr., "Electron Beam Fabricated GaAs FETs Inverter," IEEE Trans. Electron Devices, vol. ED-25, p. 1340, Nov. 1978.
24. R. C. Eden, B. M. Welch, and R. Zucca, "Low Power GaAs Digital ICs Using Schottky Diode-FET Logic," 1978 Int. Solid State Circuits Conf., Digest of Tech. Papers, pp. 68-69, Feb. 1977.
25. B. M. Welch and R. C. Eden, "Planar GaAs Integrated Circuits Fabricated by Ion Implantation," 1977 Int. Electron Device Meeting, Tech. Digest, Dec. 1977, pp. 205-208.
26. S Yanagisawa, O. Wada, and H. Takanashi, "Gigabit Rate Gunn-Effect Shift Register," 1975 Int. Electron Devices Meeting, Tech. Digest, Dec. 1975, pp 317-319.
27. M. Cathelin and G. Durand, "Logic ICs Using GaAs FETs in a Planar Technology," L'onde Electrique, vol. 58, pp. 218-221, Mar. 1978.
28. R. L. Van Tuyl, C. A. Liechti and C. A. Stolte, "Gallium Arsenide Digital Integrated Circuits," Tech. Rep. AFAL-TR-76-264, Apr. 1977.
29. M. R. Splinter, "A 2-μm Silicon Gate CMOS/SOS Technology," IEEE Trans. on Elect. Dev., vol. ED-25, pp. 996-1003 (1978).
30. M. T. Elliott, M. R. Splinter, A. B. Jones, J. P. Reekstin, "Size Effects in E-Beam Fabricated MOS Devices," 1977 Inter. Electron Device Meeting, Tech. Digest pp. 11A-B.
31. G. Nuzillat, C. Arnado and J. P. Puron, "A Subnanosecond Integrated Switching Circuit with MESFETs for LSI," IEEE J. Solid-State Circuits, vol. SC-11, pp. 385-394 (1976).
32. T. Nakamura et. al., "Punch-Through MOSFET for High-Speed Logic," 1978 Int. Solid-State Circuits Conf., Tech. Digest. pp. 22-23.
33. H. M. Darley, T. W. Houston and G. W. Taylor, "Fabrication and Performance of Submicron Silicon MESFET," 1978 Int. Electron Devices Meeting, Tech. Digest, Dec. 1978, pp. 62-65.
34. R. C. Eden, "GaAs Integrated Circuits, MSI Status and VLSI Prospects," 1978 Int. Electron Devices Meeting, Tech. Digest, Dec. 1978, pp. 6-11.
35. G. Bert, T. Pham Ngu, G. Nuzillat and M. Gloanec, "Quasi-Normally-Off MESFET Logic for High Performance GaAs ICs," Paper 7 in Research Abstracts of the First Annual Gallium Arsenide Integrated Circuit Symposium, Lake Tahoe, Sept. 17, 1979.
36. J. K. Notthoff and C. H. Vogelsang, "Gate Design for DCFL with GaAs E-JFETs," Paper 10 in Research Abstracts of First Annual Gallium Arsenide Integrated Circuit Symposium, Lake Tahoe, Sept. 27, 1979.
37. R. C. Eden, F. S. Lee, S. I. Long, B. M. Welch, and R. Zucca, "Multi-Level Logic Gate Implementation in GaAs ICs using Schottky Diode-FET Logic,"- 1980 Int. Solid State Circuits Conf., Digest of Tech. Papers, Feb. 1980.
38. S. I. Long, B. M. Welch, R. C. Eden, F. S. Lee and R. Zucca, "MSI High Speed Low Power GaAs Integrated Circuits," 1979 Int. Conf. on Solid State Devices, Digest of Tech. Papers, Tokyo, Aug. 1979, pp. 29-30; to be published in Japanese J. Appl. Phys., vol. 19, No. 2, Feb. 1980.
39. R. C. Eden, B. M. Welch, R. Zucca and S. I. Long, "The Prospects for Ultrahigh-Speed VLSI GaAs Digital Logic," IEEE J. Solid-State Circuits, vol. ED-26, pop. 299-317, Apr. 1979.

40. T. Mizutani, N. Kato, S. Ishida, K. Asai, Y. Sakakibara, K. Komatsu, and M. Ohmori, "High-Speed Enhancement-Mode GaAs MESFET Integrated Circuits," 1979 Int. Conf. on Solid State Devices, Digest of Tech. Papers, Tokyo, Aug. 1979, pp. 33-34; to be published in Japanese J. Appl. Phys., vol 19, No. 2, Feb. 1980.
41. R. H. Dennard, F. H. Gaensslen, E. J. Walker and P. W. Cook, "1 µm MOSFET VLSI Technology: Part II - Device Designs and Characteristics for High-Performance Logic Applications," IEEE J. Solid-State Circuits, vol. SC-14, pp. 247-255, Apri. 1979, and IEEE Trans. Electron Devices, vol. ED-26, pp. 325-333, April 1979.
42. G. L. Troeger, A. F. Behle, P. E. Friebertshauser, K. L. Hu and S. H. Watanabe, "Fully Ion Implanted Planar GaAs E-JFET Process," 1979- Int. Electron Devices Meeting, Tech. Digest, Dec. 1979, pp. 497-500.
43. P. I. Suciu, E. N. Fuls, and H. J. Boll, "High-Speed NMOS Circuits Made with X-Ray Lithography and Reactive Sputter Etching," Electron Device Letters, vol. EDL-1, pp. 10-11, Jan. 1980.
44. M. Gloanec, G. Nuzillat, C. Arnodo and M. Peltier, "A GaAs Integrated Edge-Triggered D-type Flip-Flop," Digest of the First Specialty Conference on Gigabit Logic for Microwave Systems, Orlando, Fla., pp. 114-119, May 1979.
45. F. E. Eisen, B. M. Welch, K. Gamo, T. Inada, H. Mueller, M. A. Nicolet and J. W. Mayer, "Sulfur, Selenium and Tellurium Implantation in GaAs," Inst. Phys. Conf. Ser. No. 28 (1976): Chap. 2.
46. B. M. Welch, F. E. Eisen and J. A. Higgins, "Gallium Arsenide Field Effect Transistors by Ion Implantation," J. Appl. Physics 45, 3685, 1974.
47. P. T. Greiling, F. S. Ozdemir, C. F. Krumm and B. F. Lohr, Jr., "Electron Beam Fabricated GaAs Integrated Circuits," 1979 IEDM Tech. Digest.
48. B. M. Welch, Y. D. Shen, R. Zucca and R. C. Eden, "Planar High Yield GaAs IC Processing Techniques," 1979 IEDM Tech. Digest.
49. B. M. Welch, Y. D. Shen, R. Zucca, R. C. Eden, and S. I. Long, "LSI Processing Technology for Planar GaAs Integrated Circuits," to be published in IEEE Trans. Electron Devices, Special Issue, June, 1980.
50. R. E. Lundgren, C. F. Krumm and R. L. Pierson, "Fast Enhancement-Mode GaAs MESFET Logic," 37th Ann. Dev. Research Conf., Boulder, Co. 25-27 June, 1979.
51. R. E. Lundgren, C. F. Krumm and R. F. John Jr., "Enhancement-Mode GaAs MESFET Logic," Presented at the IEEE GaAs IC Symposium, Lake Tahoe, NV, Sept. 1979.
52. F. E. Eisen and B. M. Welch, "Radiotracer Profiles in Sulfur Implanted GaAs," Proc. 5th International Conference on Ion Implantation in Semiconductors and Other Materials, Boulder, CO, August 1976.

Additional references (on p. 297)

6. VLSI Architecture

E. Swartzlander

With 39 Figures

In the last two decades, integrated circuit technology has evolved from primitive circuits consisting of a few transistors interconnected to implement a basic gate or flip-flop, to the current levels of Very Large Scale Integration (VLSI) where complete subsystems consisting of several thousand gates are operated at clock rates approaching 100 MHz. In spite of this phenomenal growth (or perhaps because of it), there is an expanding list of crucial functions that await the development of circuits with more gates, higher speed, or both.

Effective architecture development requires coordination of the technological constraints, internal device structure, and external interfaces to develop chips that are producible, efficient, and widely applicable. In this chapter, the technology basis is reviewed to establish the perspective for detailed architectural examination. The internal chip architecture and system architecture are then examined in detail, and a case study of VLSI architecture selection for an advanced radar digital beam former demonstrates the application of these concepts to chip development in the context of a demanding signal processing application.

6.1 VLSI Technology Basis

Circuit technology has improved dramatically in the period since the first integrated circuits as illustrated in Tab. 1. The Small Scale Integration (SSI) circuits with the order of 10 gates/chip were configured as simple logic gates and flip-flops with all signals externalized (i.e., brought out through package pins). At the Medium Scale Integration (MSI) level, more complex circuits such as 4-bit, ripple carry decimal, duo-decimal (÷ 12), and hexa-decimal (÷ 16) counters and 2- and 4-bit ripple carry full adders were developed. For these circuits, it was necessary to limit the input/output (I/O). This introduced questions of testability and necessitated the development of signal patterns to diagnose the correct operation of internal circuits for input and output signals that were not directly accessible. At the Large Scale Integration (LSI) level, functions with approximately 1000 gates were developed, such as 4-bit Arithmetic Logic Units (ALUS), bit slice microprocessors, and 1K dynamic memories. The ALUS and microprocessors could be diagnosed and tested by extending the MSI approaches, but early dynamic memory revealed pattern sensitivity such that the order of testing became as important as the patterns themselves. At the VLSI level, with over 3000 gates/chip, complete subsystems such as parallel multipliers, "Flash" Analog-to-Digital (A/D) converters, and large dynamic memories are currently available. A significant challenge lies in defining and developing additional chips of VLSI complexity.

Table 1 Integrated circuit technology

TECHNOLOGY	GATES/CHIP	FUNCTION
SSI	1-30	GATES, FLIP-FLOPS
MSI	30-300	COUNTERS, 4 BIT ADDERS
LSI	300-3K	ALUS, MICROPROCESSORS
VLSI	3K-30K	MULTIPLIERS, A/D CONVERTERS
VHSIC	>30K	FFT'S, COMPLETE COMPUTERS

Finally, no discussion of circuit evolution would be complete without mention of the U.S. Department of Defense effort in developing Very High Speed Integrated Circuit (VHSIC) technology. As shown in Tab. 1 VHSIC is aimed at increasing the gate count to levels in excess of 30,000 gates/ chip (and simultaneously increasing the performance of the individual gates). This should allow development of complete systems (e.g., FFT spectrum analysers, complete computers, etc.) on single chips. The VHSIC effort has further significance in that it is focusing on ways to reduce the cost and development time for complex chips which will result in increased opportunity for advanced chip architecture.

An important aspect of the technology evolution is that the device performance has improved along with the growth in the level of integration. This is illustrated in Fig. 1, where the speed and density are plotted as a function of design date for the triple diffused bipolar silicon technology. Part of the explanation for the close tracking in speed and density is that in most technologies both depend on the minimum feature size. Current projection photomasking techniques produce devices with 1-2 micron minimum feature size with reasonable yield. The nearly exponential growth in device performance and integration level is expected to continue as feature sizes are pushed to the submicron level by replacing optical methods with E-beam or X-ray lithography.

As is becoming widely recognized, the utility of an integrated circuit depends on the product of speed and density. This is because most systems may be implemented with a simple but very fast processor or with many slower processors. Partitioning is frequently employed in signal processing where multiple processors are configured in parallel arrays or in serial pipelines to accomplish the desired processing. A chart showing the attainable gate-rate (i.e., the density speed product) for isolated emitter coupled logic (ISO-ECL), triple diffused bipolar, and NMOS technologies is shown in Fig. 2. These speeds and densities are intended to be indicative of devices which are commercially available in 1980. The speeds refer to the maximum clock rates of large shift registers, and the densities to the number of equivalent two input gates. For each technology, the approximate speed and density limits are shown. It should be emphasized that gate-rate levels of approximately 10^{12} transitions/sec are achievable over a range of roughly two orders of magnitude in speed (i.e., 10 MHz to 1 GHz).

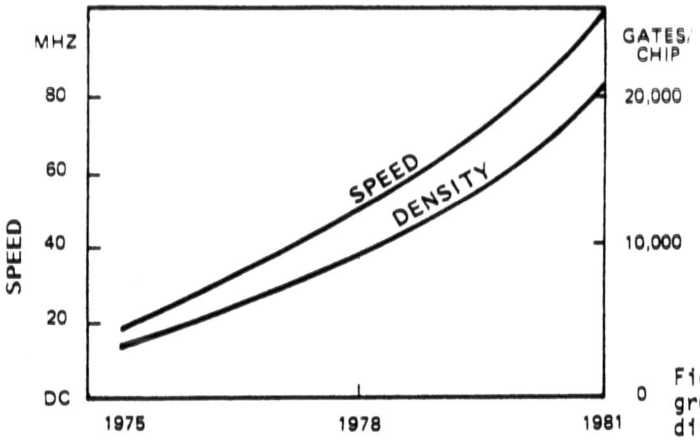

Fig.1 Speed and density growth for the triple diffused technology

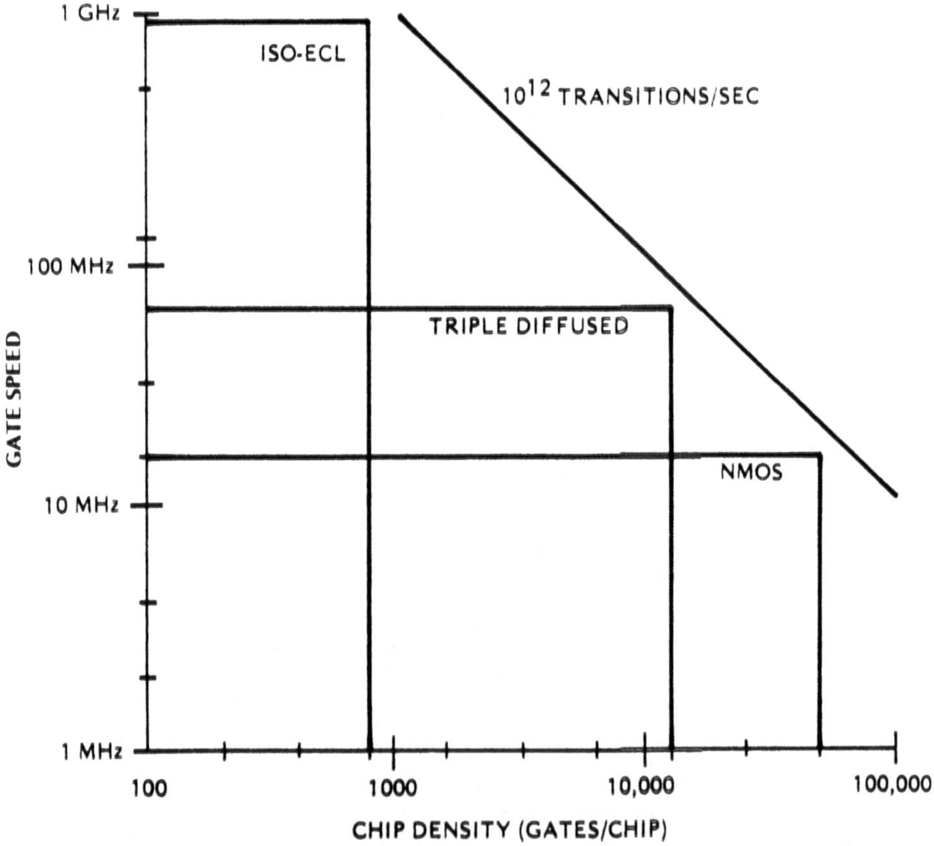

Fig.2 VLSI Gate-rate chart

Extensions to the gate rate metric are useful in characterizing memory requirements: a reasonable approach is to compute the product of the number of bits of storage with the chip access rate to arrive at the bit rate. Use of chip access rate in developing the bit rate metric gives a higher figure of merit for devices with more than 1 data port and to memories that are organized in wide (e.g., 8 bit) words. Complete processors may be characterized by their combined gate-rate and bit-rate requirements: the arithmetic section contributes to the total gate-rate, the memory contributes to the bit-rate, and the control unit contributes to both. Families of processors of roughly equivalent complexity may span a wide range of gate-rate and bit-rate as shown in Fig. 3. The curves on Fig. 3 are plotted on a relative scale and the shapes are, in fact, arbitrary since the equivalence between various levels of gate-rate and bit-rate depend on the relative complexity of logic and memory which is highly dependent on technology.

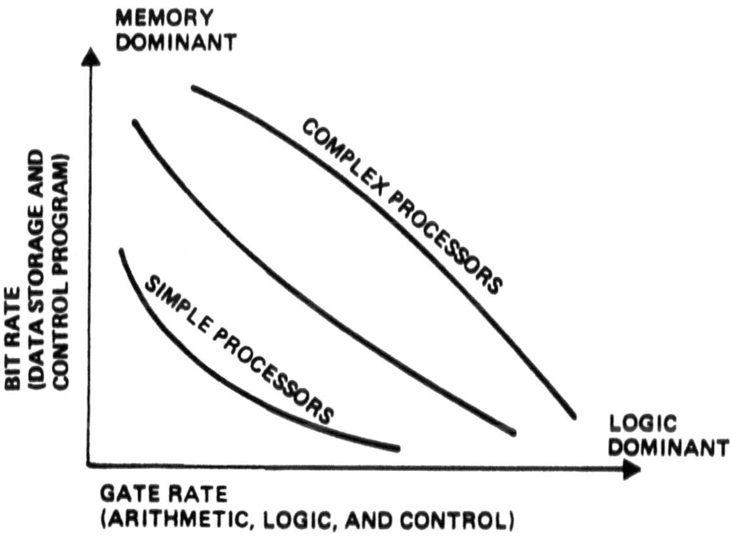

Fig.3 Processor complexity comparison

One significant constraint on the development of high complexity VLSI chips is the problem of packaging. This is due in part to the prevalence of the Dual In-Line Package (DIP) which has leads in two rows (typically on 0.3, 0.6, or 0.9 inch centers) for insertion into printed circuit boards or wire wrap socket panels. The growth in size of the largest commercially available DIP is shown in Fig. 4. As of 1980, the largest package in wide use is the 64-pin DIP. As the figure shows, the growth in size has been approximately linear with time; this suggests that packages will not become substantially larger in the near future. Part of the explanation for slow growth is the difficulty in constructing and using the large packages. Even with aluminum heat spreaders like that shown in Fig. 5 for a 64-pin EIP, the ceramic packages are fragile and require considerable care to avoid breaking the package or damaging the pins. These constraints are expected to continue, posing a significant challenge in chip architecture, as exponentially increasing chip complexity must communicate through, at most, about 100 package pins.

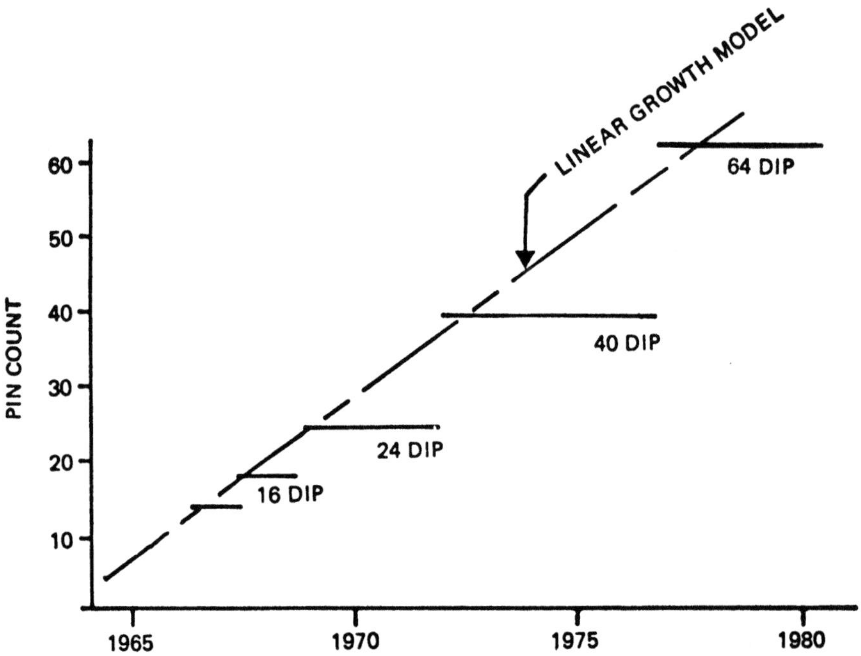

Fig.4 Commercial DIP sizes

Fig. 5 Sixty-four-pin DIP

6.2 VLSI Device Architecture

With the tremendous advance in integrated circuit technology, designers have assessed device architecture as being non-critical since "technology will fix the problems." As the technology has become increasingly constrained by packaging limitations, this attitude has begun to change. Designers now realize that it is necessary to optimize the device architecture as an explicit step in the system design process.

The design issues will be briefly considered in this section. Then an example of the architecture comparison studies implicit in the design process will illustrate how the issues impact the device development for a state-of-the-art computed tomography unit. Finally, several examples of the current state-of-the-art chips will show general characteristics of successful designs.

6.2.1 Device Architecture Issues

In device architecture it is necessary to consider ways to simplify the design process, to maintain clean functional partitions, and to minimize the amount of output from the chip.

Design Simplification

As the integration level of VLSI chips grows, the development and use of architectures which simplify the design process becomes increasingly important. A key concept in this area is the use of cellular designs in which a small logic circuit is designed and replicated in an array to realize the desired function. Cellular designs are directly applicable to arithmetic functions because many of the simple pencil and paper arithmetic approaches developed by and for humans can be directly converted into cellular logic. The resulting regular structures are easier to design than an "optimized" chip and easier to test and debug, but they may use more chip area and slightly more power.

Functional Partitioning

The second key aspect of chip architecture is the need to preserve clean functional partitions. This accomplishes two basic goals: partitioning minimizes the number of signal lines, by placing a complete function or subfunction on a chip, and it simplifies the use of the chip. Minimization of signal lines is necessary to meet package pin constraints as discussed in the previous section. It also minimizes the chip area which is devoted to bonding pads. For example, the TRW MPY-16H single chip 16-bit parallel multiplier shown in Fig. 6 devotes 17 percent of the total chip area to I/O bonding pads. Although this chip has 64 pads, it is an extremely complex chip with about 3000 equivalent gates so that the pad area occupies a smaller area than less complex chips. When the area required for I/O circuits (i.e., level shifters, buffers, drivers, 3-state interfaces, etc.) is included, a typical chip devotes 30 to 40 percent of the total area to I/O. Reduction in the number of signal lines can be accomplished by careful functional partitioning which is simplified by the ever increasing level of integration.

Fig.6 TRW MPY-16 Die photograph

Output Minimization

The third architectural issue is the desire to minimize the output (i.e., the number and speed of output lines) for VLSI circuits with an aim toward reducing the device power dissipation. Lithographic scaling of devices has reduced the power dissipation per device, but the number of devices per chip has increased at a faster rate so that power management is a significant VLSI problem. Power dissipation can be reduced by minimizing the number and speed of device output signals. In addition to eliminating or reducing the area required for output drivers, it also reduces the driver power consumption. Specifically, the driver current, i, is that required to charge a load capacitance, C_ℓ, at a given slew rate, de/dt:

$$i = C_\ell \, de/dt$$

where de/dt is the logic swing divided by the switching time. If, as is generally the case, the current is approximately constant, the voltage across the driver output transistor is a ramp and the average power during switching is:

$$p = \frac{V_{CC} - V_L + V_{CC} - V_H}{2} C_\ell \, de/dt$$

where V_L is the low output voltage and V_H is the high output voltage. For example, if $V_{CC} = 5V$, $V_L = 0$, $V_H = 3V$, $C_\ell = 50$ pf, and the switching time is 4 nsec (i.e., $de/dt = 0.75 \times 10^9$ V/sec), the power during switching is 0.13 watts. The power can be reduced by decreasing the capacitive load, C_ℓ; the supply voltage, V_{CC}; or by decreasing the rate at which the driver is switched. From an architectural viewpoint, a reduction in outputs or in the switching rate of the outputs is the only available technique. Chips could be developed with complete systems so that only a single binary output is required in many situations. However, this would require a higher level of integration than is anticipated in the foreseeable future.

In summary, the basic internal architecture features that are desired for VLSI design are: cellular logic design (to simplify design and test), clean functional partitioning (to simplify chip usage and to minimize pin count), and minimum output (to reduce power and pin count).

6.2.2 VLSI Device Design Example

To illustrate the issues in chip level architecture, the design of a multi-term multiplier-adder will be examined in detail. This device development is used for the implementation of convolutional FIR filters and circuits to evaluate inner products for high speed computed tomography; however, the analysis is applicable to many signal processing algorithms.

Performance Requirements

The performance requirements for the specific computed tomography unit, the Mayo Clinic Dynamic Spatial Reconstructor (DSR), result from the need to collect and process data sufficient to reconstruct three-dimensional images of the chest (with a resolution of approximately 240 x 240 x 240 pixels) at 60 images/sec for up to 4 sec. With a conventional single slice x-ray scanner, this would require processing up to 57,600 cross sections per 4 sec of patient time. Computation of this number of cross sections with a general purpose computer (at 30-150 sec/cross section) is clearly impractical. By use of the filtered back projection (FBP) algorithm, the required processing is decomposed into standard signal processing functions which are implemented as computation of inner products of vectors (i.e., sums of products). Filter back projection [1] consists of two distinct steps: linear filtration of each x-ray projection, followed by "back projection" of the filtered projections to form the final reconstructed images.

In linear filtration [2], the digitized samples from each x-ray projection are linearly filtered (convolved) with a single, pre-established filter kernel. Convolution consists of repeated evaluations of inner products between the filter kernel and time-shifted copies of the projection. The back projection operation [2] estimates the density values of the reconstructed image pixels. One sample from each of the projections is selected, weighted, and summed with weighted samples from the other projections to determine the x-ray density, or linear attenuation coefficient, of each pixel. Both of these operations may be implemented as an evaluation of a succession of inner products. In the case of linear filtering, one vector is the filter kernel and the other is the successive input data. For back projection, one vector is the weights, and the other is one sample from each of the projections.

In both cases, 16-bit fixed point arithmetic is adequate for the inputs to the multiplier. Multiplied outputs are summed in full precision with truncation or rounding after the accumulation. This application requires approximately 500 million multiplies/sec and a comparable number of additions. Because of the specialized nature of the system, and the low likelihood that the algorithms will be modified in the future, multiple parallel processing channels or a single high speed channel may be used.

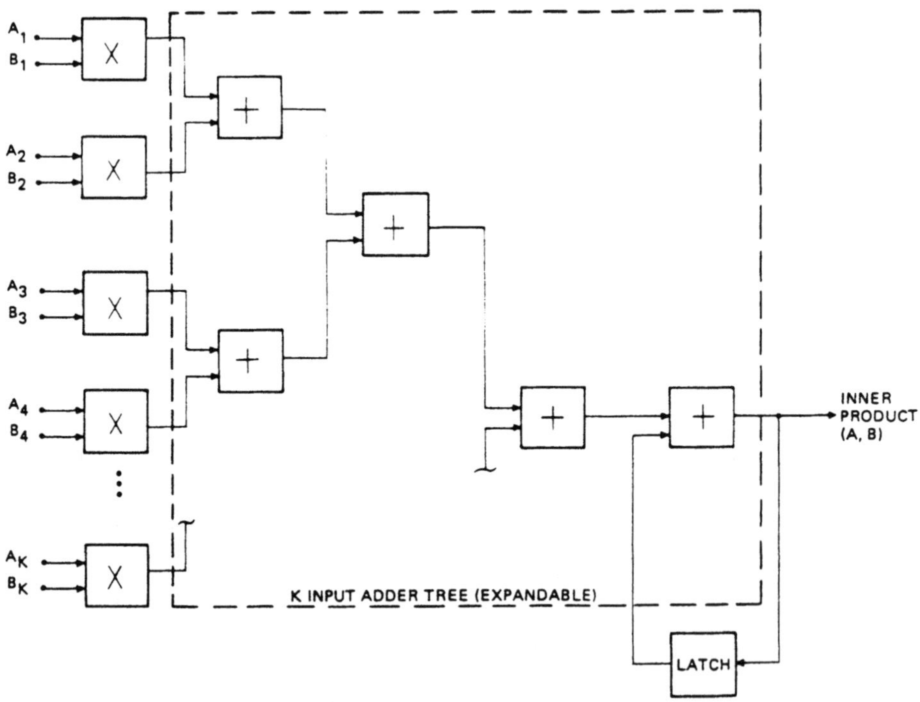

Fig.7 Multi-term inner product processor

The basic circuit structure which will be implemented is a multi-term inner product processor, as shown in Fig. 7. To form an inner product of two vectors of length m, the first k elements of the vectors are multiplied and summed; then the next k elements are multiplied, summed, and added to the previous sum. After repeating this process [(m-1)/k] + 1 times*, the inner product has been computed [3]. The circuit uses k multipliers and a (k-1) element adder tree to multiply and sum the k term. An additional adder and latch accumulate the sums. At each stage in the adder tree, adders of a sufficient width are used to avoid rounding or truncating the data. The accumulator width sets an upper limit on the vector length which can be multiplied without the possibility of overflow; however, this width may easily be made large enough to satisfy practical requirements (i.e., with ten growth bits, 1024-term inner products may be computed without overflow).

Three distinct arithmetic implementation approaches will be considered: a modular array which serves as a building block for constructing multipliers and adders; a merged arithmetic multi-term multiplier-adder; and a commercial VLSI multiplier-accumulator. All three approaches use conventional two's complement arithmetic, but due to different assumptions about the available technology, markedly different designs have been developed.

Modular Array

The modular array is based on an ultra high-speed two's complement integrated circuit which is used for addition, subtraction, and multiplication. Such a device can operate at projected clock rates of up to 500 MHz by pipelining, although the latency (the time from when the operands are entered until the result is available) will be on the order of tens of nsec. In the multi-term inner product processor, as well as many other signal processing applications, large blocks of data are processed without data-dependent branch instructions, so that achieving high processing rates outweighs minimizing the latency.

The circuit is a carry-save implementation of Booth's multiplication algorithm [4], which has been widely accepted for both software and hardware multiplication applications. This algorithm may be implemented for pipelining using a carry-save approach for the addition/subtraction of the multiplicand to the partial product [5]. For example, a 16 x 16 multiplier requires a 32 stage pipeline; the first 16 stages of the pipeline form two intermediate results (i.e., a sum word and a carry word) which sum to the product via the carry-save process; the final 16 stages perform carry assimulation. Since the pipeline clock interval need only be long enough to permit execution of a carry-save addition operation between successive clock cycles, pipeline rates on the order of 500 MHz can be attained with advanced oxide isolated ECL process technology. Booth's algorithm was selected for possible implementation as a modular array because a single VLSI circuit design can be used to multiply (by using the complete multiplier configuration), or to add/subtract by using only the carry-save adder array.

In Booth's algorithm, adjacent bits in the multiplier operand are examined in overlapping pairs, and a decision is made to add or subtract the multiplicand from the accumulated sum of partial products, or to

*[X] denotes the ENTIER function, whose value is the largest integer which is not greater than X.

retain the previous sum. Two adjacent multiplier bits are inspected at a time; if both multiplier bits are the same (either 00 or 11), the partial product is shifted one position to the right. If the multiplier bit pattern is 10, the multiplicand is subtracted from the partial product; if the bit pattern is 01, the multiplicand is added to the partial product. This procedure is repeated N-1 times for an N-bit multiplier. The requirement for physical realization of this algorithm is an array of full adders and a small amount of peripheral logic, as depicted in Fig. 8. Algebraic subtraction is performed, when required, by adding the two's complement of the multiplicand; this is implemented directly within the device by forming the one's complement, and then by adding a 1-bit at the least significant bit position.

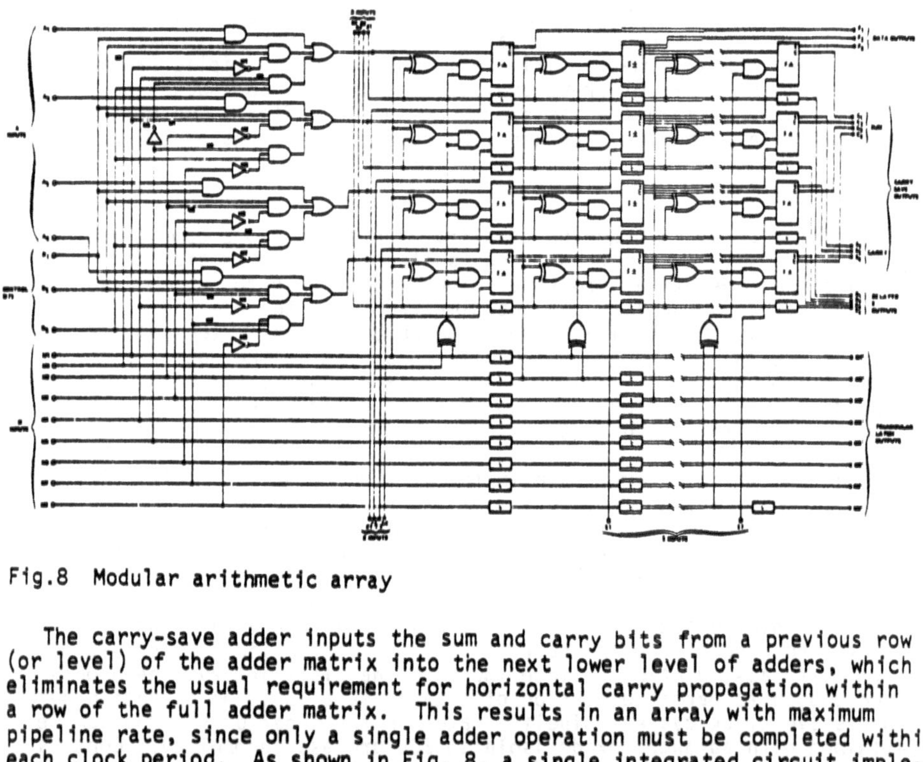

Fig.8 Modular arithmetic array

The carry-save adder inputs the sum and carry bits from a previous row (or level) of the adder matrix into the next lower level of adders, which eliminates the usual requirement for horizontal carry propagation within a row of the full adder matrix. This results in an array with maximum pipeline rate, since only a single adder operation must be completed within each clock period. As shown in Fig. 8, a single integrated circuit implementing this modified form of Booth's algorithm consists of three sections: a triangular array of (multiplier) pipelining latches; a rectangular array of (multiplicand) pipelining latches; and an array of full adders supported by a small amount of ancillary logic. The two arrays of latches have all input and ouput leads externalized for independent use as delay elements. The triangular array of latches on the left side appropriately delays the multiplier bits; this delaying action corresponds to the delay encountered by the multiplicand bits and partial products during their propagation through the carry-save adder circuit. Including pipeline latches at every level in the array ensures a 500 MHz operating speed, assuming the same gate delays achieved by an 8 x 8 ECL multiplier [6], which performs 17 add operations in one path in 25 nsec.

The sizing of the modular array has been set at 8 x 4, as shown in Fig. 8. The array requires 56 signal pins which (when augmented by the pins necessary to supply power and clock inputs) can be conveniently packaged in a 64-pin integrated circuit package. Further increases in the array size (e.g., to 16 x 8 bits) would require packages with more than 100 pins.

The modular arithmetic array may be used to implement a variety of operations, including addition, subtraction, and multiplication. The simplest designs generate the results on a "time skewed" basis; that is, the resulting bits are not all present simultaneously at the output, but emerge from the unit sequentially from the least significant to more significant bit positions as they are formed. For example, the least significant bit is available one clock period after the operands are presented to the device inputs. The next most significant bit is available one clock period later, and so on. If necessary, the skewed result can be deskewed by using the triangular latch arrays.

Table 2 summarizes the number of modular arrays required for the implementation of the various arithmetic circuits. The complexity level ranges from two devices for realization of an 8-bit adder, to thirty devices for a 40-bit adder. All entries in this table include deskewing logic.

Table 2. Modular array count for different arithmetic functions

FUNCTION	WORD SIZE	NUMBER OF 4 x 8 MODULAR ARRAYS
ADDER	8	2
ADDER	16	6
ADDER	24	12
ADDER	32	20
ADDER	40	30
MULTIPLIER	8	6
MULTIPLIER	12	12
MULTIPLIER	16	20

At first, a pipelined adder with a 40-stage latency (e.g., the output is available 40 clock periods after the operands enter the adder) appears unsuitable for use in accumulating inner products; however, highly efficient operation can be obtained by interleaving N ($\geq$ 40) inner products and computing a new component for each of the inner products every N clock cycles. This process is shown for convolutional filtering in Fig. 9. For each input datum H_i, N kernel values ($\alpha_1, \alpha_2, \ldots, \alpha_N$) are accessed and multiplied to form N intermediate products which become available after the multiplier latency time. These products are then summed in the accumulator, adding to each of the N convolutions which are in progress. During each cycle one of the accumulators will be completed and another initiated in its place.

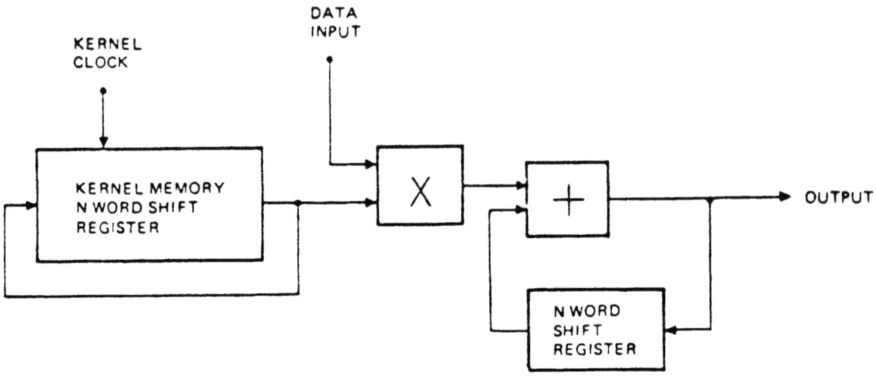

Fig.9 Modular array convolutional filter

To implement convolution and weighted summation, as required for the filtered back projection reconstruction algorithm, a single high-speed multiplier/accumulator may be used. Assuming 16-bit operands and a 40-bit result — which allows summation of 256 full precision products without overflow or roundoff — a total of 50 modular arrays are required (i.e., twenty for the multiplier and thirty for the adder). Allowing for control logic and clock buffers, the convolver implementation complexity increases to approximately 70 integrated circuits, with a computational capacity of 500 million multiply-add operations/sec (i.e., a throughput of 7 million multiply-add operations/sec per integrated circuit). It should be noted that the modular arithmetic array has not been developed, but that its complexity appears to be well within the present state of the art.

By adopting an unconventional approach to the development of a logical array which is realized via three separate (interacting) iterative functions on a single integrated circuit, a circuit design results which can be used efficiently as either an adder/subtractor or as a multiplier. This circuit is well-suited to signal processing applications in which its relatively long pipeline length and attendent latency is not a limitation. In contrast, the long output delays would pose substantial problems for general-purpose data processing applications (due to the frequent need to execute data-dependent branch instructions).

Merged Arithmetic Multi-term Multiplier Adder

The basic premise of two's complement merged arithmetic is that an optimal realization of a composite arithmetic function does not require distinct arithmetic operators (i.e., adders, subtractors, and multipliers), but instead generates the bits comprising the function directly [7]. To simplify discussion, the algorithm will be described for sign/magnitude numbers. This design can be easily modified to provide direct two's complement operation by inclusion of correction terms [8]. The sequence of operations required to compute an inner product of two K-element (each of M bits) vectors via merged arithmetic involves three basic steps: first, the bit product matrix is generated with an array of KM^2 and gates*; then

*One or more wide word addend inputs may be included as part of the composite bit product matrix to facilitate expansion.

the matrix is reduced by counting the 1-bits in each column and by performing carry processing to produce a two-row matrix; and finally, the two rows are summed in a carry lookahead adder to generate the desired inner product.

The algorithm used for the second step, matrix height reduction, is clearly the key to achieving an efficient design. Dadda's heuristic minimization procedure which seems to result in minimum complexity designs, based on use of full adders as the counters, is a simple recursion [9]:

1) Let $d_1 = 2$ and $d_j = [3d_{j-1}/2]$. Find the largest j such that at least one column of the bit matrix has more than d_j elements.

2) Use full or half adders as required to achieve a reduced matrix with no column containing more than d_j elements. Note that only columns with more than d_j elements (or those which receive carries from less significant columns) are reduced.

3) Repeat step 2) with $j = j - 1$ until a matrix with only two rows is generated (i.e., $j = 1$).

In step 2) each full adder accepts three inputs from a given column and produces a sum bit which remains in that column and a carry bit which moves to the next significant column. Thus, each use of a full adder reduces the number of elements in the composite partial product matrix by one. Similarly, half adders take in two elements from a column and produce two outputs: the sum in the original column and the carry in the next significant column.

Figure 10 shows a two term inner product reduction performed for 8-bit operand precision. At the top the two 8 x 8 trapezoidal bit products matrices are formed: one for A x B and one for C x D. The highest column is that in the middle of the figure with a height of 16. Since the d sequence is 2, 3, 4, 6, 9, 13, 19, ... the first matrix reduction is to a matrix (Matrix II) where each column has 13 or fewer columns. A total of 6 full adders are used to effect the reduction; these are shown by connecting the outputs of the adders with a line as they appear in Matrix II. Thus, a full adder is used to reduce 3 of the entries in the seventh column from the right (which has a total of 14 entries in the composite bit product matrix) to a sum bit in the seventh column of Matrix II and a carry into the eighth column. The seventh column of Matrix II has 12 entries (the 11 "unreduced" entries and the sum bit) and satisfies the desired constraint. Columns 8 and 9 each require two full adders to satisfy the constraint, and column 10 requires a single full adder. Note that column 10 in the composite bit product matrix had only 12 entries, but that the two carries that resulted from reducing column 9 would have caused violation of the constraint. Half adders are used in forming columns 7-10 of Matrix IV. They are shown as two outputs connected by a line (like full adders) except that the connecting line is "crossed". Half adders are also used to form columns 5 and 6 of Matrix V, and column 3 of Matrix VI. All of the numbered matrices on Fig. 3 have been drawn with the dots pushed to the highest possible row of each matrix (with the sole exception of the third column of Matrix V) to simplify checking each column's height.

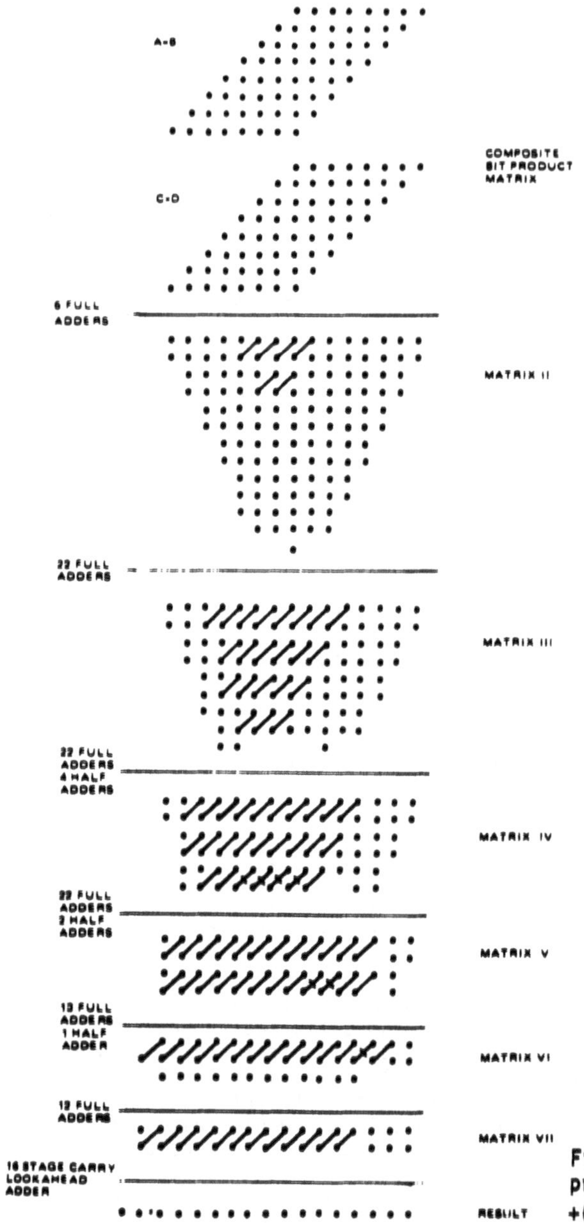

Fig. 10 Two-term, 8-bit inner product reduction example (AxB) +(CxD)

In comparison to the direct methods exemplified by the conventional convolution implementation, a merged logic implementation generates all of the bit products "en masse" using a single (i.e., merged) reduction network. This approach reduces the number of carry lookahead adders from 2K-1 to 1 (K used at the output of each multiplier, and K-1 used as an adder tree) at a cost of increasing the number of full adders in the reduction network. For the two-term 8-bit merged inner product element example, the two bit-product matrices are reduced through six stages of adders to produce a pair of intermediate operands which are summed with a single carry lookahead adder.

Comparison of the merged approach with conventional practice [7] demonstrates that for this example, merged arithmetic requires 20 more adder modules, but two fewer 15-bit carry lookahead adders. Since a 15-bit carry lookahead adder is significantly more complex than 20 adder modules [10], the conventional implementation is considerably more complex than the merged approach. In terms of timing, the merged approach incurs two additional levels of full adder delays at a saving of one 15-bit carry lookahead delay. Examination of the gate level timing indicates that the two approaches exhibit equivalent speeds.

From the viewpoint of VLSI technology implementation, the complexity reduction achieved through merged arithmetic is even greater than might be expected from gate count considerations. Since carry lookahead adders require interconnection topologies which are especially difficult to realize in VLSI circuits (because of the large number of signal crossovers), the saving of 2K-2 (K at the multiplier outputs and K-2 in the adder tree) carry lookahead adders in a K-term convolver is quite significant.

The equivalent circuit of an expandable two-term convolver building block is shown in Fig. 11. The output of the building block is the sum of the two products and the expansion input. This allows vectors of arbitrary length to be convolved by repeatedly computing two terms at a time.

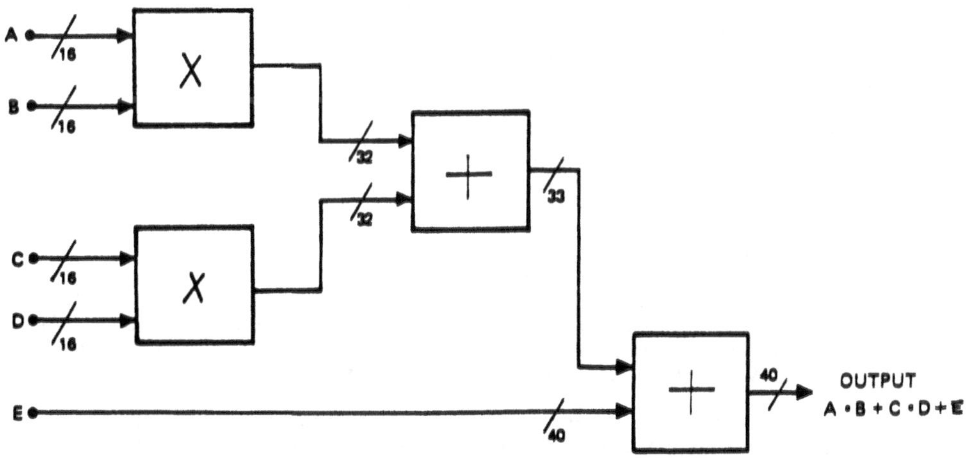

Fig.11 Two-term multiplier adder

Within current VLSI processing constraints, the implementation of a two-term convolution block is feasible as a single VLSI circuit. If implemented with the triple diffused bipolar technology [11], throughput of the merged arithmetic two-term multiplier-adder is estimated conservatively at 30 million operations/sec. This is equivalent to 60 million multiply-add operation/sec per integrated circuit. The total complexity of this circuit is approximately equivalent to 5000 two-input gates which is the current limit for this technology. It is a structure which lacks the desired regularity for VLSI design at the present time. Development of sophisticated computer-aided design techniques may allow the design and production of this device in the near future.

The large number of package pins required for the signals is an apparent problem with this circuit. A total of 144 signals are shown in Fig. 11 (2 pairs of 16-bit inputs, a 40-bit expansion input, and a 40-bit result). In fact, in most applications of large inner products (i.e., FIR filters, convolution, etc.), the vectors are elements of either a time sequence or they are fixed kernels. In the first instance, a single input port is required with shift registers to propagate the data from one multiplier to the other. In the second, the same shift register structure is used, but the kernel is loaded only at start-up or when a new kernel is required. This reduces the input pin count to 32 (excluding the expansion input). The expansion input and output still require 80 pins. A potential solution is to establish a communications protocol at the integrated circuit level which allows transfer of input data operands into the VLSI circuits in either a bit-serial or a byte-serial stream. Such a communications protocol is currently difficult to implement due to the lack of a standard bit- or byte-serial interface.

Commercial VLSI Multiplier-Accumulator

The third approach is to use a commercial multiplier-accumulator. This is exemplified by the TRW TDC 1010 16 x 16 parallel multiplier-accumulator, which has been available since 1978. The basic structure is shown on Fig. 12. It consists of two 16-bit input registers which feed a combinational multiplier array. The double precision output of the multiplier drives a 35-bit adder and an accumulator register. Three-state output drivers allow reading and loading of the accumulator register. A variety of control signals are required: selection of two's complement or unsigned arithmetic; accumulator controls (i.e., add/subtract, accumulate or passthrough, and preload); 3-state driver enables; and clocks for the input register and the output register.

This design shows the effect of package pin limitations. Since 67 signal pins would be required to accommodate the two 16-bit inputs and the 35-bit output word, significant compromise was required to use a standard package. Specifically, the least significant 16 product bits are multiplexed with the Y_{IN} input on a bidirectional port. This reduces the number of data lines to 51, and a 64-pin DIP is used.

Multiplication is performed with the successive add algorithm [11]. This is equivalent to a fully parallel realization of the Baugh and Wooley algorithm [8]. The bit product matrix is formed as shown in Fig. 13. The chip consists of a 16 x 16 array of full adders where the bits of the two operands are run across the chip perpendicularly, forming a grid. At each grid intersection, a bit product (i.e., X_jY_k) is formed and added to the carry from the adder at the grid point to the immediate right and to the sum from the adder at the grid point immediately above. The results are new sum and carry bits which propagate down and left, respectively.

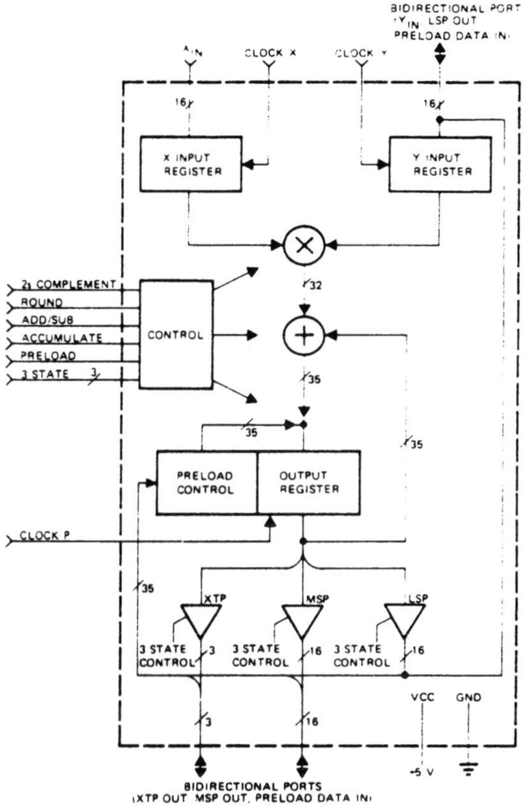

Fig.12 TDC 1010 functional design

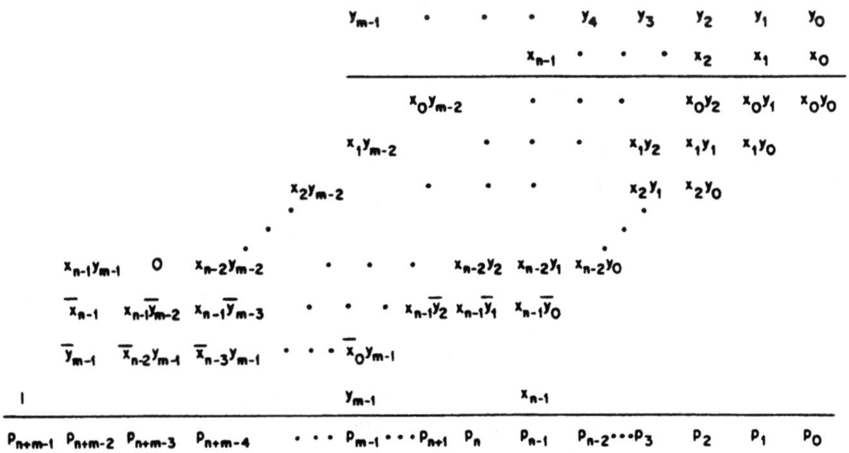

Fig.13 Baugh and Wooley multiplication algorithm [8]

This approach requires signals to travel through 32 full adders in the longest carry path, which is slower than Dadda's technique as applied to the merged multiplier-adder. This chip does not use intermediate threshold restoring buffers at each stage. As a result, the propagation delay through N cascaded stages (each with a delay of τ) is approximately $\sqrt{N}\tau$. The logic margins may degrade as the number of stages increases but, as Fig. 14 shows, there is still an adequate margin after 32 stages. Within the chip all logic signals are differential pairs (i.e., dual-rail logic) with both true and complement signals throughout. Consequently, the final detector sees twice the margin predicted from Fig. 14.

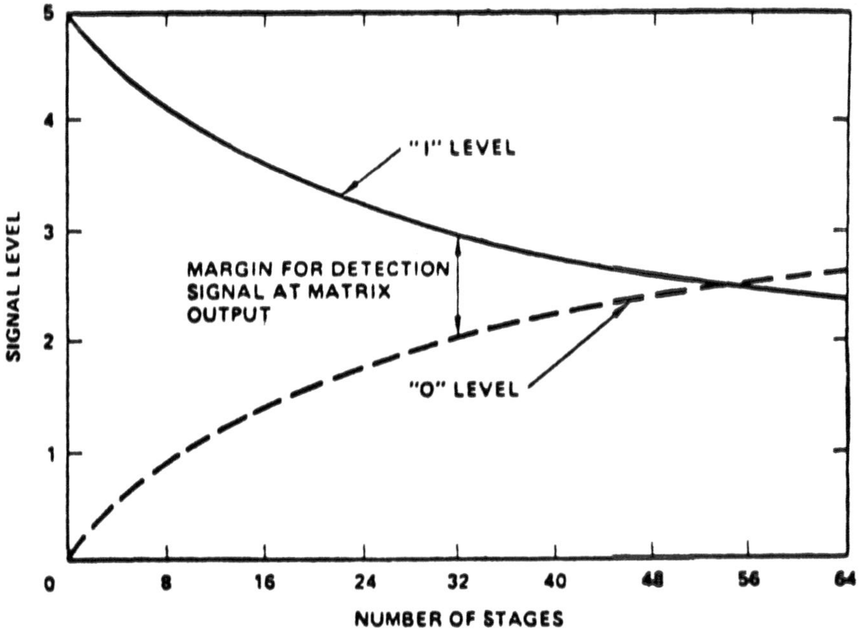

Fig.14 Logic level degradation through cascaded non restoring full adders

A microphotograph of the TDC 1010 (Fig. 15) shows the cellular array nature of the chip layout. The multiplier array is the 16 x 16 cell structure in the central section of the chip with input buffers, adder and output register, and 3-state output buffers around the periphery of the chip.

Typical delay time for the TDC 1010 is 115 nsec so that a multiply rate in excess of 8 million multiplies/sec and an equivalent accumulation rate are achieved. This is comparable to the throughput/chip count of the modular array which assumed use of a very advanced oxide isolated ECL technology. The merged arithmetic approach merit is roughly an order of magnitude higher although the merged design results in a less regular structure (and, as a result, will be much harder to design, layout, and test). It also has pin-out problems which have not yet been resolved.

Fig.15 TDC 1010 Chip microphotograph

6.2.3 Example VLSI Designs

A few examples of VLSI designs will serve to illustrate the types of functions which are presently being implemented and will further clarify some of the issues involved in developing VLSI chip architectures. These circuits are building block circuits for a variety of applications.

Multipliers

The first VLSI device is the triple diffused 16 x 16 parallel multiplier that was first sold in 1976 [12]. A photograph of the chip is shown in Fig. 6. The basic arithmetic function architecture is shown in Fig. 16. The two operands are loaded into input latches, the arithmetic function (in this case multiplication) is performed asynchronously, and the result is loaded into an output register. Use of 3-state drivers on the output simplifies interfacing in a variety of applications as shown in Fig. 17. A single bus can be used to supply operands to both data inputs of the arithmetic function by paralleling the data A and data B inputs and time-phasing the input register clocks. The result may be communicated on the same bus, if desired, since there is a 3-state output circuit. Alternatively, the function can be used in a 3-port pipelined mode. The output circuit is continuously enabled and all three registers are clocked simultaneously. On each cycle, two operands are loaded into the input register and the previous result is clocked into the output register. This provides maximum speed, yet requires only simple control (a common signal clocks the three registers and all data transfers are synchronous). Multipliers for 8-, 12-, and 16-bit word sizes are currently available, with speeds on the order of 10 MHz and power levels of 1-3 watts. For the 16-bit multiplier, the use of a 64-pin package precludes use of dedicated pins for each signal; instead, the least significant half of the product is multiplexed with the B input port. In data processing applications (where the least significant half of the product is frequently required), the multiplier is connected with all three data ports tied to a system data bus, so multiplexing the L.S. product is acceptable.

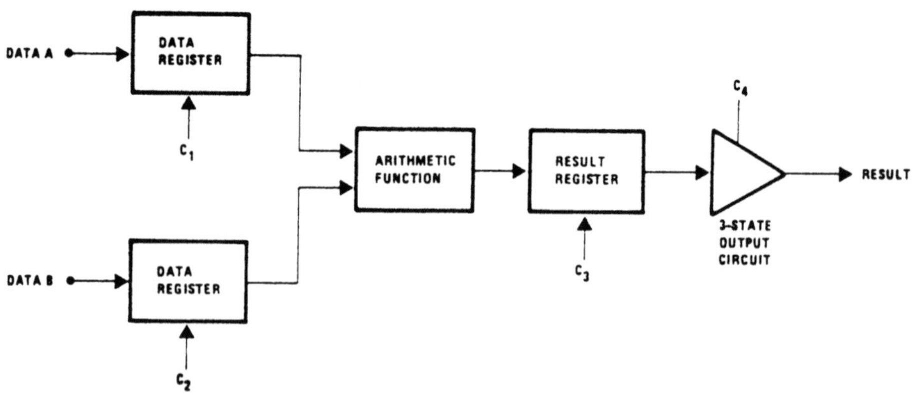

Fig.16 Arithmetic function architecture

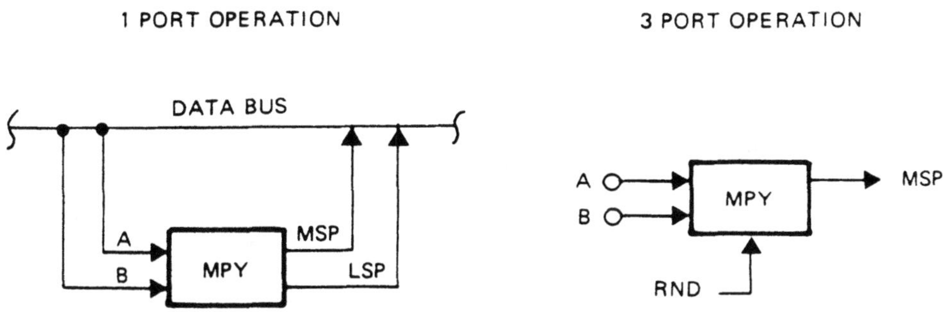

Fig.17 Multiplier interface versatility

Four-Port Memory

The second VLSI device is the 4-port memory shown in Fig. 18. It consists of a four-bit wide "slice" of 512 memory locations with four data ports (two read ports and two write ports). Each read port comprises an address input and output enable control which activates the 3-state output driver. Similarly, the write ports consists of address and data inputs and write enable strobe signals. In addition to usage as the working memory of signal processors, this circuit greatly simplifies the implementation of large first-in first-out (FIFO) buffers as shown on Fig. 19. FIFO buffers are used to couple asynchronous processes by providing what amounts to an elastic shift register. The FIFO design uses two counters, some simple comparison logic, and a 4-port memory (only two ports are used). Since only two-ports are used, the other two can be used to implement a second FIFO or may be addressed in parallel with the other two ports to give 8-bit data word width. The buffer can be used to couple one data source with two users by adding a third address counter and comparison logic. The device is currently in the early stage of development with an expected cycle time of less than 100 nsec at a 1.5 watt power level [13].

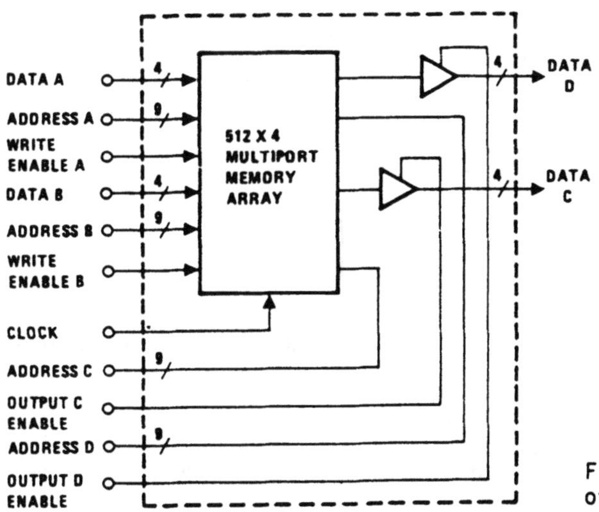

Fig.18 Block diagram of 4-port memory

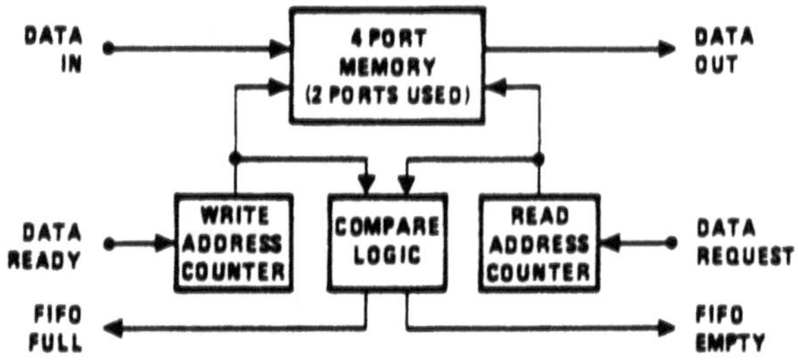

Fig.19 FIFO buffer implemented with the 4-port memory

Flash A/D Converters

The final VLSI example is an N-bit fully parallel A/D converter. The basic circuit for N = 8 consists of 255 comparators and encoding logic as shown on Fig. 20. As the chip microphotograph (Fig. 21) shows, this is a very repetitive structure, amenable to step and repeat layout.

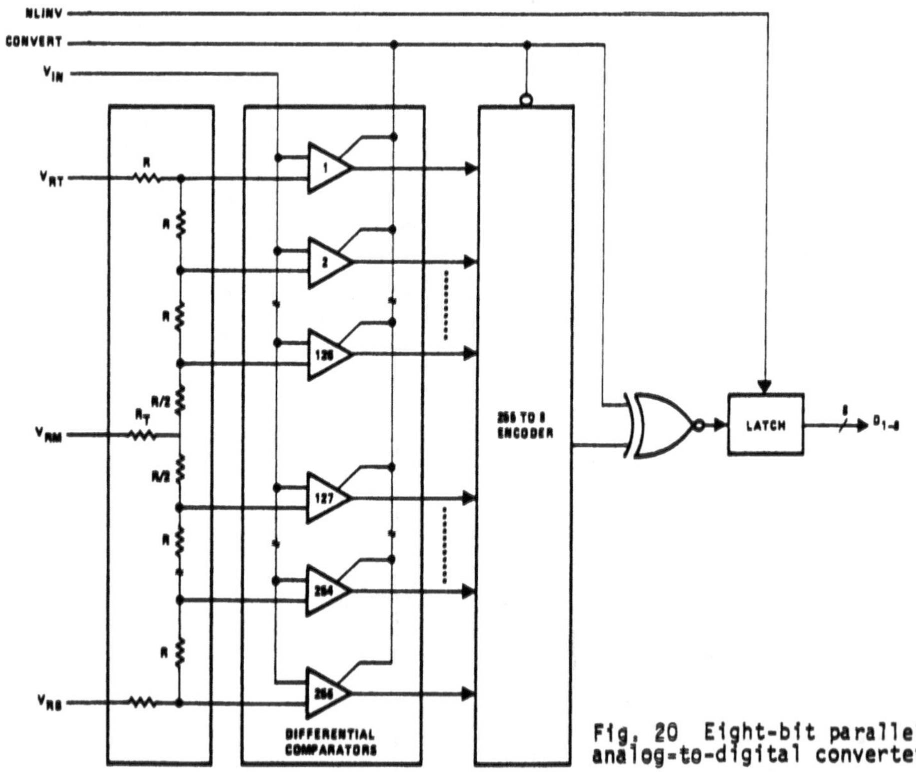

Fig. 20 Eight-bit parallel analog-to-digital converter

The analog input signal is applied in parallel to the positive inputs of the 255 comparators. The negative input to each comparator is derived from a resistor chain (with a center tap accessible to permit curvature

Fig. 21 Photograph of TRW 8-bit parallel analog-to-digital converter

adjustment) that provides 255 different voltage thresholds between V_{RB} and V_{RT}. Depending on the input voltage, all of the comparators up to some level are on, while above that point all are off. The transition point is encoded in a 255 to 8 encoder, and the output is latched. This all-parallel or "flash" approach is only feasible for VLSI technologies, since an N bit A/D requires 2^N-1 comparators. At present, coverters for 4-, 6-, and 8-bits are commercially available. A 10-bit version (with 1023 comparators) will be entering production in the near future. Because this is a fully parallel A/D converter, in many cases there is no need to use a high-complexity sample-and-hold circuit preceding the A/D. Current devices operate at sampling rates of up to 30 megasamples per second (Ms/sec). This is adequate for sampling a signal which is band-limited at 7.5 MHz without using a complex analog sample-and-hold circuit.

6.3 VLSI System Architecture

Perhaps the single most significant application of VLSI is in the implementation of digital signal processors. Until the mid-1960's most signal processing was performed with specialized analog processors because the complexity of digital systems (built with SSI) was prohibitive. Digital processors provide accurate long-term memory, eliminate calibration and adjustment, and can be readily reprogrammed to accommodate new or changing requirements.

In contrast to data processing which has been effectively implemented with microprocessors at the LSI level, the demands of many signal processing applications exceed the capability of available devices. The distinction between data processing and signal processing, which accounts for this disparity, is that data processing involves sequences of simple instructions (i.e., add, subtract, mask, compare, etc.) performed in data-dependent order; signal processing involves large functional primitives (i.e., FIR filter, FFT, CFAR detection, etc.) which are executed in a highly structured fashion.

This section gives an overview of signal processing algorithms to establish an understanding of the problem. A distributed signal processor architecture and its implementation are examined, and finally an example of an advanced sonar adaptive equalization system demonstrates application of the concepts to a real system.

6.3.1 Signal Processing Algorithms

Digital signal processing algorithms are traditionally dichotomized into two categories: time domain processing and frequency domain processing. In terms of implementation, it is more appropriate to consider inter-domain processing (i.e., spectral analysis) and intra-domain processing (i.e., filtering) as listed in Tab. 3. Inter-domain algorithms such as the Fast Fourier Transform (FFT) are used to translate between the time domain and the frequency domain or vice-versa. Other inter-domain algorithms are the Discrete Cosine Transform (DCT), the Karhunen-Loeve transform, the Walsh (Hadamard) transform, etc.

Intra-domain processing includes filtering and detection algorithms. Specific algorithms are correlation, DPCM encoding, error correcting encoding, filtering with finite impulse responses (FIR) or recursive (i.e., infinite impulse response) filters, constant false alarm rate (CFAR) detection, noise spectrum estimation (NSE), equalization, and windowing algorithms.

It is of course important to understand the speed, dynamic range, and memory requirements of these algorithms. Tab. 4 is an example of the spread in speed, dynamic range, and memory required for various FFT applications. The speed is measured in a number of Radix-2 butterflies/sec*, which is indicative of the arithmetic and control performance requirements. Most applications require 10 million or fewer butterflies/sec.

Table 3 Signal processing algorithms

INTER-DOMAIN PROCESSING
DISCRETE COSINE TRANSFORM
FAST FOURIER TRANSFORM
KARHUNEN-LOEVE TRANSFORM
WALSH (HADAMARD) TRANSFORM
WINOGRAD TRANSFORM
INTRA-DOMAIN PROCESSING
ARITHMETIC OPERATIONS: DIVIDE, SQUARE ROOT, MAGNITUDE, . . .
COORDINATE TRANSFORMATION
CORRELATION
ENCODING: DPCM, REED-MULLER, BCH, . . .
FINITE IMPULSE RESPONSE (FIR) FILTERS
RECURSIVE FILTERS
SIGNAL AVERAGING AND DETECTION: CFAR, NSE
WINDOWING

Table 4 FFT Requirements

SYSTEM PARAMETER	PROCESSOR PARAMETER	SIZE	APPLICATION
SPEED	ARITHMETIC RATE (RADIX-2 BUTTERFLIES/ SEC)	100,000	SONAR
		1,000,000	RADAR, COMMUNICATIONS
		10,000,000	RADAR
		100,000,000	SIGINT
DYNAMIC RANGE	DYNAMIC RANGE (BITS)	FIXED POINT 4 TO 8	IMAGE COMPRESSION
		FIXED POINT 12	RADAR, COMMUNICATIONS
		FIXED POINT 16	RADAR, SONAR
		FLOATING POINT 16	COMMUNICATIONS
		FLOATING POINT 32	DIGITAL VOICE
MEMORY	MEMORY SIZE (WORDS)	32 - 64	IMAGE COMPRESSION
		256 - 1024	RADAR
		4096 OR MORE	SIGINT, SONAR

The dynamic range requirements vary from 4 bits to 16 bits (or more) with fixed-point arithmetic. To satisfy future requirements, floating-point arithmetic with 16-bit or larger word sizes will become necessary.

*The FFT Butterfly consists of a complex multiply and two complex additions. It is called a butterfly because the operation flow diagram looks somewhat like a butterfly with its wings open.

The amount of memory determines the size of the largest transform which can be implemented without use of a secondary memory (which introduces a significant data transfer overhead). Most applications require on the order of 1024 to 4096 words of memory.

This estimate of FFT requirements indicates a significant proportion of signal processing systems will be satisfied with a distributed processor family where at least one of the modules is capable of 10 million butterflies/sec, 16-bit word size, and 256 to 1024 words of memory. Future implementation of floating-point arithmetic is desirable.

Intra-domain processing is largely concerned with digital filtering. S.L. Freeny has published an excellent tutorial on the hardware requirements and design considerations for filtering application [14]. These considerations apply equally to the development of the intra-domain signal processing modules.

6.3.2 Signal Processing Architectures

The first digital signal processors were hardwired digital emulations of earlier analog signal processors. These used fixed data routing (established by the interconnection) and could not be modified without extensive redesign. Tailoring the hardwired designs to implement specific signal processing algorithms minimizes the total parts count, but the need to customize each processor design and the inherent inflexibility results in high design and operational costs for many applications.

In the last few years, the programmable signal processor (PSP) concept has evolved. One or more arithmetic units and an assortment of I/O channels are operated under stored program control to provide an extremely flexible system. Although a single PSP design can be used with different software to implement signal processors for radically different applications (e.g., radar signal processing, sonar beamforming speech recognition, image enhancement, etc.) most PSP designs have special features to optimize their use within a single application area where software changes are used to implement multiple operating modes. Because of the desire to achieve efficient PSP designs, hardware is added to achieve data path flexibility. As a result, PSP hardware is more complex than a comparable hardwired processor. Also, as is becoming well-known, the wide word microcode used to program the early PSP designs is expensive to prepare, debug, modify, and maintain. The PSP is advantageous where processor flexibility is required, as for example, in a radar signal processor where software modification allows mission adaptive operation.

An alternate approach to signal processing is the distributed signal processor which uses a family of signal processing modules that are interconnected as required to optimally implement each specific processor [15]. The basic distributed signal processor system architecture is shown in Fig. 22. A control microprocessor generates a stream of "high level" signal processing commands to the various signal processors. Typically, when a command is issued to one signal processor, the control processor can begin to generate a command for another processor. Since the first signal processor requires some time (typically: 0.1 to 10 msec) to perform the task, and since there are seldom more than a dozen signal processors in a system, the control microprocessor issues less than 100,000 commands/sec. Typical rates are in the 1,000 to 10,000 command/sec range. The software although perhaps not as efficient as PSP software, is easier (and cheaper) to write and maintain.

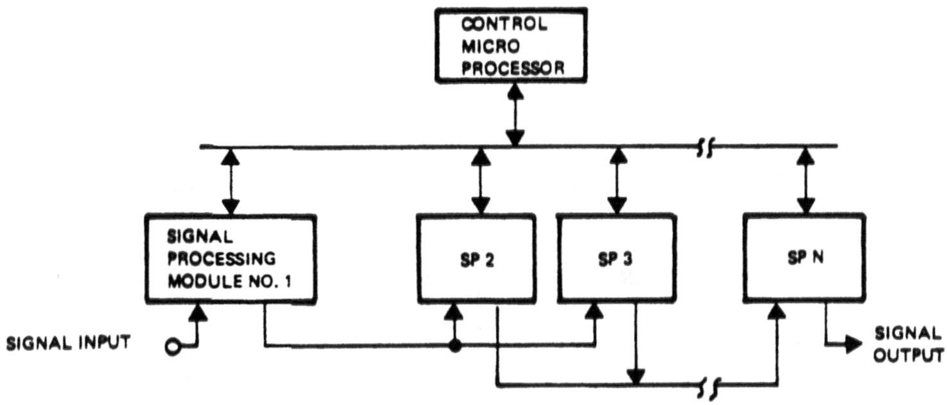

Fig.22 Multi-level control architecture for distributed signal processing

Each of the signal processing architectures is optimum for specific classes of applications. Generally, applications with fixed algorithm requirements are best suited for hardwired realization. The PSP approach is best where extreme flexibility is required or where a processor must be quickly reconfigured to accommodate changing operational modes. The distributed signal processor offers a compromise with more flexibility than hardwired processors and better implementation efficiency than PSP-based systems.

Of these three architectures, the distributed signal processor takes greatest advantage of VLSI. The hardwired approach is well matched to the SSI and MSI technologies as the available (low-level) functions were adequate for implementing each new processor design. At higher levels of integration each new processor design requires the development of custom circuits which is prohibitive for many programs. In contrast, the modules of the distributed signal processor are usable for many projects in much the same way as standard subroutines are used for a wide variety of non real-time signal processing simulations implemented on scientific computers. At the other extreme, the PSP, because of the desire to provide data path flexibility, does not use LSI or VLSI efficiently, due to packaging constraints which limit pin counts and the difficulty of providing interconnection flexibility [16]. Because the signal processing modules of the distributed signal processor perform complete primitive operations, I/O rates are relatively low and great interconnection flexibility (and efficiency) is not required.

The signal processors are interconnected as required for each specific application. For example, consider the two spectrum analyzers shown in Figs. 23 and 24. The first uses N Discrete Fourier Transform (DFT) processors to compute N points of the spectrum at a time. Data is sent to all DFT processors in parallel, each DFT computes the intensity of a different spectral component, and the results are sent to an output processor. In contrast, the second approach uses K Fast Fourier Transform (FFT) processors in a time skewed mode. The first data block is loaded into an FFT processor where all spectral components are computed (while successive data blocks are loaded into other FFT processors), and the results are output. In each case the interconnection and control sequence are optimized for the given signal processing modules.

Thus, this architecture offers interconnection flexibility, and also provides the capability for modular expansion to achieve higher speed, since additional signal processors may be added in an optimum configuration.

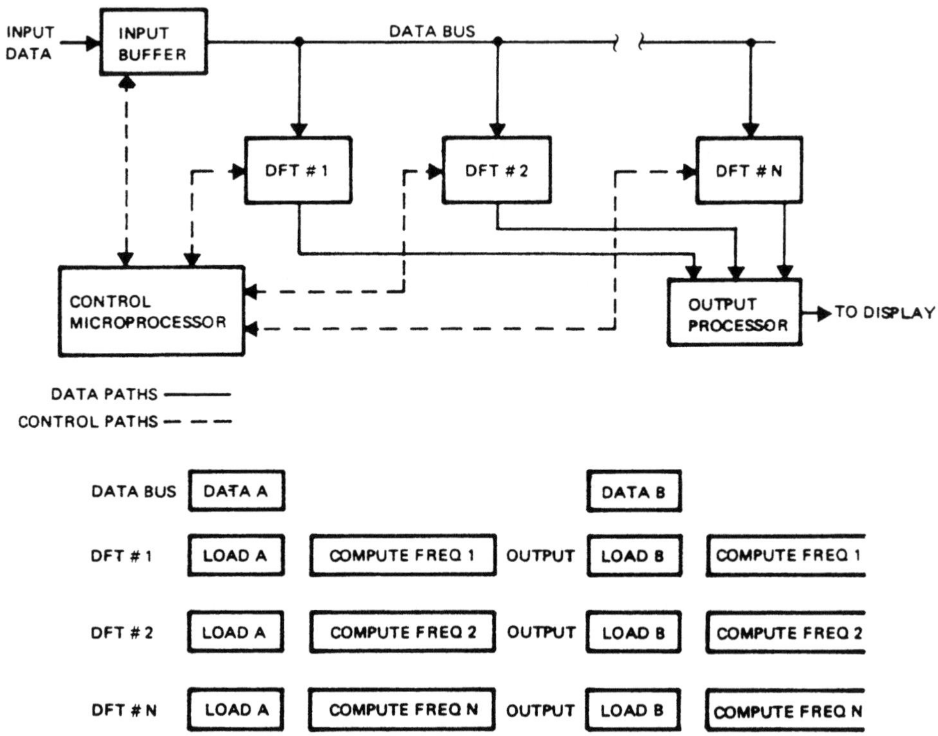

Fig.23 DFT Spectrum analyzer (parallel operation)

Control Microprocessor

The control process involves issuing sequences of commands to each signal processor to initiate the various signal processing functions. Each command contains the module address, operation code, and pertinent parameters. Several additional functions are implemented in the control microprocessor, including user interface, system monitoring, fault detection, etc.

The simplicity of the control microprocessor software is illustrated by Fig. 25, which is a portion of a control program for a synthetic aperture radar processor. The first instruction tests the status of the input processor to determine whether the first line of data is ready. When the data is ready, instruction 2 prepares the transform processor and instruction 3 commands the input processor to transfer the data.

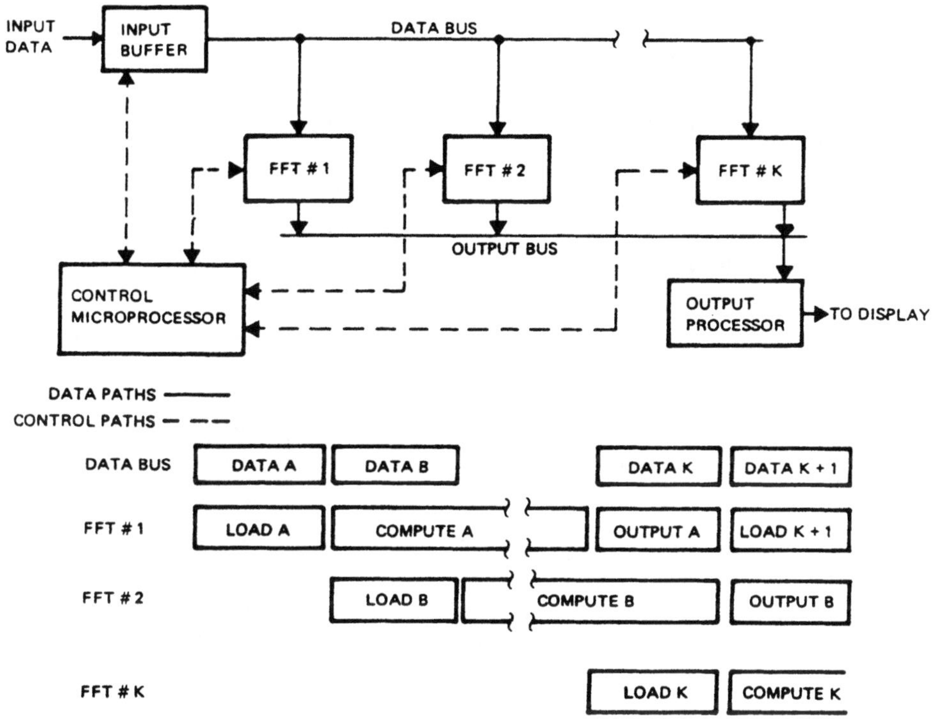

Fig.24 FFT Spectrum analyzer (time overlapped operation)

Once the transfer is completed, an FFT is performed in the transform processor. Instructions 6, 7, and 8 transfer the transformed data to the mass memory. The data is transferred in reverse order in this case to indicate the flexibility of the data transfer protocol. When the data has been transferred to the memory, it is stored via instruction 10. Instruction 11 checks the input processor to see if the next line of data is ready. If data is not ready, instructions 12 and 13 transfer a column of data from the mass memory to the second transform processor which performs azimuth filtering. Thus the control program is simple to understand (and as a direct result, easy to write, debug, modify, and maintain). A major factor in achieving this software simplicity is the use of powerful commands to the signal processors.

Signal Processors

With this introduction to the "top level" control aspects, it is appropriate to focus attention on the signal processors. There are three basic requirements for a signal processor [3]: It must be applicable to a significant signal processing function or class of functions, it must be realizable at cost commensurate with its performance, and it must not impose unreasonable I/O demands (i.e., it must provide substantial processing in relation to the I/O requirements).

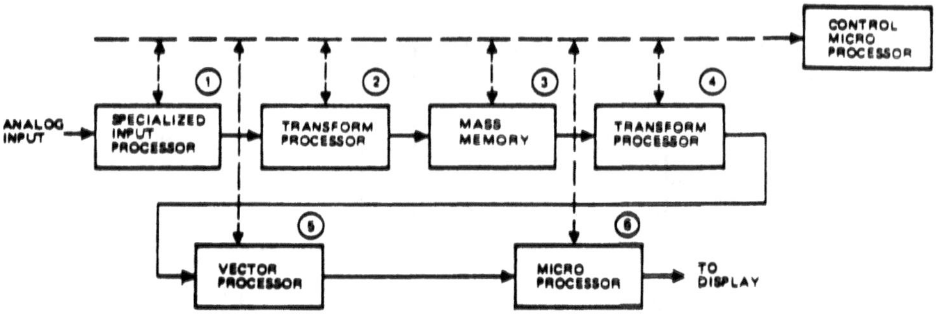

INSTRUCTION NO	MODULE	INSTRUCTION	ARGUMENTS	COMMENTS
1	1	CHECK STATUS	DONE⇒2, ELSE⇒1	CHECKS FOR DATA READY
2	2	ENABLE INPUT	512	
3	1	PERFORM OUTPUT	BASE=0, INCR=1, 512 PTS	TRANSFERS DATA TO TRANSFORM PROCESSOR
4	1	CHECK STATUS	DONE⇒5, ELSE⇒4	CHECKS FOR DATA TRANSFER DONE
5	2	PERFORM FFT	512	
6	3	ENABLE INPUT		PREPARES MEMORY TO RECEIVE FFT OUTPUT
7	2	CHECK STATUS	DONE⇒8, ELSE⇒7	CHECKS FOR FFT DONE
8	2	PERFORM OUTPUT	BASE=511, INCR=-1, 512 PTS	
9	2	CHECK STATUS	DONE⇒10, ELSE⇒9	CHECKS FOR DATA TRANSFER DONE
10	3	STORE DATA	ROW 1, BASE=0, INCR=1, 512 PTS	CHECKS FOR DATA TRANSFER DONE STORES DATA IN ROW 1
11	1	CHECK STATUS	DONE⇒15, ELSE⇒12	CHECKS FOR DATA READY
12	4	ENABLE INPUT	256	
13	3	PERFORM OUTPUT	COLUMN 1, BASE=0, INCR=1, 256 PTS	TRANSFERS DATA TO TRANSFORM PROCESSOR
14	1	CHECK STATUS	DONE⇒15, ELSE⇒14	

Fig.25 Control microprocessor software

Table 5 describes four signal processors. The first module is an inter-domain processor which implements not only the Fast Fourier Transform but also performs several alternative transforms. This module clearly satisfies the criteria of signal processing applicability and flexibility.

Table 5 Signal processors

NAME	FUNCTION
INTER-DOMAIN PROCESSOR (TRANSFORM PROCESSOR)	FFT, INVERSE FFT, DISCRETE COSINE TRANSFORM, AND WALSH (HADAMARD) TRANSFORM
INTRA-DOMAIN PROCESSOR	VECTOR ADD, SUBTRACT, MULTIPLY, DIVIDE, SCALE, SQUARE ROOT, MAGNITUDE, COMPARE, CFAR DETECTION, MOVING WINDOW SUMMATION, DATA REORDERING, INNER PRODUCTS, AND DIRECT CORRELATION
MASS MEMORY	512 x 512 x 8 MEMORY WHICH STORES OR READS ROWS OR COLUMNS OF DATA
SPECIALIZED INPUT PROCESSOR	A/D CONVERTER WITH SPECIALIZED FIR FILTER, SAMPLE INTERPOLATION, AND/OR PRF BUFFER CAPABILITY

The intra-domain processor may be viewed as a processor which operates on vectors of data. For example, a "vector add" operation is equivalent to the following simple FORTRAN program:

```
DO 1    I = 1, number of points
        J = J base + I * J increment
        K = K base + I * K increment
        L = L base + I * L increment
   1    X(L) = X (J) + X(K)
```

Since it implements the four basic arithmetic operations and also performs scaling, square rooting, magnitude computation, Contrast False Alarm Rate (CFAR) detection, direct correlation, and inner product computation, etc., this processor is clearly useful for a wide variety of intra-domain signal processing tasks.

The other signal processors in Tab. 5 are a mass memory module storing 512 x 512 words of 8-bit data (suitable for "corner turning" in radar processing, and for image buffering in video bandwidth compression systems), and a specialized input processor which is essentially a "smart" analog-to-digital converter (with PRF buffering, FIR filtering, and interpolation capability).

This, then, is the multi-level distributed signal processor architecture. A (relatively slow) control microprocessor issues commands to specialized signal processors which are each efficient hardwired or microcoded designs. The control program is easily written, debugged, and maintained because of the use of "high level" signal processing commands.

6.3.3 Implementation

Before the system is implemented it is appropriate to determine the extent to which commercially available devices can be used since this has a direct impact on the cost of system design, implementation and maintenance.

There has been much microprocessor development. Specifically, there are many single chip microprocessors (e.g., INTEL 8086, Texas Instruments TMS (SBP) 9900, Zilog Z-8000, etc.) which are relatively slow, but feature excellent support software, advanced logic analyzer, and development systems which simplify system design, checkout, and integration. There are also bipolar bit slice microprocessors such as the AMD 2900 family which provide speeds in the 1-10 million instruction/sec realm at a cost of less well developed support software, and much higher hardware and software complexity. Typical "single chip" microprocessor systems require 10-50 devices when clocking, memory, and I/O requirements are considered, while the bit slice processors require on the order of 50-200 devices.

The area of fixed memories (i.e., read only memories and programmable read only memories), present devices are relatively fast, large, and organized in fairly wide (8-bit) words. The trends in random access memories (i.e., read/write memories) indicate that they are apparently optimized for computer miniframe memory applications in that high density (i.e., total bits/chip) is obtained at a sacrifice in the features needed for signal processing (wide words and multiple data ports).

Arithmetic device development has included high speed adders, single chip multipliers, multiplier-accumulators, and analog-to-digital converters. These achieve operation rates of 10^7 operations/sec.

The control microprocessor may be implemented with either a single chip microprocessor or a bit slice microprocessor. In view of the modest processing requirements (on the order of 1,000 to 10,000 commands/sec), the single chip micros are currently adequate and will simplify software development, checkout, and maintenance.

The signal processors require significant memory, arithmetic, and control logic. New LSI devices have been defined to simplify all three areas. The goal in chip definition is to develop key functional building blocks which are effective in signal processing systems and which will also be useful in general-purpose applications. No attempt has been made to cram 100 percent of the logic of a signal processor onto custom chips since it is recognized that such devices would lack the flexibility necessary to satisfy a wide variety of present and (as yet undefined) future applications.

The single most important aspect of signal processing is achieving efficient utilization of the memory. The internal architecture of the signal processors emphasize memory efficiency. Each module has its own working memory to prevent the common system bottleneck which results from a single main memory. Although the concept of using only a single main memory per system is superficially attractive, use of a dedicated working memory in each processor is far more efficient.

In recognition of the dyadic nature of signal processing (invariably two operands are fetched, a computation performed, and the result returned to memory), a multi-port memory architecture has been developed as shown in Fig. 26. A 4-port memory serves to hold the operands. Two operands are fetched simultaneously and presented to the arithmetic unit. When the

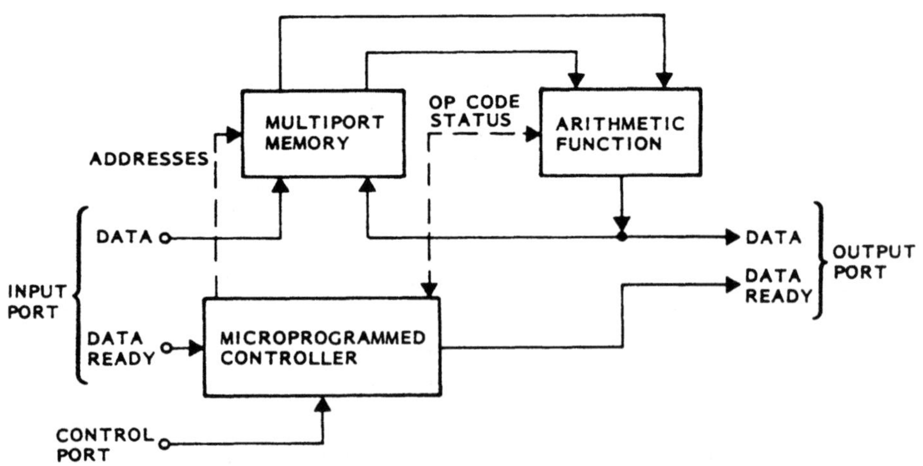

Fig.26 Typical signal processor module architecture

calculation is completed, the result (or results) is stored in the register file, while two new operands are fetched. To keep up with a 10 MHz arithmetic unit, the register file need only fetch two operands and store two results in a 100 nsec cycle, which is readily attained with current integrated circuit technology.

The control of signal processors represents another key architectural concept. Attempts to use classical data processing control concepts (e.g., the Von Neumann architecture) for signal processing have been less than fully successful. A basic problem is that data processing involves very flexible data-dependent control sequencing, whereas signal processing is highly structured with few data-dependent jump operations. In signal processing, fixed control programs (stored in ROM or PROM) with micro-programmed control sequencing generally provide adequate flexibility [17] without the complex data paths and memory and ALU interference of the Von Neumann architecture.

6.3.4 Adaptive Sonar Equalizer Design Example

To illustrate these systems architecture concepts, a breadboard model of an adaptive sonar equalizer will be examined. The adaptive equalization process is illustrated in Fig. 27. Time-varying analog input data is transformed with a spectrum analyzer. The resulting amplitude/frequency plot is smoothed to determine the average noise level as a function of frequency to estimate the relative acoustic conductance. Finally, the input spectrum is divided by the average noise level to produce the equalized spectrum.

The implementation of this process with a distributed signal processor is shown in Fig. 28. Input data is digitized and buffered to 4096 word blocks in an input processor. The basic signal processing operations are

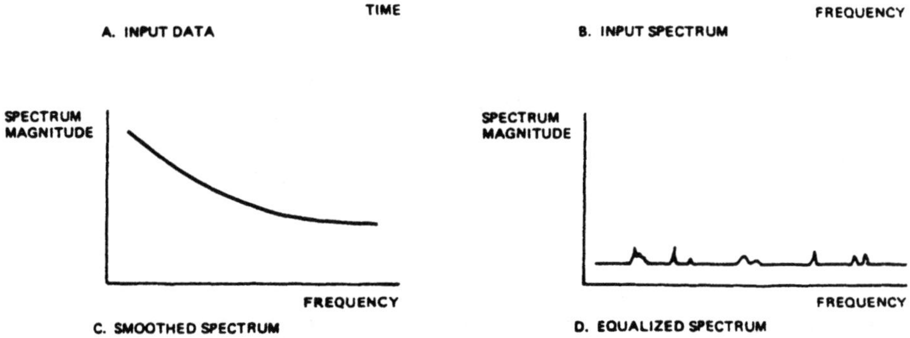

Fig.27 Adaptive sonar equalization

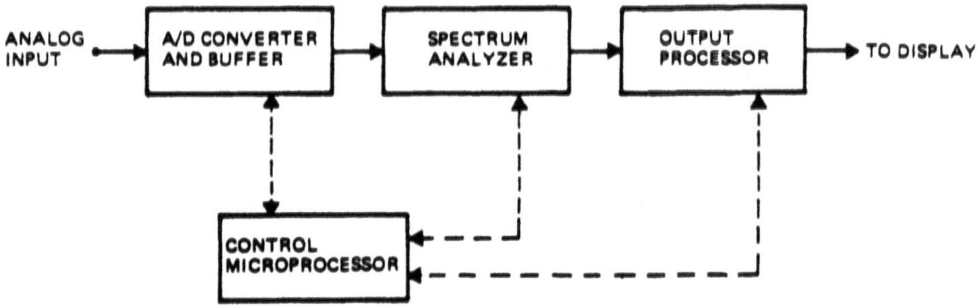

Fig.28 Adaptive spectrum analyzer implementation

implemented in a spectrum analyzer which performs both FFT and intra-domain vector operations. Finally, an output processor provides data formatting and long-term storage capability. The entire system operates under the control of an Intel 8080 microprocessor.

Superficially, it might appear that the control microprocessor is not necessary for this system, but its inclusion allowed high-level algorithm coding to be performed independently of signal processing microroutine development. The resulting breadboard processor (shown in Fig. 29) was completed in five calendar months by a three-person team (one hardware engineer, one software engineer, and one technician) at a cost of less than $100,000. This team had no previous sonar or signal processing experience, but had worked with earlier generation microprocessors. An indication of the ease of high-level coding arose during final checkout when it was learned that the wrong noise estimation algorithm had been coded; a new algorithm was coded and demonstrated in a complete system demonstration in less than two days.

Use of high-level signal processors serves to minimize the communication requirements between modules. This improves the system overhead for the present systems, but becomes even more important in the future. Specifically, VLSI chip efficiency depends upon minimizing the communication between chips [18]. As the integration levels increase, eventually each signal processor will be realized as a single VLSI chip.

6.4 VLSI Architecture Case Study

To develop a better understanding of the actual process of VLSI architecture, an example of the design and implementation of an advanced radar digital beam former (DBF) will be examined. The overall design approach is shown in Fig. 30: activities in algorithm development, architecture design, and technology selection were integrated to produce an assessment of the feasibility of real-time DBF. A series of design studies were performed to select an approach for emulation of the system. A real-time implementation has been designed based on custom VLSI circuits.

6.4.1 Basic Concepts

The basic radar beam forming concept is shown in Fig. 31. A transmitted R.F. beam is reflected from a target and received by an array of receivers. The transmitter is generally located at a different place than the receiver

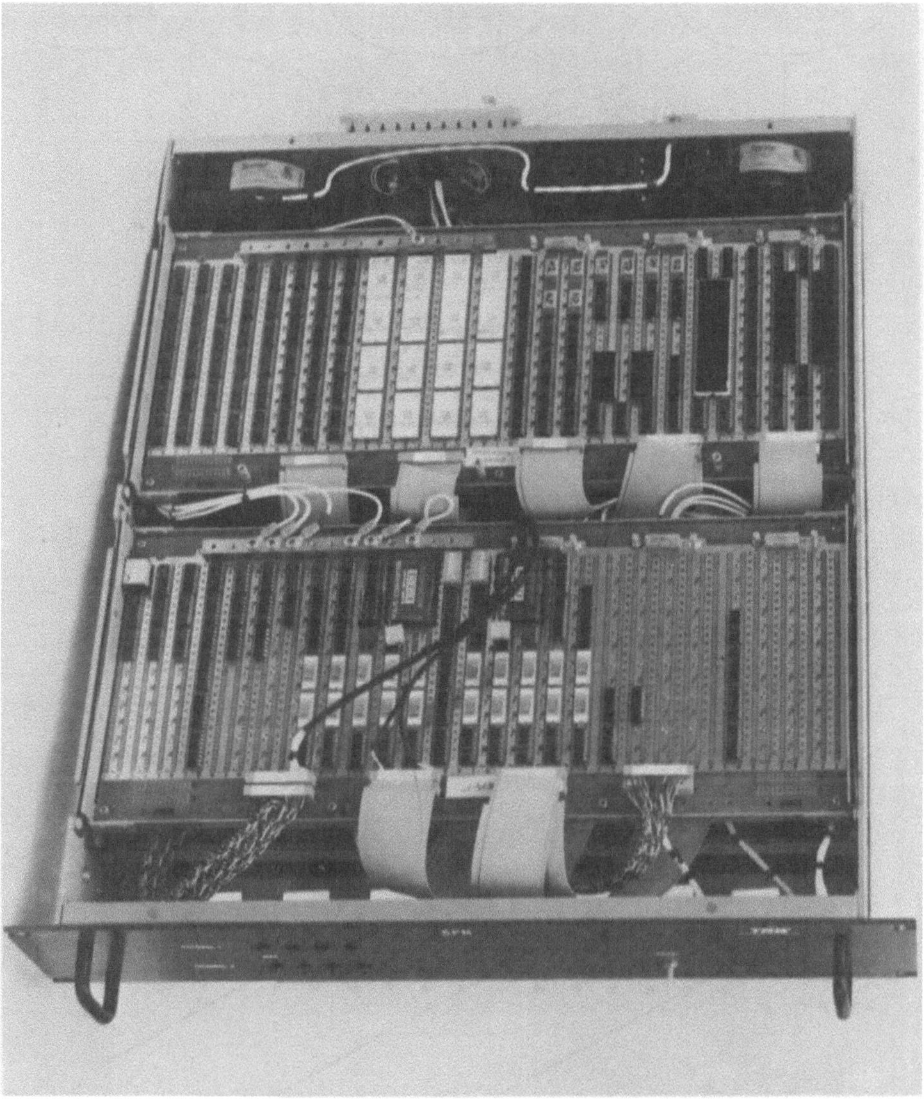

Fig.29 Adaptive sonar equalizer breadboard

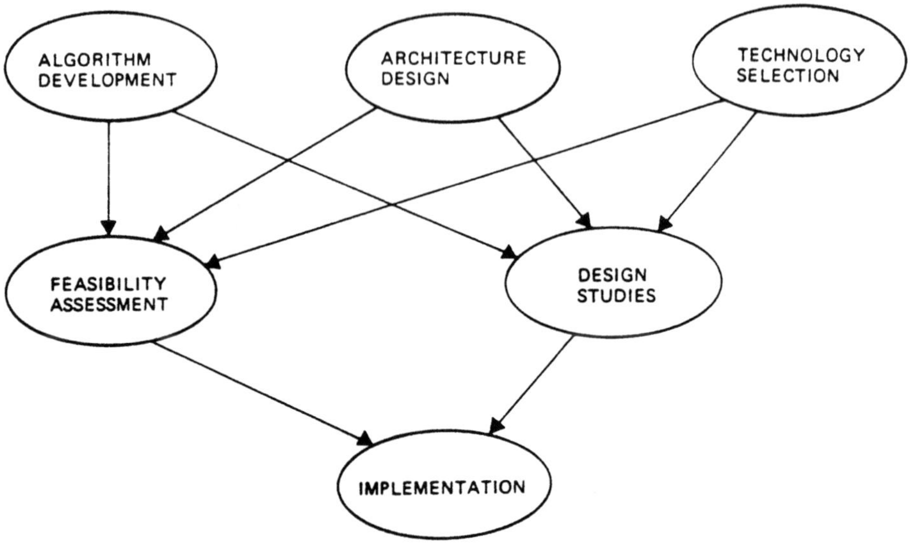

Fig.30 DBF Design approach

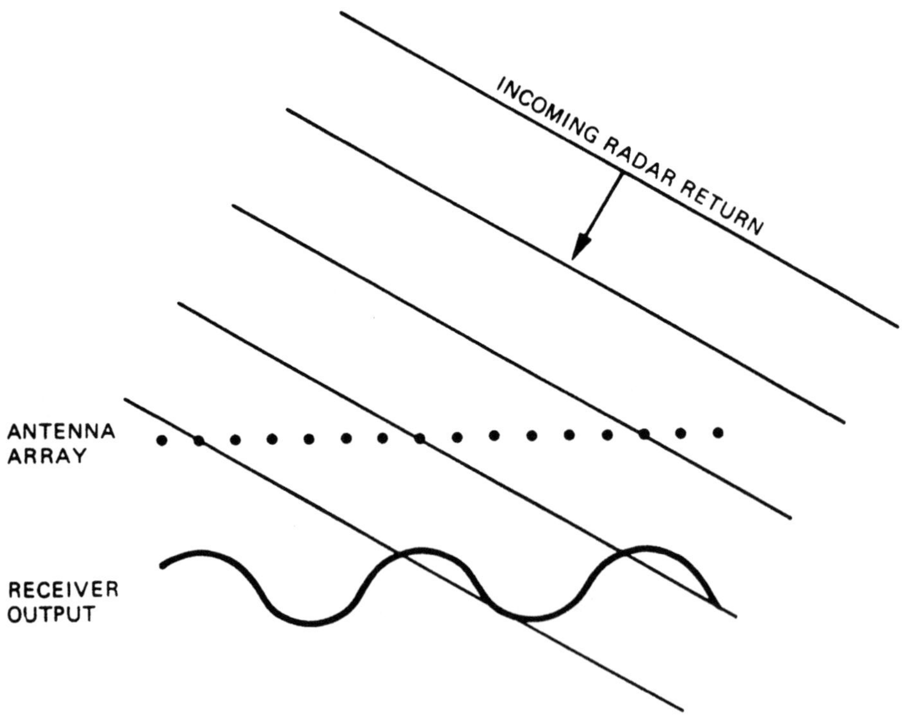

Fig.31 Beam forming concept

array. This allows a single transmitter to illuminate an area in space with several receiver sights. Such multi-static operation provides excellent object location capability. Note that although the receiver array is shown as a linear strip in Fig. 31, a two-dimensional array is used to provide azimuth and elevation information.

The DBF interfaces are shown in Fig. 32. For an N by M antenna array, NM receivers and A/D converters provide digitized baseband data to the DBF processor. The beam forming algorithms are two-dimensional spectrum analyzers because the reflections from a target form a two-dimensional sinusoid on the antenna face (as shown for the one-dimensional case on Fig. 31). Two-dimensional transforms can be computed as sets of one-dimensional transforms as shown in Fig. 33. The digitized data from the antenna array is transformed on a column-by-column basis via one-dimensional column transforms. This forms an array of intermediate results which are transformed on a row-by-row basis to form the desired beam patterns.

A variety of spectrum analyses techniques were investigated for DBF as shown in Tab. 6. The Discrete Fourier Transform (DFT) exhibits moderate arithmetic complexity (proportional to the number of beams which are formed) and low memory and control complexity. The Fast Fourier Transform (FFT) reduces the arithmetic complexity, but increases memory and control complexity. The Winograd Fourier Transform (WFT) also has low arithmetic complexity, but has high memory and control requirements.

6.4.2 DBF Emulation Model

The goal of the DBF Emulation Model development is to provide a vehicle for architecture design verification and technology selection. In architecture design selection, the critical issues include balancing of the amount of process distributuion (i.e., distributed versus centralized); selection of pipeline or parallel processing, if distributed processing is employed; and development of a bus approach.

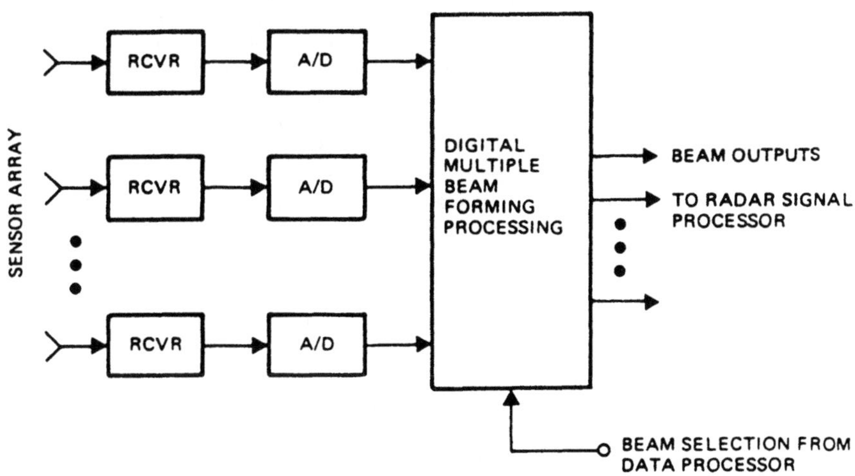

Fig.32 DBF Interfaces

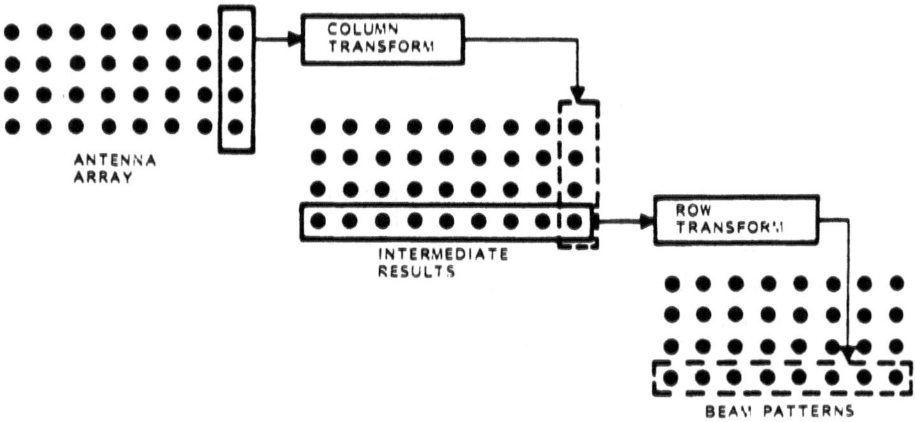

Fig.33 Two-dimensional transformation

The technology selection issues include functional verification, arithmetic implementation (i.e., word size, rounding scheme, number system, etc.), and processor sizing (i.e., memory capacity, word size, throughput, etc.).

The emulation model consists of a VAX mainframe computer, a Zilog Z-80 based controller, and N Zilog Z-8000 based beam formers as shown in Fig. 34. This architecture can be used with parallel operation of the beam formers as shown on Fig 23, where data is transferred in parallel to all beam formers so that each beam former computes one or more beam outputs in parallel, using the DFT algorithm. It may also be used in a time overlapped mode to compute beam outputs using FFTs as shown in Fig. 24.

The Beam Former Microprocessor is shown on Fig. 35. It consists of a Zilog Z-8000 16-bit microprocessor with 4K words of program memory and 16K words of data memory. To enhance high-speed beam forming, a sin/cos ROM and 16-bit multiplier-accumulator (MAC) are included. The CPU memory, sin/cos ROM, and MAC are coupled to a local bus which is coupled through a bidirectional FIRST-IN-FIRST-OUT (FIFO) Buffer to the 8-bit global bus. This is, in fact, a one-MIP microprocessor which can be used effectively

Table 6 DBF Implementation complexity comparison

	ARITHMETIC	MEMORY	CONTROL	TOTAL
DFT	MODERATE*	NONE	LOW	LOW
FFT	LOW	MODERATE	MODERATE	MODERATE
WFT	LOW	HIGH	HIGH	HIGH

*DFT ARITHMETIC PROPORTIONAL TO NUMBER OF BEAMS

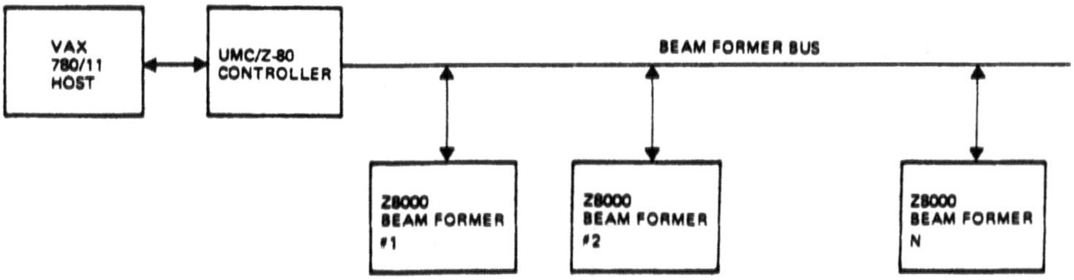

Fig.34 DBF Emulation model

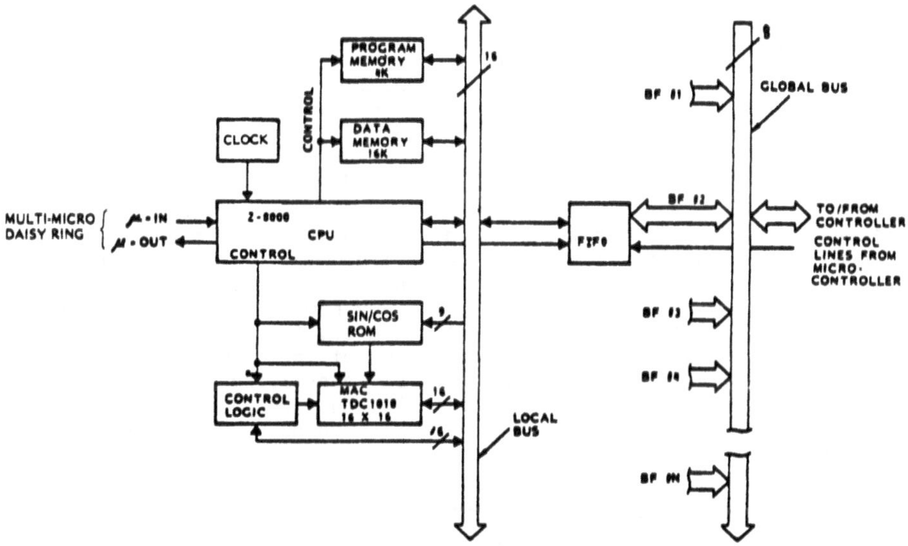

Fig.35 Beam former microprocessor

for a variety of signal and data processing applications, including the architecture and technology verification for which it was developed.

Upon completion of the emulation studies the DBF concept will have been verified and demonstrated, the arithmetic implementation will have been selected, and algorithm development will have been completed.

6.4.3 VLSI for the Real-Time DBF System

It was noted on Tab. 6 that the DFT algorithm seems to offer the lowest total system complexity for the DBF system. The emulation model beam former has verified the utility of the DFT. Accordingly, the gate-rate requirements for DFT and FFT algorithms have been evaluated in the context of specific radar systems requirements, as shown on Fig. 36, where the computational requirements have been added to the gate-rate chart of Fig. 2. It becomes apparent that roughly 10 VLSI chips/beam are required for the DFT algorithm and about 250 chips are required for the FFT

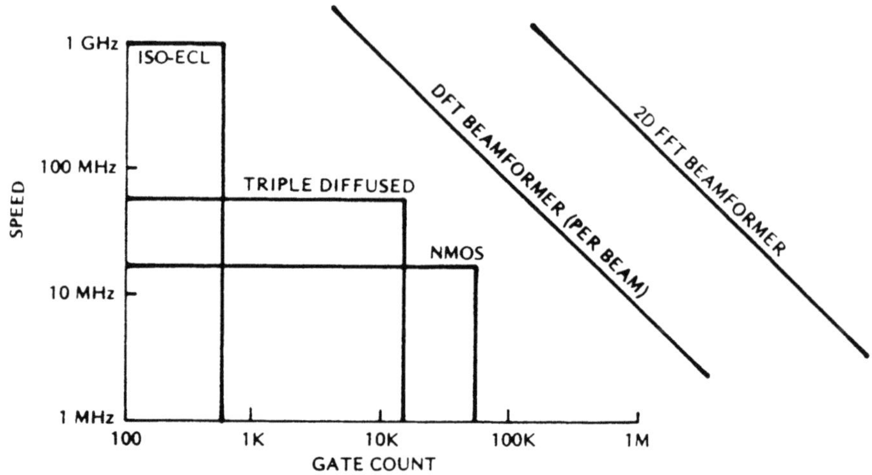

Fig.36 Gate rate requirements for digital beam forming

approach. Thus, to a first approximation, the DFT is more efficient as long as fewer than 25 beams are to be formed. Individual DFT processors may operate in a parallel mode as shown on Fig. 37. Data from the receiver array is buffered and routed to any number of DFT processors in parallel. Since there is a single data source, there is no contention on the bus. Each DFT processor is independent of the others, and the outputs may be bused or multiplexed as desired for interfacing to the radar signal processor(s). As is apparent from Fig. 37, this DFT approach provides modularity, ease of isolation for diagnosis and debug, and inherent fault tolerance.

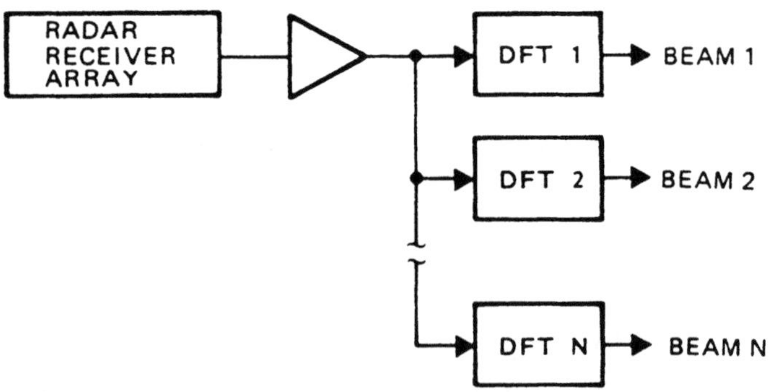

Fig.37 DFT Beam former architecture

The architecture of Fig. 37 can be effectively implemented with a VLSI chip design for DFT. The DFT chip architecture is shown on Fig. 38. It consists of a complex/real multiplier-accumulator, a sin/cos generator, and a magnitude converter. The complex/real multiplier-accumulator consists of two real multiplier-accumulators connected to form the product of a complex number (the sin/cos wave) with the real input. The sin/cos generator is implemented with a uniform step generator addressing a ROM which contains the sin/cos values. The magnitude computation involves summing the absolute value of the larger output component with 1/4 the absolute value of the smaller component (i.e., MAG = MAX (|I|,|Q|) + 1/4 MIN (|I|,|Q|)). With five data ports, this chip appears to exceed pin limitations, but the three outputs occur at a much lower rate than the X_i inputs and may be multiplexed onto a single output data port.

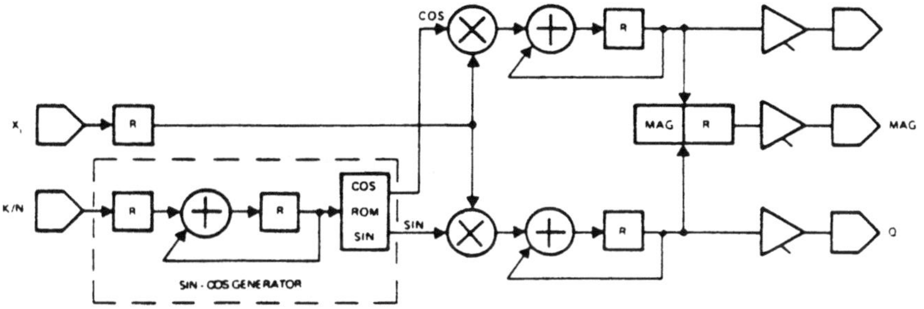

$$I = \sum_{i=1}^{N} X_i \ \text{COS (Ki/N)} \qquad Q = \sum_{i=1}^{N} X_i \ \text{SIN (Ki/N)}$$

Fig.38 VLSI Chip for DFT

To evaluate the impact of the DFT chip upon the DBF processor, a detailed cost analysis was performed. The results are shown on Fig. 39. This shows the estimated costs for DBF and (analog) phased-array Radar systems as a function of design date. These costs include Receiver and A/D converters, but do not include Antenna or Radar Signal Processor costs. By use of the DFT chip, the DBF cost is reduced by at least a factor of six in the 1978-1980 time frame. A further decrease by a factor of four is expected to occur by 1985 as a result of declining VLSI costs. In contrast, the phased array system costs exhibit minor variations and are much higher than the DFT chip-based DBF designs.

The cost advantage is, in fact, the bottom line justification for VLSI development. The DBF example clearly shows the payoff resulting from architecture optimization at the chip and system levels.

6.5 Conclusion

In this section the interaction between VLSI technology and circuit and system architecture have been examined. Although the remarkable growth in technology has solved many problems, it has exacerbated design, packaging, and interface challenges. As VLSI technology matures, these problems will most assuredly be solved and systems costs will decline precipitously.

Device architectures require development of cellular designs to minimize the effort required to design very large devices. It is also necessary to devote attention to minimizing the number of chip I/O lines (especially the

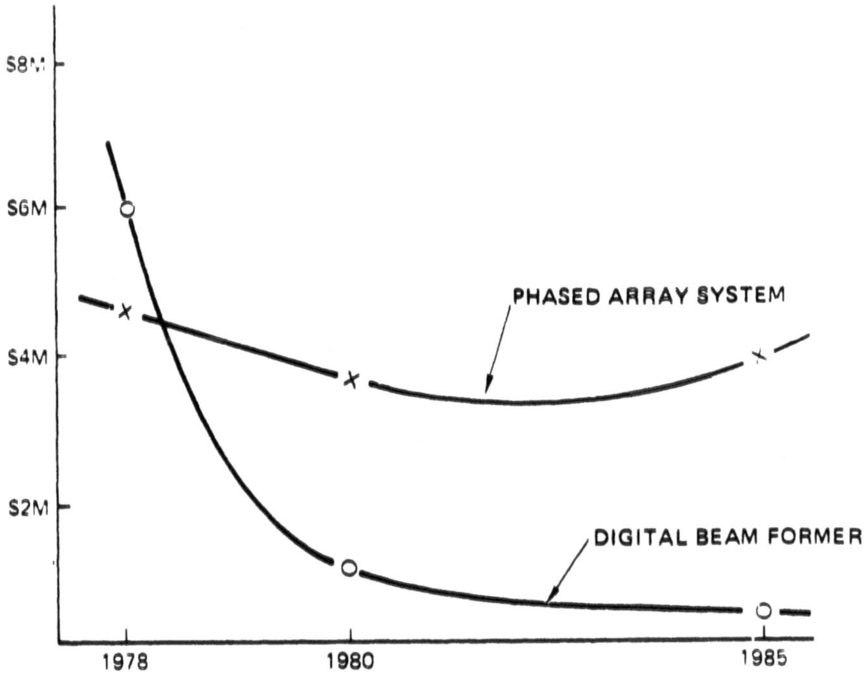

Fig.39 Beam former cost comparison

number of outputs). Such interface simplification reduces the chip area required for bonding pads, reduces package size and cost, and minimizes output driver power requirements. It requires great care, however, to insure that the application flexibility of the chip is not compromised.

As the level of integration has grown, the distinction between chip architecture and system architecture has begun to dissolve. Crucial issues remain in the design of distributed systems, especially in the areas of control, fault detection/localization, and implementation.

Examination of the design and development of an advanced radar digital beam former serves to illustrate the process used to resolve system design issues. To refine the chip design and usage characteristics, an emulation of the system was constructed using programmable microprocessors in lieu of the proposed VLSI devices. This allows reprogramming to simulate the impact of modifications to the chip architecture, arithmetic algorithms, and data wordsizes as required to develop an optimum compromise chip and system design. The final result is a beam forming system which is an order of magnitude less expensive than current (and projected) analog implementations; and at the same time, a design which offers better technical performance and simplified logistic support requirements.

REFERENCES

1. H.J. Scudder, "Introduction to Computer-Aided Tomography," Proc. IEEE, Vol. 66, pp. 628-637, 1978.

2. G.T. Herman, et al., "Rapid Computerized Tomography," Medical Data Processing Symposium (Toulouse), Taylor and Francis Ltd., London, pp. 581-598, 1976.

3. E.E. Swartzlander, Jr., B.K. Gilbert, and I.S. Reed, "Inner Product Computers," IEEE Trans. Computers, Vol. C-27, pp. 21-31, 1978.

4. Andrew D. Booth, "A Signed Binary Multiplication Technique," Quarterly Journal of Mechanics and Applied Mathematics, Vol. 4, pp. 236-240, 1951.

5. J.R. Jump and S.R. Ahuja, "Effective Pipelining of Digital Systems," IEEE Trans. Computers, Vol C-27, pp. 855-865, 1978.

6. L.A. Fletcher, "High-Speed Monlithic 8x8 Multipliers," Proc. Government Microcircuits Applications Conference, Vol 6, pp. 362-365, 1976.

7. E.E. Swartzlander, Jr., "Merged Arithmetic for Signal Processing," Fourth IEEE Symposium on Computer Arithmetic, Santa Monica, CA,1978.

8. Charles R. Baugh and Bruce A. Wooley, "A Two's Complement Parallel Array Multiplication Algorithm," IEEE Trans. Computers, Vol. C-22, pp. 1045-1047, 1971.

9. L. Dadda, "Some Schemes for Parallel Multipliers," Alta Frequenza, Vol. 34, pp. 349-356, 1965.

10. O.L. MacSorely, "High-Speed Arithmetic in Binary Computers," Proc. IRE, Vol. 49, pp. 67-91, 1961.

11. J.L. Buie and T.A. Zimmerman, "Very Large Scale Integrated Circuits for Digital Signal Processing," Circuits and Systems, pp 2-8, 1977.

12. E.E. Swartzlander, Jr., "VSLI Technology for Signal Processing," Proc. Government Microcircuits Applications Conference, Vol. 7, pp. 76-79, 1978.

13. E.E. Swartzlander, Jr., "Signal Processing Architectures with VLSI." Proc. International Conference on Acoustics, Speech, and Signal Processing, 1980.

14. S.L. Freeny, "Special Purpose Hardware for Digital Filtering," Proc. IEEE, Vol. 63, pp. 633-648, 1975.

15. E.E. Swartzlander, Jr., High Speed Micro Signal Processor Study, Air Force Avionics Laboratory Technical Report, AFAL-TR-77-63, March, 1977.

16. R.N. Noyce "Large Scale Integration: What is Yet to Come?" Science, Vol. 195, pp. 1102-1106, 1978.

17. E.E. Swartzlander, Jr., "Microprogrammed Control for Specialized Processors," IEEE Trans. Computers, Vol. C-28, pp. 930-934, 1979.

18. S.S. Patil and T.Welch, "An Approach to Using VLSI in Digital Systems," Proc. 5th Symposium on Computer Architecture, IEEE #78CH1284-9C, pp. 139-143, 1978.

7. VLSI Applications and Testing

H. H. Whalen

With 14 Figures

The subjects of VLSI applications and testing are far too broad to be treated thoroughly here. Fortunately, several excellent references are available. OPPENHEIM [1] is probably the best current compendium of digital signal processing applications, and each section contains an excellent list of references to the details of each application. That book contains chapters on telecommunications, audio, speech, imagery, radar, sonar, and geophysics. PRATT [2] is a valuable sourcebook on digital image processing. Enhancement, compression and pattern recognition are all treated in considerable depth.

VLSI Testing is still a fertile field of research and development. The best sources of ideas in this field are the proceedings (digest of papers) of the annual Semiconductor Test Conference, sponsored by the Computer Society of the IEEE. Those proceedings may be ordered from the Society, 5855 Naples Plaza, Suite 301, Long Beach, California 90803. The issues 1973-1979 are quite useful.

Also available from that source is the IEEE tutorial on LSI Testing (second ed.) edited by FEE [3]. This reference includes three reprints by MARSHALL from EDN Magazine [4,5,6]. These articles are very comprehensive and informative, as well delightfully entertaining, being written as parodies of Lewis Carroll's "Alice in Wonderland".

7.1 VLSI Applications

VLSI is a new plateau in the integration of digital semiconductor circuits. We can use 10,000 gates per chip as the threshold level of integration for VLSI. This amounts to 30,000 to 40,000 devices (transistors, resistors, and capacitors) per chip.

This level of integration will be obtained by shrinking minimum feature sizes from current dimensions of 3 to 6 microns down to dimensions of 1 to 2 microns. As a consequence of this scaling, circuit speeds can be increased by a factor of two to eight depending on the technology.

The net result is that the functional throughput per chip (number of gates times speed) can be expected to increase by a factor of 20 or so in the next three years. This could mean a 20:1 performance increase.

VLSI makes this type of performance increase possible but does not guarantee it. Corresponding improvements in packaging, testing, architecture and algorithms are required to make this 20:1 improvement a practical reality.

Assuming that the full potential of VLSI is realized, applications are not lacking. A number of digital processing solutions await VLSI for their cost effective implementation. A twenty fold increase in digital signal processing performance could be put to immediate use in Communications Processing, Image Processing, and Radar Processing.

All signal processing problems are candidate applications for VLSI. During the past twenty years, digital algorithms for solving a vast number of problems have been published. The practical realization of these algorithms depends upon the available technology. Each generation of technology has extended the feasibility of digital processing to new areas of application. Small scale integration (SSI) brought with it the computer. The major applications were in business data processing as well as scientific problem solving in physics, civil engineering, and aerospace engineering.

Medium scale integration (MSI) lowered the cost of computers and led to the minicomputer. Immediately, applications of computers increased by a factor of one hundred. New applications included small business applications, dedicated signal processing, and automatic test systems.

Large scale integration (LSI) made possible the microprocessor chip and the single chip multiplier. The impact of the microprocessor on consumer and industrial electronics (including automotive electronics) was overwhelming. In many areas, a product cannot be competitive without microprocessor control. The single chip multiplier made a similar breakthrough in signal processing. It immediately lowered the cost of digital signal processing by a factor of ten for many applications, and accelerated the development of digital signal processing applications.

Each of these semiconductor revolutions has had an economic impact far larger than the impact to the semiconductor industry, because of the large value added ratio. For example, the $20 billion dollar computer industry uses about $2 billion in semicondutors. Add to this the economic impact of software and operations and the value added approaches twenty to one.

All of us would like to be able to foretell the future given the past. Will the addition of multipliers to microprocessors really make a difference? Will VLSI lead to anything more than greater use of memory based only upon Parkinson's law? The answer to both questions is undoubtedly yes.

First of all, VLSI increases more than circuit density. It also increases speed. For example, single chip microprocessors now in development operate at four to eight times the clock rate of commercially available microprocessors. This is a direct result of transistor scaling laws. In addition, functional efficiency increases dramatically when the multiplier function is incorporated on a chip. Current single chip microprocessors are poorly matched to high speed multipliers, both in speed and in input-output flexibility. For example, one commercial microprocessor coupled to a 200 nanosecond multiplier requires 18 microseconds to execute a multiplication. A VLSI signal processor chip described in this chapter executes the multiplication in 250 nanoseconds and accumulates the result. This chip has all the programming flexibility of a microprocessor but can execute a second order recursive filter section fifty times faster. VLSI signal processing chips are to signal processing what the microprocessor chip is to industrial control.

In order to guess what types of VLSI chips will be important we can divide signal processing applications into four types of problems: Serial low speed problems (single channel, 100 kilosamples per second or less); Serial high speed problems (single channel, greater than 100 kilosamples per second); Parallel low speed (multiple channels, each 100 kilosamples per second or less); and Parallel high speed (multiple channels, each greater than 100 kilosamples per second).

We shall denote the four cases as SL, SH, PL, and PH, respectively. Assume that each channel requires 10 to 100 operations per sample. This is a reasonable assumption in many cases. For example, a 1025 point FFT takes about twenty operations per sample. A second order recursive filter section requires nine operations per sample.

Examples of SL applications are voice channel communications, digital audio signal processing, speech compression and synthesis and facsimile processing.

Examples of SH applications are radar pulse compression, digital video processing for television and video recording, and digital pulse compression for radar systems.

Examples of PL applications are array sonar processing, multichannel narrowband communication systems, private automatic branch exchanges (PABX), computer aided tomography systems (CAT), and low rate video data compression systems.

Examples of PH applications are synthetic aperture radar processing, wideband digital communications processing, and high resolution image processing systems.

For all SL problems, a ten million operation per second processor would suffice. With VLSI, it will be feasible to put such a processor on a single chip. (A four million operation per second single chip processor has already been announced.)

A group of single chip processors can solve PL type problems. A single high speed processor could also be programmed to solve the PL type of problem but both hardware and software costs would be more expensive because multiplexing, demultiplexing and pipeline overhead are required.

The SH type of problem is a more difficult tradeoff. A high speed processor could be constructed from a group of single chip processors (multiple instruction multiple data, or MIMD machine). The software cost for this type of machine is substantially higher, because either the processors must be synchronized, or a complex real time data management and resource allocation system must be implemented to provide for data dependent branching. In some applications, data dependent branching does not exist, and a rigidly connected pipeline process is a practical solution. A matched filter for pulse compression in a radar processor is such an application. The only operation required is the convolution of the input data with a fixed sequence of coefficients.

We now describe a number of specific applications to illustrate the impact of VLSI on signal processing. The examples presented are drawn primarily from communications and radar. The communications applications described here are all continuous, or pipeline problems. The data arrive and are processed as an uninterrupted stream. The solutions are applicable to many problems of this type, such as speech processing and facsimile.

The radar problem discussed is synthetic aperture radar. It has much in common with array processing, image processing and computer aided tomography. Because two dimensional processing is required in these applications, it is convenient to block the data into arrays. This permits almost arbitrary channelization of the processing problem, but it requires careful attention to the organization of memory. In multidimensional problems, the data are often accessed in a different manner than that in which they are stored.

For convenience, one dimensional problems are often converted into multi-dimensional problems. For example, one can compute the discrete fourier transform (DFT) of a sequence of data of length N = KL by arranging the data into a K by L matrix. The DFT of the rows can be computed, and then the DFT of the columns of the result. A permutation of the second set of DFTs is the DFT of the original sequence. This technique is valid for any circular convolution, with certain restrictions on K and L, and by repeated application can be used to derive the fast Fourier transform (FFT) the fast Hadamard transform (FHT), the Winograd Fourier transform (WFT) and other so-called fast transformation techniques.

A memory in which the data are stored by rows and read by columns or vice versa is called a corner turning memory. A generalized corner turning memory would be useful in a number of signal processing applications. This will be evident from the discussion on synthetic aperture radar.

We have selected communications applications as examples because the communications industry will probably surpass the computer industry as the largest single user of integrated circuits in the next twenty years. VLSI is the reason for this change. VLSI is necessary to make digital techniques reliable and cost effective for a number of communications applications. The reliability issue is extremely important in communication systems and concern for the reliability of digital systems has somewhat slowed their introduction. Given that reliability can be assured, there will be undoubtedly rapid introduction of digital circuits into the telephone system.

VLSI circuits should be more reliable in a per gate sense than LSI for three reasons. First there has been a maturing of silicon processing technology. Failure mechanisms are better understood and techniques to reduce failures are known and practiced. Second, the impact of design on reliability and testability is better understood. Third, the gate to pin ratio is increased substantially in VLSI. A large portion of the past failures in integrated circuits have been pin or package related [2]. With VLSI, the total number of pins and packages is greatly reduced.

7.1.1 Telephone Echo Canceller

Digital cancellation of telephone echoes is an outstanding example of a VLSI digital signal processing application. This application has the following attributes:

1) Cost Effectiveness of VLSI. The solution can be reduced to practice in a single VLSI chip.

2) High Volume. Many millions of cancellers are needed (one per 12 or 24 telephones).

3) System Impact. Use of echo cancellers throughput the system would enable the use of communication satellite links for local telephone traffic.

Telephone systems use two wire circuits (both send and receive on one wire pair) for local subscriber loops and four wire circuits (separate send and receive circuits) for interswitch traffic. The delay in a synchronous satellite relay link is about 270 milliseconds. This is slightly greater than the syllable rate in normal speech.

At the subscriber switches, hybrid coils are used to terminate the four wire circuits and connect them to the two wire subscriber lines. If the hybrid is perfectly balanced and matched to the subscriber line impedance, signals transmitted into the receiver are not returned through the sender path. Such return signals are echoes, and occur in practice because the hybrid is fixed in its characteristics to match an average subscriber line. The echo is annoying when it becomes a fraction of a syllabic period (about 225 milliseconds). The two-way delay on transcontinental calls ranges from 30 to 60 milliseconds, depending on the transmission medium and path. The echo is usually suppressed by echo suppressors in the four wire circuit which interrupt the return path when the send path is on. Echo suppressors are not very satisfactory because they interrupt the conversation when the talkers are not synchronized and often the first or last parts of words are lost. The return path interruption technique becomes hopeless when satellite links are used, for then the two-way echo delay is 540 milliseconds.

The echo canceller solves this problem by removing the echo at the point it returns from the hybrid on the distant end of the line, as seen in Fig. 1.

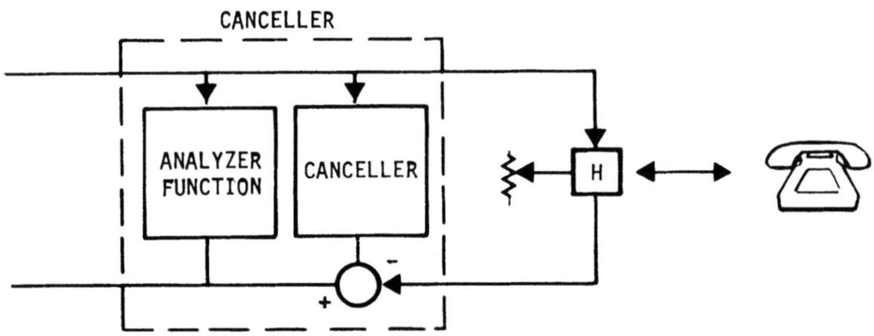

Fig.1 Telephone Echo Canceller

The subscribers can be denoted as A and B. "A"'s echo must be cancelled at "B"'s hybrid. At the B hybrid there is an A send line in, an A receive line out and a B subscriber line on the two wire port. The function of the echo canceller at that point is to construct a filter which will duplicate the leakage path which the A send signal sees as it traverses the hybrid to the A receiver line port.

The output of the filter is an estimate of the echo and is subtracted from the A receive line at that point. When the echo is cancelled at this point, the long distance, long delay path need not be considered. The filter response therefore needs only to be a few milliseconds (16 to 25 milliseconds is commonly used).

A typical canceller uses an 8 kilosample A-D converter and a 128 point digital correlator to cross correlate the A send signal with the A receive signal [8].

This correlation function is used to generate the tap weights for a 128 point transversal filter. A number of algorithms have been proposed for deriving the tap weights from the cross correlation function. In most of these the computation is dominated by the computation of the cross correlation function and the implementation of the filter. Both filter and correlator are implemented by 128 point convolutions requiring 255 multiplications each. Thus, the signal processing load is approximately 8,000 x 510 or about four million multiplications per second per channel.

Additional computations are required for the adaptation and for the interpolation required prior to D-A conversion. These could be performed in a microprocessor. Alternatively, a single five megamultiply per second processor could be designed to perform the entire echo cancellation function. Both alternatives are feasible with VLSI. A single chip echo canceller using custom logic is described in [8].

7.1.2 Narrow Band Voice Channel Processing

The echo canceller is typical of many problems in voice channel processing. These problems involve a 8000 per second sample rate and a filter or data window length of 80 to 320 samples (10 to 40 milliseconds). The window length is determined by either the channel characteristics or the rates of change in the speech production mechanism. Equalizers, voice pitch extractors, linear predictive speech encoders and modems, touch tone receivers and other functions all require about the same amount of signal processing.

One approach is then to build a general purpose VLSI chip which can be programmed for multiple applications in narrow band voice processing. At least two such single chip voice band processors have already been developed. Not surprisingly, the developers of these processors are Nippon Electric Co. Ltd, [9] and Bell Telephone Laboratories, [10].

The NEC chip contains a RAM, a coefficient ROM and a program ROM. It contains a program stack, a 32 bit Arithmetic Logic Unit (ALU) and a 16 by 16 bit parallel multiplier. Data words are 16 bits and coefficient words are 13 bits.

The NEC processor instruction cycle is 250 nanoseconds. This chip computes a second order filter section (four multiplies, four adds and two moves) in 2.25 microseconds. Address generation, memory access and instruction decoding are done in parallel with computation. This parallel operation is achieved through direct memory access ports and an ALU Bus Isolator.

The result is a single chip signal processor which performs four million operations per second. This capability exceeds that of any commercial minicomputer, and far exceeds the signal processing capability of any commercial microcomputer. It is an excellent example of the potential of VLSI. The chip area is 0.28 cm^2 and the technology is 3μ N channel MOS. The chip contains 40,000 transistors.

The BTL chip is also N MOS but uses 5μ linewidths, has 45,000 transistors and an area of 0.68 cm^2. The chip has a 16 bit coefficient word, a 20 bit data word, and a 40 bit accumulator. The architecture is similar to the NEC chip, except that the BTL chip does not have a direct memory access to the multiplier. The computational efficiency is about the same, however. The clock for the NEC chip is 37% faster and its computation rate for 2nd order recursive filter sections is 42% faster (2.25 μsec vs. 3.2 μsec). If the BTL chip were designed in similar technology, the throughputs would be comparable.

These two signal processor chips portend a breakthrough in narrowband communication processing. They demonstrate the feasibility of single chip programmable digital processors for voice channel applications. This not only will permit cost effective digital processing for a number of applications, but also permits the assignment of processors on a per channel basis. In the past, digital processors could only be cost effective if they served groups of twelve or twenty-four channels. Cost effective "per channel" processors may greatly simplify the system architecture.

7.1.3 Dual Tone Multifrequency (DTMF) Receiver

Pushbutton telephone systems present an excellent application for the single chip VLSI processors just discussed. The signaling system uses two groups of four frequencies, as shown in Fig. 2. Each number keys two simultaneous tones, one from each group of four. In order to decode the signal at the subscriber switch, the receiver divides the voiceband into two sub bands, each of which contains a group of four frequencies. The low band and high band filters can each be constructed from three second order recursive filter sections. After limiting, each band is fed to four separate matched tone filters, each of which is a second order recursive filter. Each of the eight tone filters is low pass filtered by another filter section and a decision logic decodes the result. If two additional filter sections are used to eliminate noise and dial tones at the input, a total of 24 filter sections is required [9]. The BTL chip can execute 39 filter sections in the 125 μsecond sampling interval, and the NEC chip can execute 55 filter sections in the same interval. Thus, the NEC chip can be programmed as a one or two channel DTMF receiver and the BTL chip as a single channel receiver. Additionally, each channel would require an analog antialiasing filter (low pass at 3.2 kilohertz) and an A-D converter.

Two other methods for digitally implementing the DTMF receiver have been suggested. The first is to FFT the band between 100 Hz and 1800 Hz and put detectors on the appropriate FFT taps. However, this is not quite so simple. First the signal must be bandpass filtered to prevent aliasing. Then the FFT must be windowed. The tones, as can be seen from Fig. 2, are not evenly spaced. They are purposely spaced aperiodically in order to minimize a phenomenom called aliased harmonic distortion [1, page 15] and intermodulation effects. The additional processing required and the large number of frequency bins required (32), cause the FFT approach to be less attractive than the digital tuned filter approach described above.

The second alternate method is to bandpass the signal, limit it and count zero crossings. This method is fairly easy to implement with special digital logic, but is more susceptible to noise and split bauds than is the tuned filter approach.

Another practical problem in DTMF receivers is "talk off". This is the effect of voice on the line being interpreted as a tone signal, leading to incorrect decoding of the dialed number. With the tuned filter approach, the thresholds of the detectors can be precisely matched to the limiter output level. This is an advantage of the tuned filter approach.

One DTMF receiver can serve a number of channels, since the average line will be in dialing status less than ten percent of the time. The switching system will use a number of these receivers on an "as available" basis. Even at the ratio of one DTMF receiver per twenty subscriber lines, the market for digital DTMF receivers is in the millions of units.

	HZ		HZ
ROW 1	697	COLUMN 1	1209
ROW 2	770	COLUMN 2	1336
ROW 3	852	COLUMN 3	1477
ROW 4	941	COLUMN 4	1633

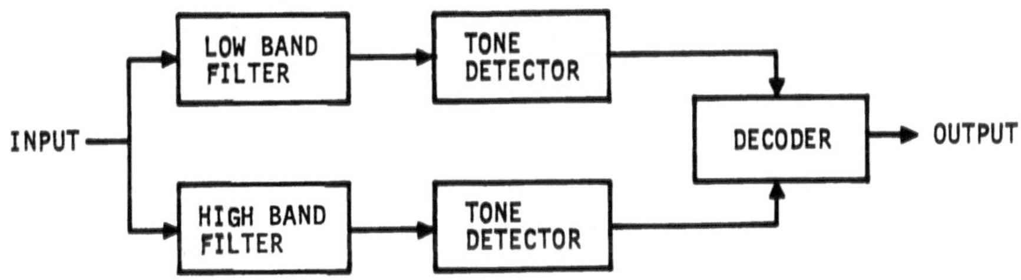

Fig 2. Dual tone multifrequency receiver

7.1.4 TDM-FDM Transmultiplexer

Two very different multiplexing systems are used throughout the telephone system. Because of the enormous investment in equipment, the older frequency division multiplexed single side band (FDM-SSB) system will continue in use through this century. However, the newer pulse code modulation time division multiplexed techniques used in AT&T's electronic switching system (ESS) represent a substantial fraction of the total system. The process of interfacing the two different multiplexing systems is called transmultiplexing.

We shall consider a relatively simple case, the digital modulation of 24 channel PCM group into two 12 channel FDM SSB groups.

The ATT D2 format multiplexes 24 8 kilosample voice channels into a 1.544 megabit per second bit stream. Each sample is companded by the μ-law into a 255 level 8 element code. Each data frame consists of 8 x 24 = 192 data bits plus a frame marker bit [11].

Both the NEC and the BTL processor chips have the speed and the decoding circuitry to demultiplex and decompand the D2 bit stream back into twenty four parallel channels, each of which is linearly encoded in 16 bit parallel data words. A smaller, custom VLSI chip could also perform this function easily. A possible advantage of a custom chip is that larger output buffers could be added to the custom chip.

After demultiplexing and decompanding, the transmultiplexer shown in Fig. 3 must perform the following operations:

- Baseband generation. Each channel is mixed with an in phase sine wave at the mid band frequency of 2000 hertz and a quadrature phase (delayed 90 degrees) sine wave at the same frequency. Each signal is then separately low pass filtered to remove signals outside of 200 hertz. Each channel requires 16,000 multiplies per second for the mixers and two to eight filter sections for the low pass filters. This requires 20% or less of the capacity of either the NEC or the BTL chip.

- FDM group band generation. Twelve channels are to be frequency division multiplexed into 4 kilohertz slots in the group band from 60 KHz to 108 KHz.

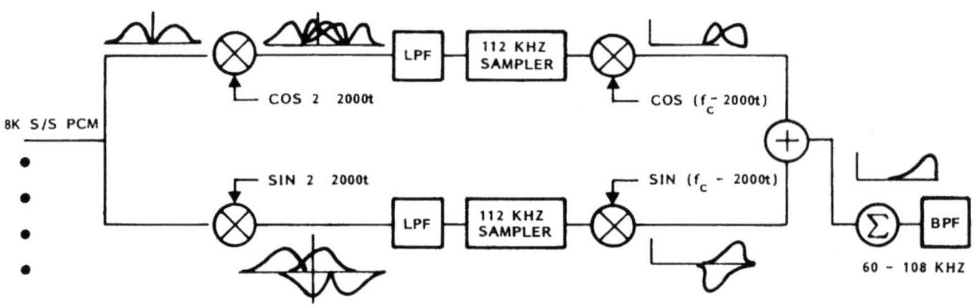

Fig.3 Weaver digital TDM-FDM transmultiplexer (12 channel)

To accomplish this each I and Q signal is interpolated to 112 kilosamples per second [12] and digitally mixed by multiplying by a reference centered at the middle of its frequency slot. The reference sine waves in the two mixers must be 90 degrees apart in phase. The outputs of the digital mixers are then added and the lower side bands will cancel. The interpolation and mixing requires about two million operations per second. One chip can then perform baseband and group band generation for one channel, and perhaps for two channels, depending upon the implementation.

To complete the group band generation, the channel outputs are added, passed through a digital to analog converter and then low pass filtered to remove the mixing products above 112 KHz. A possible implementation would use one demultiplexor chip, twelve voice band processor chips and two D-A converter chips, plus the analog low pass filter. A similar configuration could be employed to reverse the process and to convert two 12 channel FDM group bands to a 24 channel D2 (or D1) TDM group. An alternative approach to the FDM to TDM transmultiplexor is to use FFT techniques to demultiplex a 60 channel super group or even a 300 channel master group.

7.1.5 Synthetic Aperture Radar (SAR)

Synthetic aperture radar is an excellent example of a parallel high speed processing problem. If direct techniques are used, the number of operations increases as the cube of the resolving power. The number of resolution elements (pixels) increases as the square and each resolution element requires the processing of a set of data points which increases linearly with the resolution.

Synthetic aperture techniques are used to generate high resolution radar maps by coeherently processing the stored returns from a radar which is flown along a flight path parallel to the area which is to be mapped. The stored returns can be thought of as a rectangular array, or image. The SAR processor performs a two dimensional transformation of this array into a radar image which resembles a photograph. The number of picture elements in the image is potentially equal to the number of points in the array of stored returns.

The resolving capability is not caused by doppler effects and so is independent of the aircraft velocity. It is due to the displacement of the physical antenna to a different position at the time of each transmitted pulse.

Synthetic aperture radar is a special type of antenna array processing. We shall derive the algorithm from this point of view because many other array processing problems are similar and also excellent candidates for digital signal processing.

Consider first a plane wave impinging on a linear array which has N+1 equidistant elements separated by distance d, as shown in Fig. 4. The phase shift of the received signal between adjacent elements is equal to

$$2\pi\ d \sin \Theta/\lambda \text{ in radians} \tag{1}$$

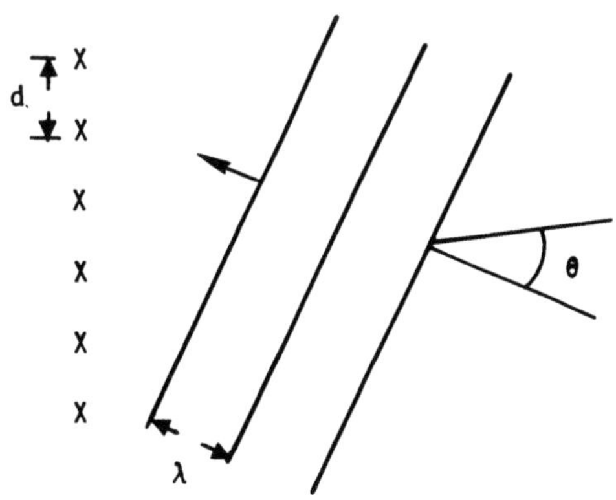

Fig.4 Plane waves impinging on a linear array

where Θ is the angle between the normal to the wave and the normal to the array. We can represent the signal at the outer element as a complex sinusoid:

$$A\exp(i2\pi ct/\lambda) \tag{2}$$

where A is the amplitude and c is the propagation velocity.

Now if we sum the received signals over all the array elements we obtain

$$A\exp(i2\pi ct/\lambda) \sum_{K=-N/2}^{N/2} \exp(i2\pi kd\sin\Theta/\lambda) \tag{3}$$

This is the sum of N+1 complex numbers whose arguments vary by $\frac{2\pi d\sin\Theta}{\lambda}$ and whose modulus is A, as depicted in Fig. 5. As we add more elements to array, the modulus of the sum grows until the total phase shift exceeds 90 degrees or $\frac{\pi}{4}$ radians. At that point the modulus decreases so that when the total phase shift is 180 degrees or $\frac{\pi}{2}$, the sum is equal to zero.

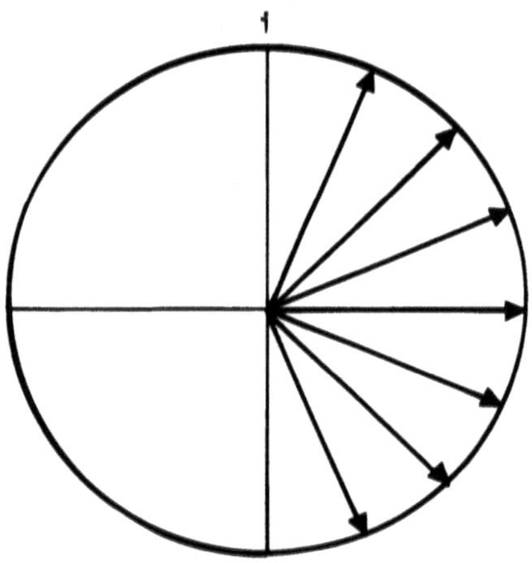

Fig.5 Complex phasor representation of signals received at the elements of a linear array

The sum can be reduced analytically by noting that

$$1-a^{N+1} = (1-a)\,(1+a+a^2+\,..\,+a^N) \tag{4}$$

then Eq. (3) becomes

$$A\exp(i2\pi ct/\lambda)\ \frac{\sin[(N+1)\pi d\sin\Theta/\lambda]}{\sin(\pi d\sin\Theta/\lambda)} \tag{5}$$

The antenna pattern of the linear array is then

$$\frac{\sin[(N+1)\pi d\sin\Theta/\lambda]}{\sin(\pi d\sin\Theta/\lambda)} \tag{6}$$

The half power antenna beamwidth can be computed by evaluating this function and is approximately $\frac{Nd}{\lambda}$, or $\frac{L}{\lambda}$ where L is the length of the array.

Now notice that the antenna pattern can be shifted merely by inserting a phase shift at each antenna element. If the difference in phase shift between successive elements is $2\pi d\sin\phi/\lambda$, then the antenna pattern becomes

$$\frac{\sin[(N+1)\pi d(\sin\Theta+\sin\emptyset)/\lambda]}{\sin\ [\pi d(\sin\Theta+\sin\emptyset)/\lambda]} \tag{7}$$

Then the pattern of (6) is shifted so that it is centered at an angle ϕ from the broadside direction. By inserting a linear progression of phase shifts, then one can generate a "beam" in any direction. A number of different beams can be generated simply by dividing the signal and using a number of different summing networks each with a different linear phase shift pattern.

In order to form the beams digitally, the complex waveform must be sampled at the Nyquist frequency at each element. The phase shift is then implemented by multiplying by a complex number whose magnitude is unity. If beam shaping to reduce sidelobes is desired, the magnitudes of the complex multipliers can be adjusted. This adjustment is done once for each beam. For a complex wave, the Nyquist frequency is the highest frequency in the input spectrum. For real sampling, suc h as occurs in acoustic pressure transducers, the sampling rate must at least be twice the highest input frequency. If the bandwidth of the sensor exceeds the bandwidth of the signal, the sensor element outputs must be prefiltered prior to sampling in order to prevent aliasing of noise and spurious signals into the sampled data.

For each beam, the number of operations (multiplications and additions) required is 2N times the sampling rate.

So far we have assumed plane waves. This is appropriate for a linear array because the array is focused at infinity. However, we have already shown that when the phase shift at an element exceeds $\lambda/4$, the contribution of that element is negative. If the waves originate at a distance R, the maximum useful aperture for an unfocused array is that which sees a maximum phase shift (for a wave normal to the array) of $\lambda/4$. This length is easily calculated from

$$(L/\lambda)^2 + R^2 = (R+\lambda/4)^2 \tag{8}$$

or

$$L \approx \sqrt{2\lambda R}$$

An array longer than $\sqrt{2\lambda R}$ must be focused. However, the focusing is accomplished by a set of phase adjustments, just as in beam forming. If ψ is the angle between the direction from the source to the array element and the normal to the array (assuming the source is on an array normal), then the phase shift at that element is $2\pi D(1-\cos\psi)/\lambda$ where D is the source to element distance. D can be approximated by R, and $1-\cos\psi$ can be approximate by $\frac{\psi^2}{2}$. In addition, we will make the approximation that

$$\psi \approx \frac{kd}{R} \tag{9}$$

Then the phase adjustment in radians at the Kth element (from the array center) is

$$\emptyset_k = \pi R k^2 d^2/\lambda R^2 = \pi k^2 d^2/\lambda R \tag{10}$$

This is accomplished by multiplying the signal samples at element K by the complex multiplier:

$$\exp(i\emptyset_K)$$

If it is desired to shape as well as focus the beam, a non unitary multiplier,

$$W_K \exp(i\emptyset_K)$$

can be used.

It is apparent that the arithmetic operations required for beam focusing are the same as the operation for beam forming. In fact, beam forming, beam focusing and beam shaping can all be accomplished simultaneously by multiplying each sample at each element by a complex number.

The above discussion assumed that the source is at the same place for all of the array elements. This is true for a physical array such as an active or passive sonar or a phased array radar. However, in the usual synthetic aperture array, both the transmitter and receiver elements are moved together. The defocusing effect of the linear array is doubled because the phase shift must be calculated from the round trip from element to source and back to the element.

The maximum effective unfocused aperture is then calculated from

$$(L/2)^2 + K^2 = (R^2 + \lambda/8)^2 \tag{11}$$

or

$$L_{max} \approx \sqrt{\lambda R} \tag{12}$$

Additionally, we must take the round trip phase shift into account in calculating the beamwidth. Recall that the antenna pattern calculation assumed a phase difference between elements of

$$2\pi d \sin \Theta/\lambda \tag{13}$$

However, in the synthetic aperture radar, the phase shift between successive elements must include the round trip phase differnce, which is

$$4\pi d \sin\Theta/\lambda \tag{14}$$

The antenna pattern is then

$$\frac{\sin[(N+1)2\pi d(\sin\Theta + \sin \phi)/\lambda]}{\sin[2\pi d(\sin\Theta + \sin\phi)/\lambda]} \tag{15}$$

Note that the beam forming phase factor, $\sin\phi$ has also been doubled.

The antenna pattern for the synthetic array is seen to be twice as narrow as that of physical array. The beamwidth is then given by

$$\beta_{SA} = \lambda/2L \tag{16}$$

In the unfocused case, using (12), the azimuth resolution is

$$\text{resolution} = \beta R = R\lambda/2L_{max} = \sqrt{R\,\lambda/2} \tag{17}$$

For the focused synthetic array, the resolution is $\lambda/2L$ where L can be made as large as the beamwidth of an individual element. Since the individual element beamwidth is λ/S, where S is the element length, the beamwidth of a focused synthetic array is $\lambda S/2\lambda$ or S/2.

The synthetic array is usually obtained by flying an aircraft mounted coherent pulsed radar in a straight line at a velocity V. The aircraft uses a single antenna of length S, and emits pulses at a periodic interval T. The interval T must be chosen so that there is at least one pulse per azimuth resolution element. This means that T must satisfy

$$T < S/2V \tag{18}$$

Each pulse is match filtered in the receiver to provide a complex waveform. This waveform is sampled and digitized. Each digital word represents a linear combination of all of the returns from a given range for a single transmitted pulse. These words are stored in a memory in what may be thought of as the columns of a matrix. Each successive pulse results in a new column of the matrix.

The rows of this matrix represent the sequence of returns from a fixed range for a sequence of pulses.

In order to obtain the maximum resolution of S/2, the aperture must be

$$L = 2\lambda R/S^2 \tag{19}$$

The total number of words in memory is then

$$N_R\ (2\lambda R/S^2) \tag{20}$$

where N_R is the number of range resolution elements.

Consider the following example

An aircraft at 200 meters per second
5 = 2M
λ = .1M
R = 20KM

If we desire a 1 meter^2 resolution then

L = 1000 meters
T = .01 seconds
N = 1000 samples

Assume a swath 5000 meters deep is to be mapped.

Focusing is required, because the maximum unfocused aperture is

$$L_{MAX} \approx \sqrt{\lambda R} \approx 45 \text{ meters}$$

The focusing operation takes the form

$$e_k(x_k,y_\ell) = \sum_{x_k - N/2}^{x_k + N/2} s(x_j - x_k, y_\ell)\, W(x_j,y_\ell) \exp[i\phi(x_j,y_\ell)] \qquad (21)$$

Where $e(x_k,y_\ell)$ is the estimate of the reflectivity at point whose coordinates are x_k, y_ℓ (x is azimuth, y is range). $W(x_j,y_\ell)$ is an amplitude weighting factor use to shape the beam to improve dynamic range. S(x,y) is the word stored in the x,y position of the stored return matrix.

In the general case $W(x_j, y_\ell)\exp[i\phi(x_j,y_\ell)]$ is variable in both j and ℓ, and therefore requires an amount of storage equal to the stored return matrix. However, in most cases this function can be held constant for many values of j, and the coefficient storage requirement is proportionately less.

The stored return matrix in this example is 5000 by 1000, or 5×10^6 complex words. The computation rate, counting additions and multiplications is 10^7 per pulse, or 10^9 per second. Since the range channels can be processed independently the problem can be made as parallel as desired.

One hundred processors, each capable of ten million operations per second could perform the basic operations. The independence of range channels is valid if the range variation from the imaged point does not exceed one range cell (one meter in this example).

This occurs when

$$(x - x_k)^2 + R^2 = (1 + R)^2$$

or (22)

$$x - x_k \approx \sqrt{2R} = 200 \text{ meters}$$

The consequence of this result is that the azimuth correlation cannot be performed over an entire row of the stored return matrix but must instead be performed over pieces of several rows, and a very precise algorithm would use some degree of interpolation.

This does not preclude the parallel processing approach however, since 100 parallel processors each can work on 50 rows of the stored return matrix. The total range migration effect over the synthetic array in this example is 12.5 meters. Therefore memory contention should not be a serious problem.

It is clear, however, that both the size and the organization of the memory are important considerations. The requirement for 5 million words amounts to 80 megabits of storage. VLSI memory chips can be expected to have between 64 kilobits of storage. Using 256 kilobit chips, the memory would require more than three hundred memory chips.

So far we have ignored doppler effects. Since the radar is moving with respect to the ground, the returns will have a doppler shift

$$\Delta f = 2Vf_o \sin\psi/c = 2V\sin\psi/\lambda \qquad (24)$$

where Δf is the doppler shift, f_0 is the carrier frequency, c is the propagation velocity, λ is the wavelength and ψ is the angle between the array normal and the target direction. Notice that the doppler shift changes sign as the angle ψ goes through zero, while the array focusing term does not. The maximum phase shift in the return is for a target at the edge of the aperture,

$$\Delta\psi = 2\pi\Delta f T_{max} = (4\pi VR/c\lambda)\sin\Theta max \qquad (25)$$

where $\Delta\psi$ is the net doppler phase change, R is the range T is the return time from the target and Θmax is the angle to the target from the array normal at the edge of the beam. Since

$$\Theta max = \lambda/2S$$

we have, for our example

$$\Delta\psi \approx \frac{2VR}{cS} = \frac{200 \times 2 \times 10^4}{3 \times 10^8}$$

$$= .0133 \text{ meters or } .133 \text{ wavelengths.}$$

The maximum doppler phase effect is seen to be about 1/8 wavelength in this example and since the doppler phase is linear, across the array, it can be ignored in most applications. We have already seen that a linear phase shift across the array merely shifts the angle of the beam. Thus the doppler effect slightly distorts the radar map but does not affect resolution.

If doppler correction is desired, the doppler phase correction is added to the focusing phase correction prior to convolving with the row of the stored return matrix.

Other Array Processing Problems

We have presented a large amount of detail on the synthetic array radar problem not only because it is a major challenge for VLSI technology, but because its solution can be extended to a number of other important problems.

As discussed earlier, the problem of beamforming with either a physical or synthetic array can be solved by multiplying the array outputs with a phase adjustment function. The only difference is that the phase function is linear rather than quadratic. In a physical array, the columns of the return matrix are collected in parallel rather than one at a time. The columns represent time samples rather than range samples. After convolution, the columns of the results matrix are the inputs to channel processors for each beam.

In the case of a sonar, the SAR processor above could perform beam forming for a 1000 element array with 100 beams at a sample rate of 5000 per second. Alternatively, a Fourier transform processor could form the beams with much less computational effort.

7.1.6 VLSI Memories

The most obvious application of VLSI is in memories. The organization of memories within the chip is relatively straightforward, except for some testing considerations to be discussed later. The demand for larger and larger memories is increasing. HODGES [13] presents an excellent discussion of large semiconductor memories.

The impact of VLSI on memories will be more than simply bigger, cheaper, and faster memories, although that will certainly occur. The architecture, both internal chip architecture and external memory architecture, is likely to change.

At present, memory chips use single bit words to minimize the number of pins required. A memory with k bit words is configured from k single bit chips. Each time the memory capacity is doubled, only one more pin (for address) is required. However, this simple organization is not the best from the standpoint of interconnect, reliability, and testability. With VLSI these considerations assume greater importance.

Hieararchial arrays have been suggested by MEAD [14] in order to minimize chip interconnect and access time. Byte oriented memories would require more input/output pins, but would increase gate utilization (a lower proportion of chips would be addressed), improve access and permit on chip error correction. Many signal processing applications are memory access limited, so that there are system benefits to independently addressable "short-wide" memories (e.g., 2K by 32 bits versus 64K by 1 bit).

In VLSI memories, the amount of charge stored in each bit is so small that errors can be caused by alpha particles generated in the package or as secondary effects of cosmic rays. ZIEGLER and LANFORD [15] have shown that these effects are significant.

Error correcting codes can correct both these "soft" errors and "hard" errors which result from chip defects. A brute force method now in vogue for correcting hard errors is to use half of a 32 kilobit memory chip as a 16 kilobit memory if the other half is nonfunctional. Error correcting coding would further improve "yield" while at the same time providing protection against soft errors. Reference [16] contains a thorough and up to date discussion of both error correcting memory and error correcting logic.

Sophisticated memories which can perform simultaneous access and limited search or "content addressing" are excellent applications for VLSI. These memories would have a major impact on the computer industry.

7.2 VLSI Testing

As VLSI increases the number of logic gates per chip, testing and design for testing become increasingly important.

SSI chips could be tested easily because the pin to gate ratio was high and the number of gates was small. The state of the chip could be controlled and observed with sequences of a few hundred bits or less. A chip that was either uncontrollable or unobservable could be redesigned eaily, when that fact became known. Systems and boards were testable because they were accessible - the nodes of the system could be probed by "bed of nails" testers which could drive the input and monitor the output of each chip. The chip itself might have twenty or fewer logic gates.

Testing of LSI has evolved to a generally accepted set of procedures [17]. The test patterns are based on both the logical design and the detailed circuit design, as illustrated in Fig. 6. With VLSI, all of the steps shown in Fig. 6 become more difficult. Probe requirements are more difficult because of pin limitations. Fault analysis is more time consuming. Pattern tests are more complex and more numerous. Structural or logical tests grow exponentially with the logical complexity. For example, if a network has k inputs and ℓ flipflops, $2^{k+\ell}$ different patterns are required in order to perform a complete functional test. In fact, it may be that there is no sequence of patterns which can test the network.

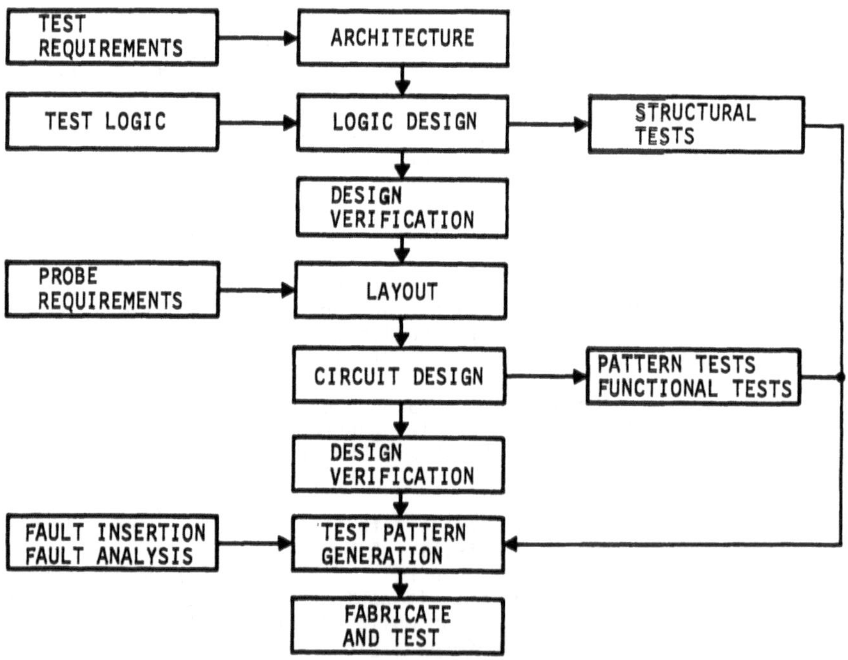

Fig. 6 The Top Down Testing Process

It is common to borrow two terms, controllability and observability, from control theory in order to qualitatively describe the testability of logic nets.

We shall use a more precise definition of these terms:

A network is controllable if that network can be brought to any of its possible states by an input sequence of finite length.

A network is observable if any state of the network can be deduced by observing a finite length result of a finite length test sequence.

A network is accessible if its storage elements can be written and read directly, without traversing the network.

The attributes of controllability, observability and accessibility are absolute. When they are present, the network is not only testable, but easily testable.

7.2.1 Design for Testing

With LSI and VLSI, these three attributes - controllability, observability and accessibility are difficult to attain. One reason for this is pin-to-gate ratio. Because chips and boards are planar devices, their interconnection capability is a linear function of the reciprocal of the circuit dimension, while their complexity varies inversely as the square. If we reduce the linewidths and linespacings at both the board and chip level by a factor of ten, we get one hundred times as many gates per chip, but only ten times as many pins per chip. The pin to gate ratio is then reduced by a factor of ten. Great care must be exercised in the electrical design, the logical organization, and the testability design, because of this lower pin-to-gate ratio. The relationship between design and testability is shown in Fig. 7.

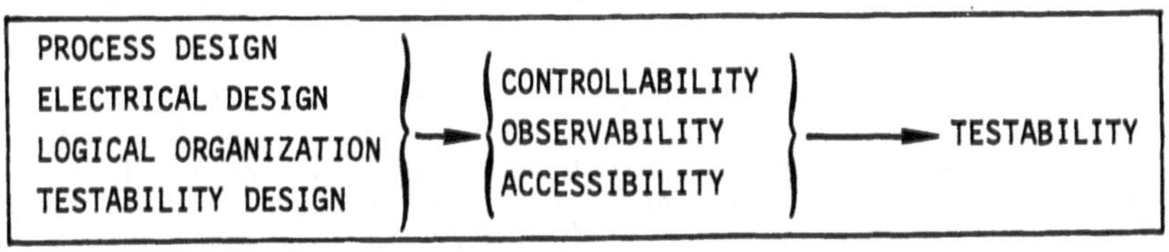

Fig. 7 Testability Comes From Design

Many discussions of testing consider only two attributes for testing - controllability and observability, i.e., can the device be put in every state from external inputs and can the state be deduced from the output? I have added the attribute of accessibility because it is a measure of the ease of controlling and observing the state of the device.

The design of semiconductor chips begins at the process level. This level is key to both reliability and testing. As the number of devices on chip increases, the probability of double faults increases. Unless the process is engineered so that double faulting is improbable, the test generation procedure must assume double faulting or even multiple faulting. Four types of faults are usually considered: gate stuck at one, gate stuck at zero, short circuit and open circuit.

A 10,000 gate chip has roughly 20,000 "stuck at" fault opportunities. In addition it has roughly 20,000 open circuit opportunities. Short circuit, or bridging opportunities depend upon the chip layout, but can exceed 20,000. Thus, approximately 60,000 different faults are possible. The number of possible double faults is 3.6×10^9. Analytic estimates of fault possibilities are given in GOLDSTEIN [18].

With existing techniques, the computer run time required for fault simulation and test generation increases as the cube of gate count. For a 10,000 gate system the time required is about 100 hours on a large computer such as the Univac 1108 [19]. This is for only single faults. Obviously double faults are unmanageable. Oxide formation techniques, metallization techniques, and lithography techniques all affect the probability of shorts and opens, as well as gate failures. Cleanliness is a critical process parameter, particularly with VLSI, because several devices can be affected by a single particle induced surface defect. Therefore, process design is important to VLSI testing, because it is a means of reducing the occurrence of multiple faults.

In addition, the wearout mechanisms for chips are mechanical or chemical in nature. Chemical stability is very process dependent. Mechanical stability is a function of the package design and the bonding technique [20, 21].

Electrical design is critical for VLSI chips. Pattern sensitivity must be held to an absolute minimum, and pattern sensitivity is basically an electrical problem. "Soft" errors are often a result of pattern sensitivity. A number of papers have dealt with this subject. Because speeds will be greater in VLSI and interconnect lines more closely spaced, capacitive coupling will be greater.

A number of suggestions have been made for reducing the pattern sensitivity of LSI circuits. In particular, much is known about "robust" memory design [22, 23]. A set of design principles for reducing pattern sensitivity of general logic circuits is badly needed.

Logical organization is the reduction of the VLSI chip to subfunctions, or macrocells, each of which is individually accessible and testable [24, 25]. This assumes that the electrical design is such that a chip's behavior is actually represented by its logic, i.e., that pattern sensitivity between macrocells is nonexistent. A microprocessor is a good example of a functionally organized chip. The RAM, the ROM, the input-output circuitry, the ALU, the registers, the shift network and the instruction decoder are usually physically separated. In many microprocessors the various functions are individually accessible. If they are not, the testing is virtually impossible.

The fourth and last design area of significance to controllability, observability, and accessibility is testability design. This is the design of specific additional features which enhance these attributes. These design features include set/scan logic (Level Sensitive Scan), continutity testers, built-in signature analyzers, and built-in pattern generators. An excellent discussion of these techniques is presented in WILLIAMS and PARKER [26].

Set/scan logic or level sensitive scan design (LSSD) is the concept of designing every latch circuit as a switchable port device. That is, it can receive its input from either of two sources, and send its output to either of two sources. Let the input ports be labled D_1 and D_2 and the outputs labeled Q. Then if the devices are daisy chained as in Fig. 8 by connecting an output of one device to the D_2 input of its nearest neighbor they form a shift register when the D_2 ports are active. When the D_1 ports are active, the device acts as a conventional latch. IBM calls such a device a "shift register latch", because an array of such devices connected as above functions either as an n-bit shift register or an n-bit latch. If the two ends of the shift register are connected to a pair of input/output pins, the states of all of the latches are easily accessible.

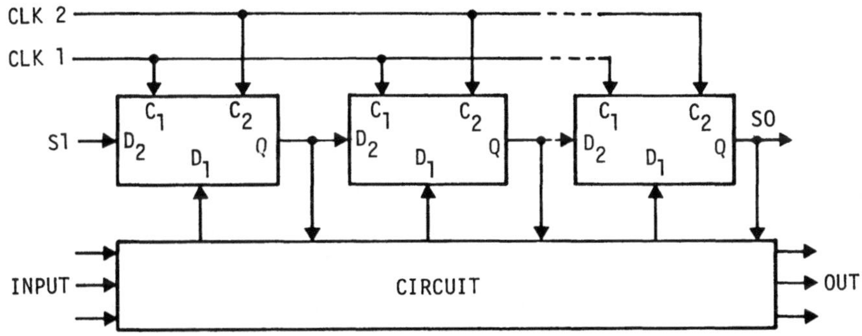

Fig. 8 Shift Register Latch Using Three Port Devices

IBM's concept of level sensitive scan design (LSSD) goes a step further. In this approach, all logic, except for RAM and ROM is partitioned into combinational (memoryless) networks and (latch) registers as shown in Fig. 9. The combinational networks are decomposed into smaller networks separated by intermediate registers. All registers are designed from four port devices and are interconnected into one or more externally connected shift register loops [26, 27].

With LSSD, each combinational net is directly accessible. Therefore, the problem of controllability and observability is reduced to ensuring that each such net is controllable and observable. The design of controllable and observable combinational nets is treated by EL-ZIQ [28].

One drawback to LSSD or set/scan is that it only provides for static or functional testing. Because the test inputs must be read in serially the device under test (DUT) cannot be clocked at the normal operating rate. One method of dynamic testing with set/scan logic is to combine the concept of set/scan with that of linear feedback shift registers (LFS).

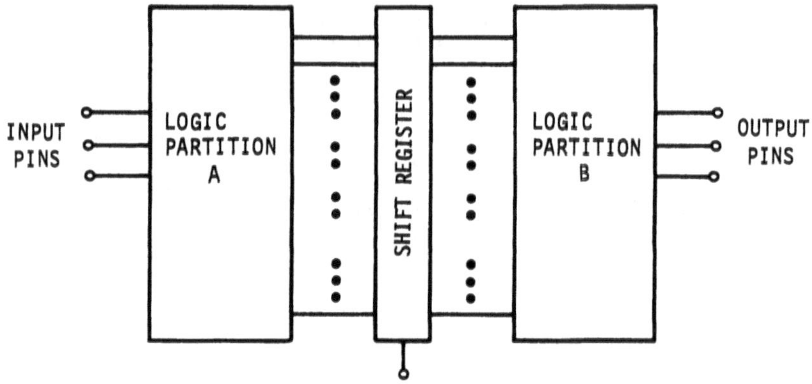

Fig. 9 Level Sensitive Scan Design Partitions Logic into Combinational Nets and Shift Register Latches

A linear feedback shift register using modulo two arithmetic can be used to generate a repeatable pseudorandom binary sequence. This sequence can be clocked into a shift register latch at the input which is toggled between shift operation and latch operation at the operating clock frequency. This provides a pseudorandom input at the operating rate. The output shift register latch can be toggled at the same rate and the sequence that is shifted out input to another linear feedback shift register. The length of this latter register can be several times the data wordlength. Given a known state of the input LFS and a known number of clocks, the contents of the output LFS is known and is called a device signature.

If the DUT is defective, the signature will differ from that of a known good device (KGD). This technique of encoding the outputs is called signature analysis. It may be used with or without set/scan techniques, but the two concepts combined provide a powerful means of built-in testing. The technique described above is illustrated in Fig. 10. This technique is described further by KONEMANN, MUCHA, and ZWIEHOFF [29].

The design of shift register latches may be implemented in various ways. Two designs, using three port devices (two in, one out) and shown in Fig. 11 have been suggested by FUNATSU, WAKATSUKI, and YAMUDA [30]. The implementation of an entire computer, the IBM System 38, in LSSD is described by STOLTE and BERGLUND [31]. Signature analysis is described in detail by NADIG [32].

Testing VLSI Memories

Memories divide into two major classes, RAMS and ROMS. In some ways ROMS are more difficult to test because they lack the attribute of controllability. Addressing errors cannot be detected because write-read tests are impossible. For that reason, signature analysis is appropriate to ROM testing [33].

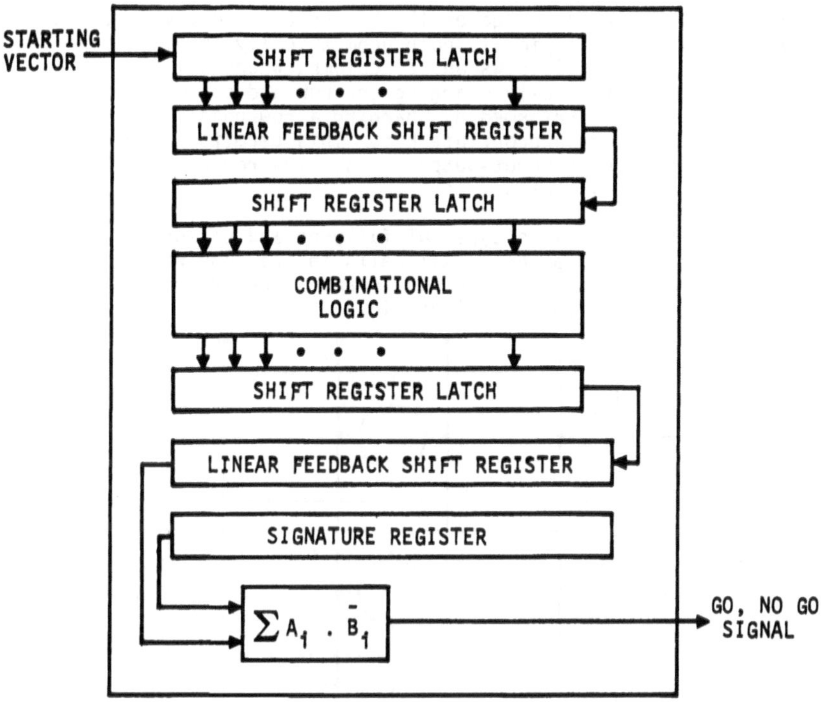

Fig. 10 Dynamic Built-In Test Concept Using, Set/Scan and Signature Analysis

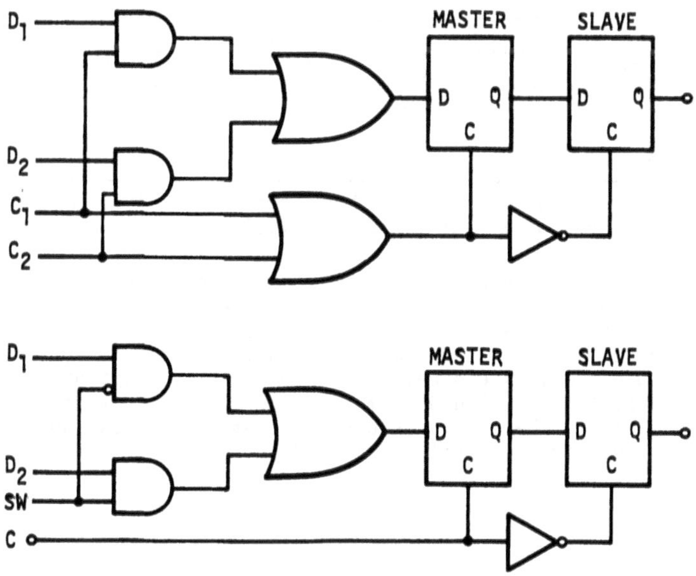

Fig. 11 Alternate Scan Set Logics for Three Port Devices

At the present time RAM's are generally bit by 2^m word devices. Their layouts are usually two dimensional, both geometrically and functionally. The 2^m bits are arranged in 2^k rows and 2^ℓ columns where $k + \ell = m$. A cell is written by simultaneously charging the row and column write lines for that bit. Reading is performed by charging a row read line and sensing a column sense line for the appropriate bit. This permits a very compact design, and efficient use of input-output pins. The row address takes k pins and the column address takes ℓ pins.

Given this organization, some types of failures and potential pattern sensitivities are readily apparent. All cells in the same row and all cells in the same column share some circuits and therefore row and column pattern sensitivities are typical. (See Fig.12)

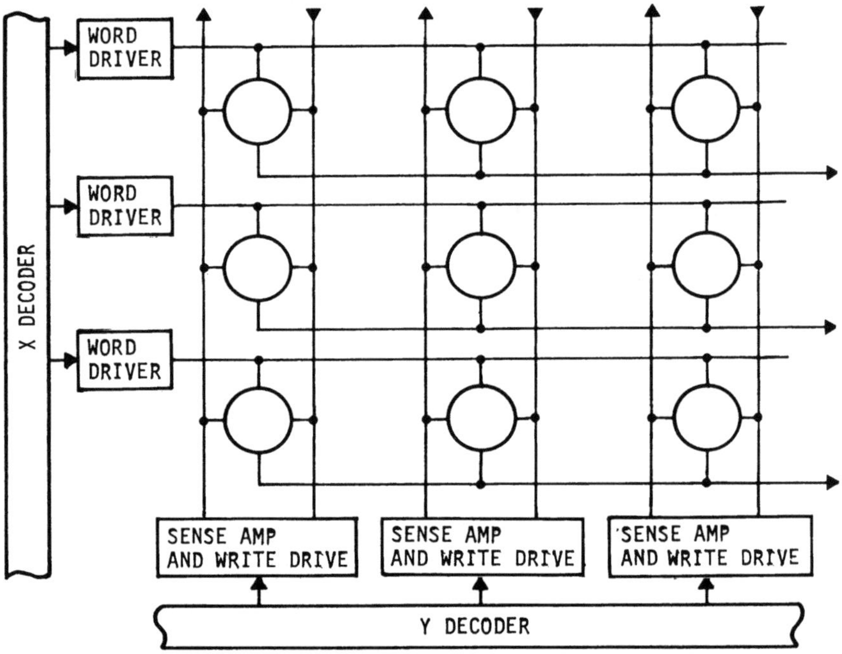

Fig. 12 Typical VLSI Memory Organization Illustrating Row/Column Pattern Sensitivities.

Other likely failures are bits stuck at zero or one, address decoder shorts or opens (more or less than 1 bit accessed with one address), and surrounding cell disturbances. The latter can be caused by capacitive coupling between cells. With the smaller dimensions and higher speeds of VLSI, the disturbance may extend beyond the immediately adjacent cells. However, the adjacent cells should suffer the greatest effect and testing than may be sufficient. We shall discuss next some well known and useful tests, and their relevance to VLSI testing [25].

Marching One-Zero

Write all cells to zero. Read all cells. Write the first cell to one and read it. Repeat for all cells. Read all cells. Write the first cell to zero and read it. Repeat for all cells. Read all cells. This test requires 8N operations. It tests functionality of all cells and some static interaction, but not dynamic interaction.

Walking One-Zero

Write the first cell to one, read all cells and restore the first cell to zero. Repeat for all cells. This test requires $2N^2$ operations. It tests the cell addressing and stresses the sense amplifier, thus testing its recovery time.

Galloping Ones and Zeros

Read the first cell and then write it (change it) and reread it. Read the next cell and reread the first cell. Repeat the alternating read for all cells. Then repeat the entire sequence for all cells. Repeat the whole process and the memory should then be restored to its original state. This test requires $4N^2$ operations. It tests functionality, all read read transitions, and all address decoding.

Galloping Write Recovery

Read cell i, write it to one, read cell j, read the cell i, read the cell j again and rewrite cell i to zero. Repeat for all i leaving j fixed. Repeat the entire process for all j. Write the memory to all ones and repeat the test, interchanging zero and one. This test requires $12N^2$ operations. It tests all write read transitions, all cells, and all addressing.

The galloping one-zero test and the galloping write recovery test together take $16N^2$ operations and make a fairly effective memory test. If the tests are performed at 100 nanoseconds per cycle, the test time for a 4096 bit RAM is about 26 seconds per chip. A 16 kilobit RAM would take about 7.2 minutes and a 64 kilobit RAM takes almost two hours. Assuming a test cost of $40 per hour, the cost of testing 64 kilobit memory chips would be $80 each.

There are three approaches to lowering this test cost. One is to partition the chip into a number of smaller memories, each with its own addressing. This multiplies the number of address pins by the partitioning factor. A second approach is to partition the chip into a number of memories which share the same addressing logic. This is a byte slice memory. The byte slice memory requires additional data pins. The number of additional data pins is equal to the partitioning factor. A good type of test for byte oriented memories is called moving inversion (MOV I) and will be discussed later.

The third approach to the VLSI test cost problem is to find effective tests which require fewer than kN^2 cycles. A number of such tests have been proposed. All are based on the assumption that the pattern sensitivities of chips are expected to be either local (adjacent cells), or row oriented, or column oriented. If that assumption is in fact true, then tests requiring on the order of $N^{3/2}$ cycles should be effective. Some examples are described below.

Walking Columns

Write the first column of cells to one. Read all cells. Repeat for all columns. This test requires $2N^{(3/2)}$ operations. It checks the column address decoder and the sense amplifier.

Galloping Columns

Read the first cell and then write it and reread it. Read the next cell in the same row and reread the current cell. Continue alternating between the written cell and all of the cells in its row. Repeat for all cells. This test requires $6N^{3/2}$ operations (if rows, columns, and diagonals are used).

Galloping Column Write Recovery

Same as Galloping Write Recovery, but only varies cell i over the same row, column and diagonal as cell j. This test requires $36N^{3/2}$ operations.

In the case of a 64K RAM, the test time for the latter test would be less than a minute and a half.

Two other useful RAM tests are surround disturb and moving inversion. Surround disturb makes use of the fact that the physically adjacent cells are the most likely (other than those sharing a row or column) to be affected by a cell write or read.

Surround Disturb

Read the first cell and then write it. Alternate the read of each neighbor cell with a read of the written cell. Repeat for all cells. Repeat, interchanging ones and zeros. This test requires 32N operations. A write recovery version of this test is similar to the galloping write recovery test but with the alternating write read loop restricted to the nearest neighbors.

Moving Inversion (byte oriented memory) [3, p.130]

Load the memory with zeros. Read the first word. Write a one in the first bit position. Reread. Repeat for each bit position. Repeat for all memory words by incrementing the least significant address bit. Repeat the process, writing ones instead of zeros. Repeat the entire process, reversing the order of addressing. This algorithm is then repeated n times, where n is $\log_2 N$. Each repetition uses a different bit as the least significant bit for incrementing purposes. The next address is generated by adding a word which is all zeros except for a one in the ith bit position. This addition is carried out using end around carry, so that the all the addresses will be generated. The total number of operations for this test is $12 WN\log_2 N$, where W is the word length and N is the number of words. Note that if the total number of bits is N, then the number of words in the memory is N/W and the number of operations is $12N(\log_2 N-\log_2 W)$.

Testing VLSI Microprocessors

VLSI Microprocessors will contain a variety of functional units or macro-cells, each of which must use a different test approach. Internal RAMS can be tested by the techniques just discussed if they are externally accessible. However, some designs may employ internal address generators in order to save pins and increase efficiency. In this case, the test algorithm must be included in the chip design. Internal ROM's can be tested by signature analysis and this can also be performed by built-in test circuits. Register stacks can be tested by techniques similar to RAM tests.

Input-output circuits and continuity can be tested by the set/scan technique. This requires a shift register latch at the input and the output of the chip. This is probably the first type of built-in test which should be implemented. The design impact is not great, and interconnection failures are the most common failures in both boards and chips.

The testing of the other parts of a microprocessor depend on its logical and physical organization. It is certainly possible to design an untestable, pattern sensitive microprocessor.

MC CASKILL[34] has suggested a sensible test methodology for AMD 2901 byte slice microprocessors. The major functional blocks in a 2901 chip are shown in Fig.13. The suggested sequence of tests is:

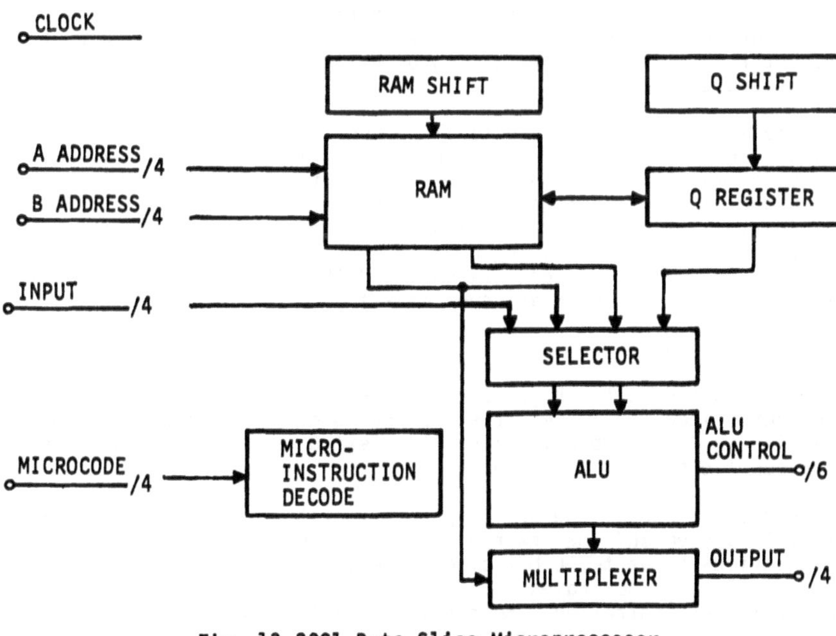

Fig. 13 2901 Byte Slice Microprocessor Architecture Main Functional Blocks

1) Test RAM with external address port A and external input output.

2) Test RAM with external address port A and input output through ALU.

3) Test RAM with external address port B and input output through ALU.

4) Test Q register.

5) Test all decoding possibilities of the ALU source selector.

6) Perform a functional test of the ALU with a fixed instruction sensitivity.

7) Test the RAM shift and Q register shift for all number combinations.

The particular tests used by McCaskey result in a total of about 12,000 test words. The 2901 is a very flexible processor, but is only four bits wide and contains only 64 bits of memory in 16 registers. VLSI microprocessors will have orders of magnitude more memory and will be 16 to 32 bits wide. Obviously, the testing problem will increase dramatically. One possibility might be to functionally and electrically partition the chip into four or eight bit slices. The RAM might also be partitioned into non-interacting "pages". To be non-interacting, they must use independent read write electronics and be physically separated.

In the past it has been necessary to understand the design and layout of a chip in order to generate an effective and efficient test program [34]. In the future it will be necessary to understand the test program in order to design the chip.

Unfortunately, design for testability does not come free. Set/scan techniques are estimated to require between 4% and 20% additional chip area. The precise amount of this penalty depends on the technology (bipolar or MOS) and the ratio of storage logic (registers, latches, and flip flops) to combinational logic. Partitioning adds chip area and subtracts from throughput performance. The reasons for this performance penalty include the delays caused by additional latches and the delays caused by partitioning fast carry logic in arithmetic devices.

VLSI Test Equipment

A number of sophisticated high speed testers have been developed for the semiconductor industry. These testers are functionally similar. The block diagram in Fig.14 is typical. Fairchild, Tektronix, and Macrodata are leading suppliers of this type of equipment.

The current state of the art is probably best represented by the Fairchild Sentry VIII system and the Tektronix 3250 system. The Sentry VIII handles 120 pin devices at up to 10 MHz clock rates. The 3250 handles 64 pin devices at 20 MHz clock rates. Both have programmable pattern generators and are computer controlled. Both have sophisticated data logging and data analysis capabilities. Both have sophisticated multiphase timing systems which permit the dynamic testing of complex memory and microprocessor chips.

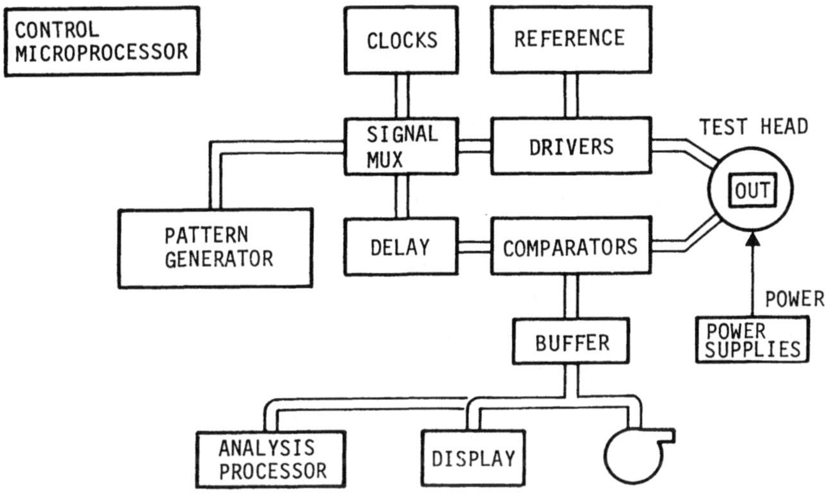

Fig. 14 Block Diagram of a Typical VLSI Tester

These testers normally operate on packaged chips and require a special test card for each package type. Some IC manufacturers believe that packaging costs for VLSI chips (up to 120 and possibly 200 pins per package) will be so great that dynamic testing at wafer probe is desirable. This is difficult for two reasons. One reason is mechanical, the large number of contacts in such a small area presents a problem in the design and fabrication of probe cards. A 160 pin chip 300 mils square has bonding pads on seven mil centers.

The second reason is electrical. To perform at clock rates of ten to thirty megahertz, the contacts must have very low impedance. Contacts at the bonding pad are one problem, inductance in the leads is a problem, and ground of the reverse side of the chip is a problem. Special preparation of the reverse side of the wafer may be required to facilitate high speed wafer probe.

The cost of current high performance test systems ranges from $200,000 to $1,200,000, depending upon the model, the features, and the options. This high cost of test equipment impacts both the supplier and the user of VLSI chips. The capital investment for incoming part testing is escalating rapidly. On the other hand, the cost of repair for failed parts is also escalating. This situation increases the incentive to design for testability and to provide built-in test for VLSI chips.

7.3 Conclusion

In this chapter the subjects of VLSI applications and VLSI chip testing were discussed. Because the subject of VLSI application is so broad, specific applications were chosen as examples to illustrate the impact of VLSI on signal processing. In the second part of the chapter, the well known and useful testing techniques were summarized and their relevance to VLSI testing was discussed.

REFERENCES

1. A.V. Oppenheim, Digital Signal Processing, Prentice Hall, Englewood Cliffs, N.J. (1978).

2. W.K. Pratt, Digital Image Processing, Wiley-Interscience, New York, (1978).

3. W.G. Fee, LSI Testing, Second Edition, IEEE Computer Society, New York, (1978).

4. M. Marshall, "Logic in Testingland - A Tale of Measurement Techniques", EDN Magazine, (Jan, 1976).

5. M. Marshall, "Through the Memory Cells - Further Explorations of IC's in Testingland", EDN Magazine, (Feb, 1976).

6. M. Marshall, "Microprocessors in Testingland - The Jury Receives Its Instructions", EDN Magazine, (Mar, 1976).

7. A.G. Glaser and G.E. Subak-Sharpe, Integrated Circuit Engineering, Addison Wesley, Menlo Park, CA, 746-799, (1977).

8. Y.S. Chen and D.L. Dutweiler, "A 35,000 Transistor Chip Echo Canceller", Proc. 1980 International Solid State Circuits Conference, IEEE, New York, 42-43, (1980).

9. Y. Kawatami, et al, "A Single-Chip Digital Signal Processor for Voice-band Applications", Proc. 1980 International Solid State Circuits Conference, IEEE, New York, 40-41, (1980).

10. J.R. Boddie, et al, "A Digital Signal Processor for Telecommunications Applications", Proc. 1980 International Solid State Circuits Conference, IEEE, New York, 44-45, (1980).

11. R.L. Freeman, Telecommunications Transmission Handbook, John Wiley, New York, 82-130 and 448-481, (1975).

12. S.L. Freeny, J.F. Kaiser, and H.F. McDonald, "Some Applications of Digital Signal Processing in Telecommunications", in Applications of Digital Signal Processing, A.V. Oppenheim, Ed., Prentice Hall, Englewood Cliffs, N.J. 1-28 (1978).

13. D.A. Hodges, "A Review and Projection of Semiconductor Components for Digital Storage", Proc. IEEE, Vol. 63, No. 8, 1136-1147, (Aug, 1975).

14. C. Mead and L. Conway, Introduction to LSI Systems, Addision Wesley, Menlo Park, CA, 317 (1980).

15. J.F. Ziegler and W.A. Lanford, "The Effect of Sea Level Cosmic Rays on Electronic Devices", Digest of Papers, 1980 International Solid State Circuits Conference, IEEE, New York 70-71, (Feb. 1980).

16. D.K. Pradham and J.J. Stiffler, "Error-Correcting Codes and Self-Checking Circuits, Computer, Vol. 13, No. 3, 27-37, (May 1980).

17. E.I. Muehldorf, "Test Pattern Generation as a Part of the Total Design Process", Proc. 1978 Semiconductor Test Conference, IEEE Computer Society, Long Beach, CA, 4-7, (Nov, 1978).

18. L.H. Goldstein, "Computational Complexity/Confidence Level Tradeoffs in LSI Testing", Proc. 1978 Semiconductor Test Conference, IEEE Computer Society, Long Beach, CA, 50-58, (Nov, 1978).

19. J.H. Stewart, "Application of Scan/Set for Error Detection and Diagnostics", Proc. 1978 Semiconductor Test Conference, IEEE Computer Society, Long Beach, CA, 152-158, (Nov, 1978).

20. E.R. Huatek, "Semiconductor Memory Attrition Summary", Proc. 1976 Semiconductor Test Conference, IEEE Computer Society, Long Beach, CA, 35-40, (Oct, 1976).

21. T.L. Palfi, "MOS Memory System Reliability", Proc. 1975 Semiconductor Test Conference, IEEE Computer Society, Long Beach, CA, (Oct, 1975).

22. R.C. Foss and R. Harland, "MOS Dynamic RAM-Design for Testability, Proc. 1976 Semiconductor Test Conference, IEEE Computer Society, Long Beach, CA, 9-15, (Oct, 1976).

23. P. Nissen, "Some Memory Testing Paradoxes", Proc. 1978 Semiconductor Test Conference, IEEE Computer Society, Long Beach, CA, 91-95 (Nov, 1978).

24. E.I. Muehldorf, "Designing LSI Logic for Testability", Proc. 1976 Semiconductor Test Conference, IEEE Computer Society, Long Beach, CA, 45-49, (Oct, 1976).

25. D.W. Bray, "Partitioning for Test Generation of Large Synchronous Sequential Circuits", Proc. 1976 Semiconductor Test Conference, IEEE Computer Society, Long Beach, CA, 50-53, (Oct, 1976).

26. T.W. Williams and K.P. Parker, "Testing Logic Networks and Designing for Testability", Computer Magazine, Vol. 12, No. 10, 9-21, (Oct, 1979).

27. F. Hsu, P. Solecky, and L. Zobniw, "Selective Controllability: A Proposal for Testing and Diagnosis", Proc. 1978 Semiconductor for Test Conference, IEEE Computer Society, Long Beach, CA, 170-179, (Nov, 1978).

28. Y.M. El-Ziq, "A New Approach for Designing Testable Combinational Networks", Proc. 1978 Semiconductor Test Conference, IEEE Computer Society, Long Beach, CA, 103-108, (Nov, 1978).

29. B. Konemann, J. Mucha, and G. Zwiehoff, "Built in Logic Block Observation Techniques", Proc. 1979 Semiconductor Test Conference, IEEE Computer Society, Long Beach, CA, 37-41, (Oct, 1979).

30. S. Funatsu, N. Wakatsuki, and A. Yamada, "Designing Digital Circuits with Easily Testable Considerations", Proc. 1978 Semiconductor Test Conference, IEEE Computer Society, Long Beach, CA, 98-102, (Nov, 1978).

31. L.A. Stolte and N.C. Bergland, "Design for Testability of the IBM System/38", Proc. 1979 Semiconductor Test Conference, IEEE Computer Society, Long Beach, CA, 29-36, (Oct, 1979).

32. H.J. Nadig, "Testing a Microprocessor Product Using a Signature Analysis", Proc. 1978 Semiconductor Test Conference, IEEE Computer Society, Long Beach, CA, 159-169, (Nov, 1978).

33. H. Thaler, "Pattern Verification and Address Sensitivity of ROM's by Signature Testing", Proc. 1978 Semiconductor Test Conference, IEEE Computer Society, Long Beach, CA, 84-85 (Nov, 1978).

34. R. McCaskill, "Test Approaches for Four Bit Microprocessor Slices", Proc. 1976 Semiconductor Test Conference, IEEE Computer Society Long Beach, CA, 22-26, (Oct, 1976).

35. B.C.S, Liu, "Microprocessor's Algorithmic Test Pattern Generation in the Higher Level Tester Language", Proc. 1979 Semiconductor Test Conference, IEEE Computer Society, Long Beach, CA, 268-290, (Oct, 1979).

8. VHSIC Technology and Systems [1]

D.F. Barbe and E.C. Urban

With 17 Figures

The principal objective of the VHSIC thrust – based on 20 years of industrial, academic, and defense experience – is to insert very high-speed integrated circuits into defense systems. To achieve this objective, certain technological barriers must be overcome and subsystems must be built to demonstrate the improved capability. These subsystems, along with the technological challenges, are discussed in this chapter.

8.1 IC Progress

From 1959 – when the first integrated circuit contained a few transistors per chip – until the present time, the density of IC components has increased at the rate of a factor of two per year (see Table 1). The highest density per chip has been achieved in memory because of its very regular structure. Since logic or signal processing functions require more random interconnections, the levels of density and rates of increase have been somewhat less – but, nevertheless, quite remarkable – in logic chips. This rapid increase in circuit density is due in part to the rapid decrease in circuit dimensions; improvements in optical lithography and chip fabrication have yielded circuits with dimensions in the 2.5- to 5.0-micrometer range as shown in Fig.1.

Table 1. Progress in IC Development

Level of integration	Number of logic gates/chip	Time period of development
Small-scale integration	3-30	Early 1960's
Medium-sclae integration	30-300	Mid-late 1960's
Large-scale integration	300-3000	Early-mid 1970's
Very large scale integration	> 3000	Beyond late 1970's

Figure 2 graphically illustrates the trends in (1) the number of gates per integrated-circuit chip, (2) the number of gates needed to implement common systems functions and (3) the number of IC chips needed to implement common systems functions. Although system complexity is continually increasing, it is not doing so as rapidly as the number of gates per chip. The result is

1 This chapter is a modified version of an article from COMPUTER, Vol.13, pp.13-22, February 1981. Reprinted with permission.

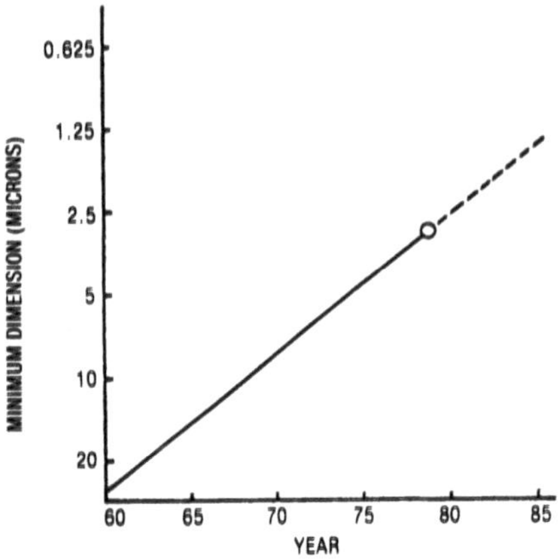

Fig.1. Minimum feature size as a function of time from 1960 to 1980 and beyond

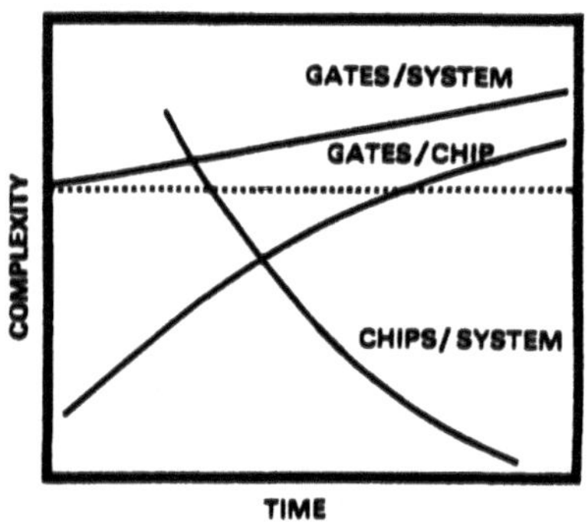

Fig.2. Chip and system complexity growth

that the number of chips necessary to implement a system function is decreasing with time. The combination of the economics of batch processing and higher speed capability is allowing digital electronics to replace electromechanical and analog functions and to therefore encompass a greater portion of a system design. In the near future, systems such as digital computers will contain all the required functions (CPU, RAM, ROM, I/O, A/D) on a single chip. More complex systems will be built using a small number of subsystem chips. As system sophistication continues to grow, the use of greater numbers of chips for increased performance will become possible.

In the context of current methods by which ICs are designed and applied, the capability to achieve significant increases in integration levels has led to problems in other areas. The man hours required to design an IC has increased with chip complexity, resulting in a concomitant increase in chip cost. Also, the specificity required to custom design a device for a particular system tends to limit the market over which high design costs can be amortized. These problems may be overcome by greater standardization and increased use of programmable devices. A higher degree of standardization can be achieved through increased use of improved computer-aided design techniques, availability of standard chip families, and by utilizing uncommitted logic arrays. This approach could also reduce the complexity of the testing problem. A greater effort in programmable and standardized system devices will be required to offset the decreasing commonality caused by increasing complexity as shown in Fig.3. Microprocessors have already accomplished this for LSI types of complexity.

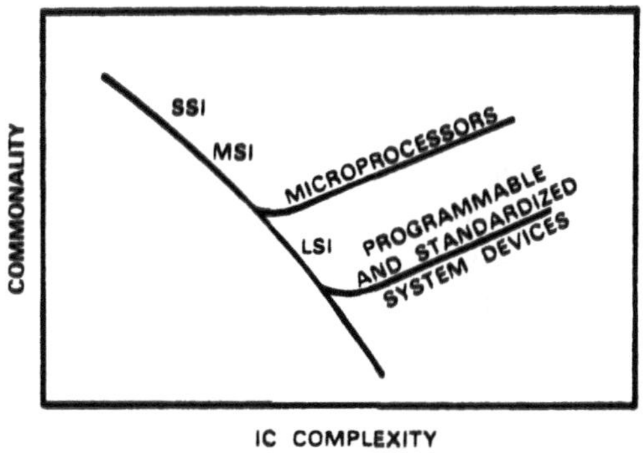

Fig.3
Commonality vs IC complexity

To achieve the requirements for defense systems, significant improvements in speed, size, weight, power and cost are necessary. New system design capabilities made practical through the use of standard functions and/or programmable devices will tend to merge design talent of integrated circuits and systems and place a greater reliance on vertically integrated industrial facilities. The United States government is addressing these problems with a systems-driven Very High Speed Integrated Circuits (VHSIC) Program, which is the subject of this chapter.

8.2 Throughput Capacity

A parameter adopted to describe the signal processing capacity of very high-speed ICs, throughput capacity is the product of the number of gates on an IC chip times the clock frequency in Hertz, so the units are gate Hertz (the clock period is taken to be four gate delays).

Figure 4 shows the throughput capacity for several systems. Microprocessors have throughput capacities on the order of 10^{10}, voice processors and the global positioning satellite processor have throughput capacities on the order of 10^{11}, the AIM-54 phoenix seeker processor and the AN/UYK 20 Navy standard computer have throughput capacities on the order of 10^{12}. JTIDS, the joint tactical information distribution system, has a throughput capacity between 10^{12} and 10^{13}. These are throughput capacities for systems composed, in most cases, of many chips. Cost, weight, and power dissipation are high, and reliability tends to be a problem because of numerous interconnections between chips. As a result, there is great incentive for fabricating chips with more and more signal processing capabilities – that is, chips with higher throughput capacities.

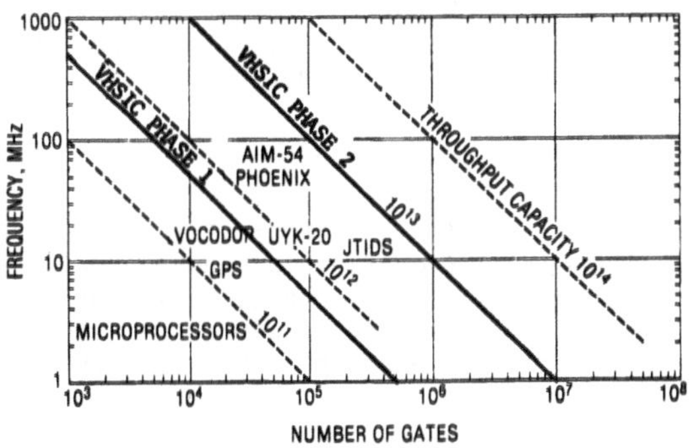

Fig.4. Performance of several systems on a plot of clock frequency vs number of gates with throughput capacity as a parameter

Table 2. Defense IC throughput needs

Application	Throughput capacities (gate Hz)
Command, control, and communications	10^{13}
Radar	10^{14}
Electronic warfare	10^{14}
Image processors	10^{13}
Computers	10^{13}

Table 2 lists projected throughput needs for several systems. As points of reference for ICs, chip throughput rates on the order of 10^{10} can be achieved with 5-micrometer technology, and chip throughput capacities between 10^{12} and 10^{13} are projected with 1-micrometer technology [1].

The power dissipation and weight of systems having various throughput capacities are a strong function of feature size (Fig.5). For example, a system requiring 2×10^{13} gate Hz and using 5-micrometer IC technology could be expected to weigh slightly more than 1000 kg and dissipate over 100 watts. The same system capability based on 1-micrometer technology could be expected to weigh about 10 kg and dissipate about one watt. Thus, very large reductions in power dissipation and system weight can be achieved by concentrating on the reduction of IC feature size.

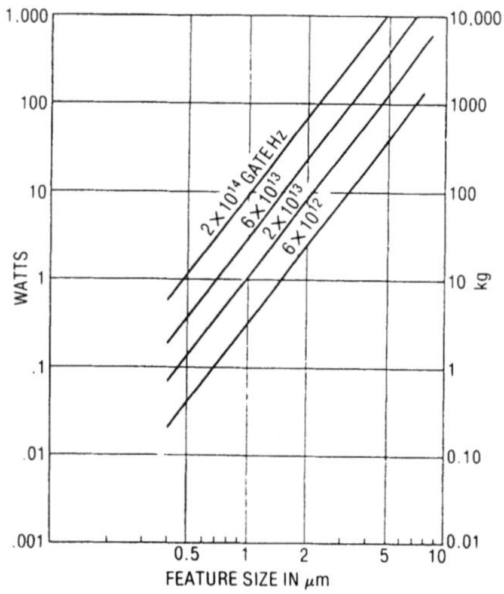

Fig.5. Power and weight of IC assemblies vs IC feature size, with throughput capacity as a parameter

8.3 Defense Systems

Several observations can be made regarding defense systems. First, their complexity and high life-cycle costs prevent us from fielding the desired number of systems. Second, performance requirements – for example, very high throughput capacities – tend to be different from those of the commercial sector. Third, the defense environment is different. The environment experienced by a signal processor on the deck of a ship – where it may be subjected to transmission from radar, for example – is very different from that of most commercial systems. Fourth, defense applications tend to require custom ICs. This is due to the fact that the budget for weight, power dissipation, and size on defense platforms tends to be very restricted, and space, power, and weight can be saved by designing custom chips specifically to do the job that is needed. Fifth, the leverage of defense in the IC market is not nearly as great as it used to be. The absolute defense need has remained essentially constant while the commercial market has increased greatly, thereby reducing, in relative terms, the military's market leverage.

Finally, defense-system logistics must be improved. This includes the need for built-in test and for systems with better fault tolerance, reliability, and repairability. System complexity, coupled with the decrease in trained technical personnel, clearly indicates a need to vector toward systems which are inherently more reliable. Also, built-in test techniques indicating fault location, so that technicians without a great deal of training could effect repairs in an efficient manner, would be highly useful.

8.4 The VHSIC Program

8.4.1 Approach

The VHSIC program [2] is an IC technology effort driven by a systems approach – that is, the technology is pushed in the direction dictated by systems needs for the mid-1980's and beyond. As shown in Fig.6, systems requirements drive the development of systems brassboards which then impact design, architecture, software, and test technology. They, in turn, guide fabrication, lithography, and CAD development, culminating in production lines which provide the ICs to meet systems requirements.

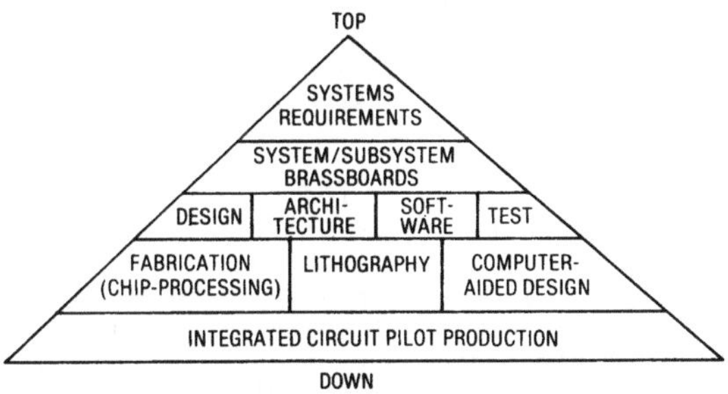

Fig.6. The VHSIC program is driven by sytems requirements

Planned program outputs include (1) a family of chips to perform the most commonly occurring signal processing functions across systems in the three military services; (2) higher order computer-aided design techniques for efficient design of very complex ICs in terms of time, cost, and manpower; (3) pilot production lines to manufacture the family of chips as well as custom chips; and (4) systems brassboards to demonstrate VHSIC capabilities.

Thus, the overall objective of the VHSIC program is the direct development of advanced ICs for introduction into future systems in a timely, affordable manner. Expected benefits include advanced capability, high return of investment including life-cycle costs, and increasing the lead in ICs by several years. The program expects not only to advance IC technology (measured in terms of computational speed and size, weight, and power requirements), but also to achieve this advance directly in circuits with high reliability, ease of testability and diagnosis, and high tolerance to high-stress environments. Each phase of the VHSIC program is intended to play a specific role in reaching these objectives.

8.4.2 Schedule

The VHSIC program is divided conceptually into four phases: Phases 0, I, and II will be carried out consecutively, but Phase III is carried out in parallel with the other three. The program began in March 1980 and is scheduled to continue into 1986.

Phase 0 was a study phase to define the necessary work, the detailed approach, and the plans for achieving the ultimate objectives of the VHSIC program. This analysis proceeded in a top-down fashion, starting with the selection and analysis of at least three projected defense systems or subsystems to determine their signal and data-processing requirements and to identify broadly applicable VHSIC chips.

In Phase 0, architectures and design approaches were selected and investigated to implement VHSIC chips with 1.25-micrometer and submicrometer minimum feature sizes using a minimum clock rate of 25 MHz and providing a minimum throughput capacity of 5×10^{11} gate Hz per cm^2 for the 1.25-micrometer features and approximately 10^{13} gate Hz per cm^2 for the submicrometer features. In addition, chip technology and processing techniques, including device technologies, interconnects, and metallizations appropriate for both feature sizes, were studied along with chip fabrication and manufacturing techniques. The studies also include packaging tradeoffs, CAD techniques, and lithographic requirements and techniques – all necessary to achieve the overall VHSIC program goals. In addition, susceptibility to thermal stress over the full defense temperature range from -55°C to +125°C, to radiation doses to 10^4 rads (Si), to electromagnetic interference, and to other environmental factors met in military applications were considered.

Phase I is subdivided into two parallel efforts. Using 1.25-micrometer IC technology, Phase Ia is directed toward developing complete electronic brassboard subsystems within about three years. These brassboards may consist of several VHSIC chips operating at a minimum clock rate of 25 MHz. It is anticipated that a pilot production-line capability will be established for this technology. Minimum requirements of reliability, testability, and environmental immunity will be demanded.

Phase Ib consists of initial efforts to extend IC technology to submicrometer feature sizes and corresponding circuit complexities. This will include high-resolution lithography and replication techniques, submicrometer device design and modeling, substrate and epitaxial growth improvements and analysis, metallization reliability and interconnect analysis, appropriate CAD techniques, and architecture and systems considerations. A feature of both Phases Ia and Ib is to design for testability and fault location.

Similarly, Phase II is divided into two parallel programs. Phase IIa will provide subsystem demonstrations based on Phase Ia brassboards, and Phase IIb will continue the Phase Ib submicrometer development effort. Phases Ib and IIb are directed at developing all aspects of IC technology necessary to cross the so-called one-micrometer barrier, considered to be the practical limit of conventional optical lithography. The end goal of these efforts is the development of production capability for advanced chips. This includes not only lithography and fabrication but also design, architecture, software, and testing technologies. Subsystem demonstrations of Phase IIb technology are expected at the end of the program.

The Phase III VHSIC support program runs parallel with the main program efforts. Whereas the Phase I and II program contracts are large, vertically integrated efforts with each contractor covering all aspects of VHSIC development, Phase III consists of many smaller, shorter term efforts in key technology areas designed to feed into the main program. Part of the motivation for establishing Phase III was to permit and encourage the participation of universities and small business in the program. Phase III efforts focus on high-resolution lithographic equipment and processing technology; advanced architecture and design concepts for reducing custom fabrication; increasing chip utilization and system reliability through fault tolerance and improving system testability through on-chip testing; advanced CAD techniques; improved silicon materials and fabrication processes; analytical methods for determination of substrate and fabrication-induced defects at the submicrometer level; methods for improving radiation, thermal, and mechanical stress tolerance; establishment of design standards and interface requirements; new device, gate, and circuit structures; and techniques for documentation and methods for improved and simplified utilization and testing.

8.5 Example Brassboard Systems

The systems demonstration units – that is, the brassboards – are the initial mechanism by which VHSIC technology will be transferred into operational equipment. For discussion, example brassboard systems [3] are organized in five categories: (1) command, control, and communications, (2) radar, (3) electronic warfare, (4) image processing, and (5) general-purpose computers as illustrated in Fig.7. These example brassboards are considered indicative of current and near-term defense signal processing requirements.

Communications, command, and control	*Radar*	*Electronic warfare*	*Image processors*	*Computers*
• JTIDS • A-J MODEM • NATO ID	• SURV • TAC • MRSP • AMRAAM	• APMS • ESM	• IMAGE PROC • EOSP • CM	• GP • USSP

Fig.7. Example VHSIC systems brassboards

8.5.1 Command, Control and Communications

Example brassboard systems in the command, control and communications category are described below.

JTIDS. The joint tactical information distribution system is a program to develop a family of terminals and interfaces for integrated communications, navigation, and identification in tactical and air defense operations. Initial implementation of JTIDS is planned using the Air Force's E-3A airborne warning and control system and F-15 and F-16 fighters, as well as Navy ships and aircraft. The JTIDS network is intended to connect many different types of weapons systems, dispersed over a wide area, with sources of surveillance, intelligence and support information, weapon controllers, and decision-making commanders. The system architecture is based on time division multiple access enhanced with an advanced TDMA for time slot accessing modes and multinet operations. Basic capabilities are derived through the use of spread-spectrum TDMA technology, and signal transmission is at microwave frequencies using cryptographic encoding for security. The role of the VHSIC components will be to provide modules which perform one or more of the various processing functions. This will simplify system integration through a standard interface.

Anti-Jam Modem. A VHSIC chip set is being sought to replace the specialized modems of the past and eliminate costly redesign. The ideal all-purpose modem would stress programmability so that the same circuitry could serve many different sets of data modulation schemes (AM, FM, SSB, FSK, PN, etc.), frequency-hopping rates, and spread-spectrum formats by simply changing the microprogram in a read-only memory. The resulting hardware is to be a black box add-on to existing communications systems and will be compatible with advanced satellite communications equipment and JTIDS. Production quantities of at least 100,000 modems are estimated.

NATO IFF. Last year, the Navy was designated the lead service to develop a triservice NATO identification system to be used by all military ships and aircraft throughout NATO by the late 1980's. The new system must be compatible with all existing identification systems (Mark V, Mark XII, and DABS — discrete address beacon system). The market is immense, in the $7 billion bracket.

8.5.2 Radar

The second category, radar, includes the following brassboard systems.

Surveillance Radar. Functions of this proposed airborne system to detect and help counter enemy missiles include adaptive beamforming combined with adaptive Doppler processing (side lobe cancellation plus moving-target indication). It will have a programmable matched filter, signal synthesizer, and detector. Throughput rate is projected at 4×10^{13} gate Hz. The radar's weight will be limited to approximately 460 kg.

Tactical Air Radar Signal Processor. VHSIC-based subsystems will be needed for signal processors in the next generation of air defense aircraft. This new processor is expected to be installed initially in the F-14 and later in the F/A-18. Its main functions include time domain correlation, 100 MHz analog-to-digital conversion, and fast Fourier transforms. It uses digital filters, a 256×16 random-access memory with 4-ns access, a complex 16-bit ALU, and a high-speed input/output controller.

MRSP. The multifunction radar signal processor is to be mounted within the nose radar of high-performance tactical aircraft. It will be used for the delivery of air-to-ground precision weapons with modes requiring the execution of large, complex fast Fourier transforms at rates greater than 20 MHz. The same system will be used for future air-to-air radar processing requirements, including threat assessment.

AMRAAM. The advanced medium-range air-to-air system, a next-generation weapon for tactical aircraft, uses an I/J-band coherent pulsed-Doppler radar with a pulse repetition frequency variable from 15 kHz to 1 MHz. By increasing throughput rates, VHSIC components will provide larger numbers of range gates and improved Kalman filters. AMRAAM also needs a high-frequency, high-stability frequency synthesizer capable of operating over the I/J-band at 100-kHz intervals with ±25-kHz accuracy.

8.5.3 Electronic Warfare

The following brassboard systems are examples in the electronic warfare category.

APMS. An advanced power management system is needed to improve airborne electronic countermeasures. ECM performance can be improved by better management of jammer resources in spectral, temporal, and spatial domains for any given receiver/transmitter configuration. VHSIC components are being considered to bring down weight to 55 kg, size to 0.074 m^3 (2.6 ft^3), power to 600 W, and cost to $180,000 in production quantities.

ESM Signal Sorter. At present, electronic-support-measures signal sorters use 15000 LSI chips; VHSIC technology is aimed at reducing that number to 30. Furthermore, the electromagnetic threat is expected to increase both in density and complexity, with sophisticated emitters producing waveforms that include both frequency and time agility. Better signal-sorting-systems will be needed on aircraft, surface ships, and submarines, with content-addressable memories improved over the current 64-bit density with 300-ns delays to 1K-bit densities with 30-ns delays.

8.5.4 Image Processing

Three systems comprise the image-processing brassboards.

Image Processor. The use of focal-plane array technology for surveillance, weapon control, and signal processing is advancing faster than processing technology. Extraction of signal information from the optical/electro-optical energy presented to the detectors of a focal-plane array exceeds the ability of conventional analog hardware, such as charge-coupled devices, to handle the high data rates. The Navy is seeking to apply VHSIC components in a modular approach for processing 64-96 multipled channels in a single chip from a sensor with up to 1000 channels. The processor will perform gain and offset correction, two-dimensional adaptive spatial filtering, multispectral ratioing, two-dimensional adaptive thresholding, and edge detection. The vector processor will sample data at 1 MHz and perform programmable algorithms, using external reference data and the main data stream, at 50 operations between data samples. Requirements for bit rates ranging from 10 to 100M bit/s and processing rates of 10^{12} to 10^{13} gate Hz are expected.

EOSP. Under a cooperative program with DARPA, the Defense Advanced Research Projects Agency, the Air Force is developing new families of space-based surveillance sensors and communications systems for the late 1980's and early 1990's. The objective of the electro-optical signal processor program is to perform the signal processing functions on board the satellites within the appropriate power, life, radiation, size, and weight constraints. VHSIC components are planned to meet the processing requirements of multimission radar, electro-optical sensor, and advanced communications systems.

Cruise Missile. Future versions of the cruise missile will require more accurate guidance, and the Air Force is looking at VHSIC components for an autonomous precision navigation and guidance system that will reduce weight, increase maneuverability, and improve survivability by allowing the missile to fly at lower altitudes and at greater speed. This effort is being conducted in conjunction with DARPA's autonomous terminal homing program. The Air Force estimates that high-velocity terminal guidance will require pattern matching of up to 256 × 256 pixel images at 5 frames/s with 4 bits/pixel. These may be divided into subarrays for parallel processing.

8.5.5 General-Purpose Computers

There are three general-purpose computer brassboards examples.

General-Purpose Computer. The Navy is considering the use of VHSIC components to produce smaller versions of existing computers. These new models will emulate the Navy's AN/AYK-14 airborne computer and AN/UYK 20 shipboard computer (which are already software-compatible with each other). The goal is to develope a chip set to build a 32-bit microprogrammable processor capable of executing 50 million instructions per second.

Universal Sensor Signal Processor. This is a program to use VHSIC technology to create a general-purpose signal processor for demonstration with the E-3A AWACS radar. The E-3A was chosen because its radar exhibits many of the needs common to existing military systems and because the aircraft is expected to have a 30-year life, during which time the threat to be dealt with is likely to change considerably. Once developed, the universal signal processor is expected to be able to support the sensors used in other programs such as Pave Paws, advanced tactical radar, space radar, space/airborne electro-optical search systems, and unattended radar. Immediate requirements are for an improved ECCM capability, but ESM is also included in plans for the future.

AN/AYK-15A. This is a program to "shrink" a general-purpose airborne computer onto a VHSIC chip. The Air Force already has two candidate processors — a microcomputer with a 300K operations-per-second throughput rate and a larger processor with a 400 KOPS rate. Both use core memory. This program is aimed at developing a smaller, more powerful computer for missile applications where space is severely limited.

8.6 Scaling to Smaller Dimensions: Benefits and Barriers

Achieving the throughput capacities discussed above requires a fundamental change in the IC transistor — that of scaling down its dimensions. Thus, a number of technological barriers must be overcome before the benefits of the VHSIC program can be realized.

8.6.1 Benefits

As an example of the benefits of scaling to smaller dimensions, consider the MOS transistor — composed of a source, a drain, and a gate — as shown in Fig.8. The gate is separated from the silicon substrate by the oxide thickness T_{ox}. The source and drain spacing is L, and the channel width is W. The scale factor, K, is defined as the old dimension divided by the new dimension. L, W, and T_{ox} are all reduced by the factor K. The doping density of the channel must increase by the factor K as the minimum feature size is decreased, and the gate voltage must be decreased by the factor K. Thus,

- device area scales by the factor K^{-2},
- delay time scales by the factor K^{-1},
- power dissipation scales by the factor K^{-2}, and
- radiation hardness increases by the factor K^{-2}.

These are the significant benefits of scaling.

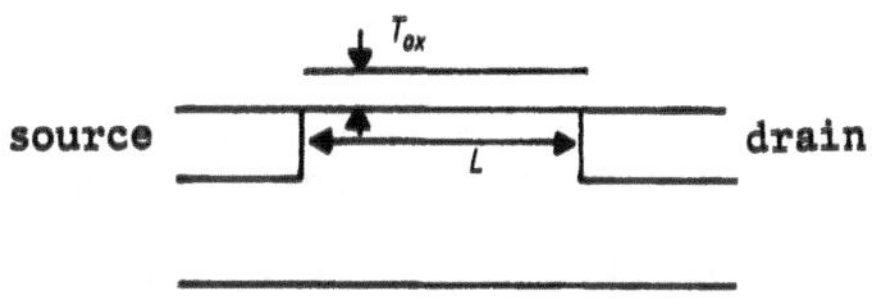

K = SCALE FACTOR = OLD DIMENSION/NEW DIMENSION

DEVICE PARAMETER	SCALING FACTOR
DEVICE DIMENSION T_{ox}, L, W	$1/K$
DOPING DENSITY N_A	K
VOLTAGE V	$1/K$
CURRENT I	$1/K$

Fig.8. The basic assumptions used in the scaling theory of MOS transistors

8.6.2 Negative Effects

Turning now to the barriers that must be overcome to achieve these very beneficial effects, first consider the negative effects of scaling [4]:

- the line resistance increases by the square of the scale factor,
- current density increases linearly with scale factor, and
- contact resistance increases by K^2.

As a result of higher current density, the median time to failure of the interconnect decreases. For example, Fig.9 shows that the medium time to failure in an aluminum interconnect line is orders of magnitude larger in a line carrying 10^5 A/cm^2 than in a line carrying 10^6 A/cm^2. This is primarily due to electron migration. Interconnect systems have been devised which improve this situation, but it is mentioned here to indicate one of the problems that results, and must be overcome in a practical way, as circuits are scaled to smaller dimensions.

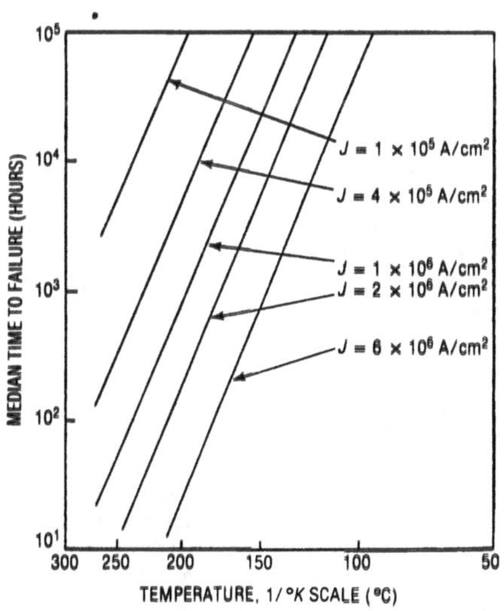

Fig.9. Median time to failure vs reciprocal temperature for aluminum interconnects

8.6.3 Noise Margins

Another detrimental effect of scaling MOS transistors is reduced noise margins. The root mean square noise of MOS transistors, primarily $1/f$ noise, is inversely proportional to the square root of the transistor area. The signal is proportional to the voltage swing, which scales as K^{-1}. Therefore, the signal-to-noise ratio in the MOS transistor decreases as the square of the scale factor, and the noise margin is less in small transistors.

8.6.4 Electromagnetic Interference

The electromagnetic interference is a function of bipolar transistor perimeter, as illustrated in Fig.10. As the perimeter decreases, the susceptibility to electromagnetic interference increases [5]. This is especially important to Navy systems, such as shipboard systems, where signal processors may be exposed to radiation due to radar antennas.

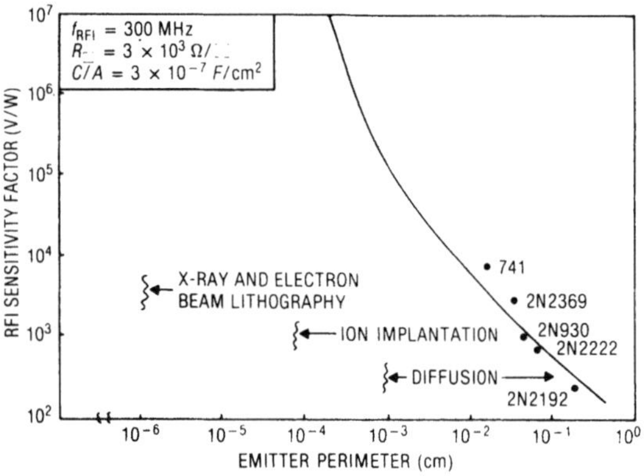

Fig.10. Radio frequency sensitivity factor vs bipolar transistor emitter perimeter

8.6.5 Soft Errors

Another problem which becomes more serious with scaling is the soft error problem in dynamic random-access memories. Loss of memory bits can result from alpha particle emissions from packages containing minute amounts of radioactive materials. As shown in Fig.11, the error rate of charge-coupled dynamic RAMs increases as the critical charge, or the size of the storage cell, decreases. This may force VHSIC designs to use static memory, which is not as susceptible to soft error problems as dynamic memory.

8.6.6 Radiation Hardness

Although scaling laws indicate a gain in radiation hardness with scaling, the trend as several technologies have evolved from MSI toward VLSI (for which no effort has been made to radiation harden) shows an alarming tendency for devices having smaller feature sizes to be more susceptible to ionizing radiation. This situation is illustrated in Fig.12. In many cases the factors causing the susceptibility to radiation have been identified and corrective measures taken, thereby increasing radiation hardness to rather high levels. But the general trend is for smaller MOS devices to be more susceptible to radiation, and care must be taken to identify inherently radiation-soft devices and techniques so that inherently radiation-hard technologies can be selected.

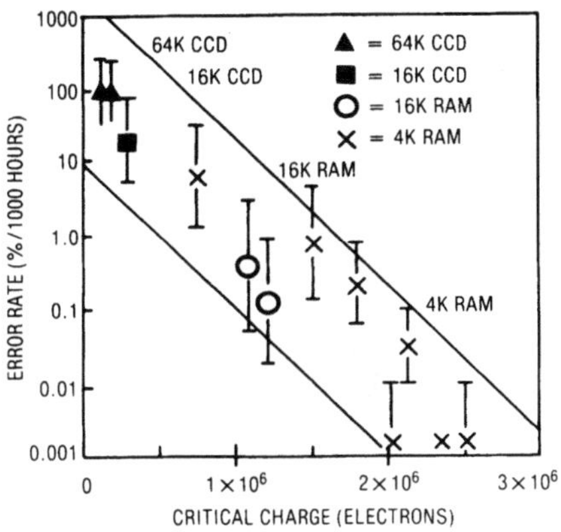

Fig.11. Error rate vs critical charge for charge-coupled RAM cell

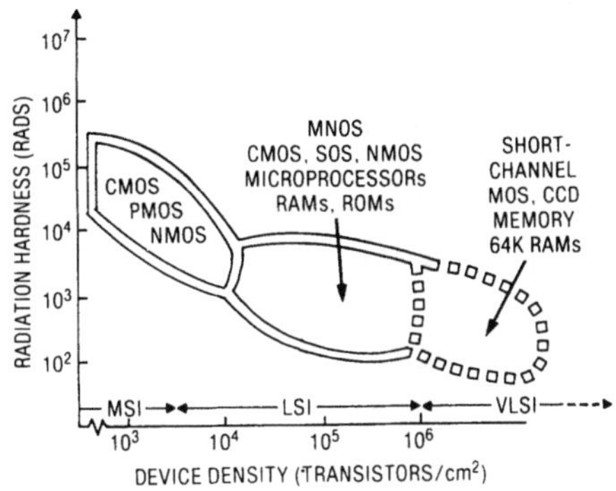

Fig.12. Radiation hardness vs device density for several technologies [6]

8.6.7 Economical Lithography

With optical lithography — the lithographic technique used during the 20-year history of ICs — whole wafers are exposed at once. This allows delineation of a larger number of chips at one time and turns out to be a very economical way of defining features. The problem arises as the feature size is reduced; a very fundamental factor — the wavelength of light — limits feature sizes to approximately one micrometer. Steps have been taken to use short-wavelength visible light, that is, ultraviolet, in lithography, but this has been taken to the point of diminishing returns.

It seems clear that achieving feature sizes well below one micrometer — for example, 0.5 micrometer — will require a fundamentally different lithographic technique. A leading contender is electron-beam lithography, which makes use of a computer-controlled scanning electron microscope system as illustrated in Fig.13. Applying digital voltages to the system causes the electron beam to scan over the wafer, exposing electron resists in the appropriate places and thereby delineating one level of the IC. This can be done in a raster-scan fashion, illustrated in Fig.14, with the electron beam turned on and off at appropriate places as it scans the wafer. Alternate schemes, shown in Fig.15, include a vector-scan system whereby the beam is directed to and moved in only the area where exposure is desired. A more advanced technique uses an extended-area square aperture to expose — at one time — areas many times the inherent beam diameter. In an even more advanced technique, a variable aperture provides rectangles of various dimensions to eliminate scanning the minimum-size beam over a large area.

The problem with these electron-beam lithography systems is one of economics. Chips must be exposed one at a time, that is, in series rather than in parallel as with the whole-wafer optical lithography techniques. However, considerably progress has been made, and definition of feature sizes necessary for VHSIC has been demonstrated. It is the economics of the situation which must be addressed by the VHSIC program.

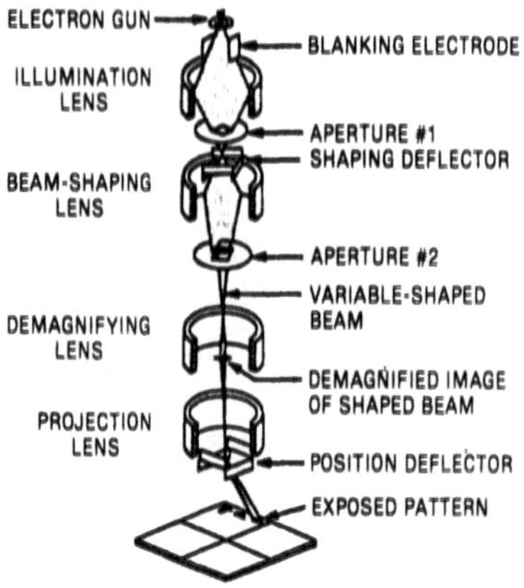

Fig.13. Schematic diagram of an electron-beam lithography column

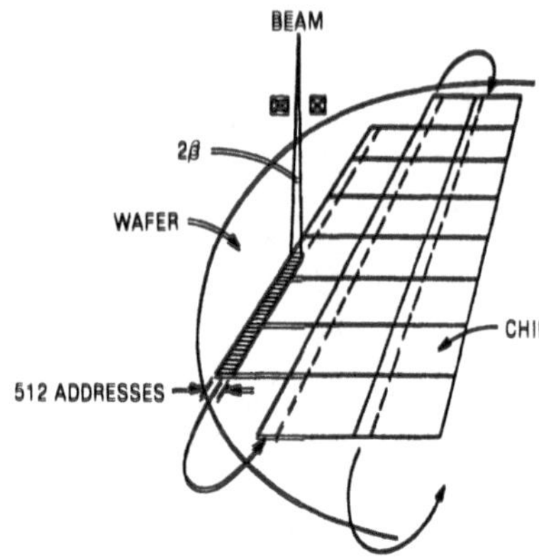

Fig.14. Schematic diagram of a raster-scan electron-beam pattern

8.7 Computer-Aided Design

Because defense applications require a good deal of customality and because of the high cost of design for large-scale and very large scale ICs, the success of the VHSIC program depends heavily on improved computer-aided design techniques.

8.7.1 The Increasing Design Problem

Consider the following points. The Intel 8086 microprocessor required 13 man-years for layout, and the Motorola 68000 microcomputer chip required 52 man-years design time [7]. The design cost of ICs, as we move into the VLSI era, is becoming very large. Large design costs can be accommodated in chips which have a very broad application base, such as memory, but justifying custom chips becomes very difficult for less extensive applications.

To illustrate the magnitude of the design problem, imagine each layer of a chip drawn on quadrille paper with the minimum IC feature as a 1/4-inch division. In the late 60's, when we were dealing with feature sizes between 15 and 20 micrometers, a feature had to be magnified by a factor of 350 for it to equal 1/4 inch, and a complete mask level would have required a 7-foot piece of quadrille paper. This would be manageable. But, moving into feature sizes achievable through the early and mid-70's and into the late 70's, such chip layouts would require pieces of quadrille paper the size of a living room floor and two tennis courts, respectively. Continuing through VHSIC I and II – that is, 1.25 and 0.5 micrometers, respectively – would require scale factors on the order of 5000 and 12,500 and paper sizes on the order of two basketball courts and two football fields.

This example, summarized in Table 3, illustrates the magnitude of the design problem as we achieve micrometer and submicrometer feature sizes. Clearly, computer aids on a large scale will be mandatory to design these chips economically.

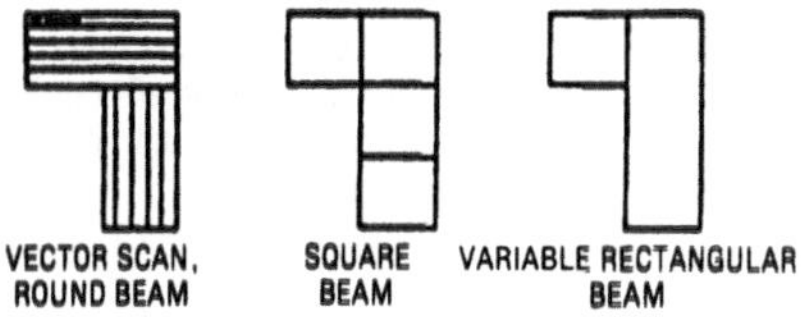

Fig.15. Diagram showing non-raster-scanned electron-beam patterns

Table 3. The design problem

Development time period	Minimum feature size: line and space	Scale factor to magnify to 1/4 inch	Paper size to lay out 250-mil chip (square)
Late 60's	17.5 μm	350	(7 ft) wall hanging
Early 70's	8	800	(17 ft) large living room floor
Late 70's	2	3200	(67 ft) two tennis courts
VSHIC I	1.25	5000	(100 ft) two basketball courts
VHSIC II	0.5	12500	(260 ft) two football fields

8.7.2 The Macrocell Approach

One of the methods under consideration for economical chip design is the macrocell approach. The design for each macrocell – a functional block, such as a multiplier or a memory, which could be equivalent to an LSI chip – would be stored in a macrocell library. Chips would then be designed by arranging macrocells into an interconnection grid structure as illustrated in Fig.16. Once a macrocell was designed, the design – that is, the macrocell – could be used over and over again and its cost amortized over many chip designs.

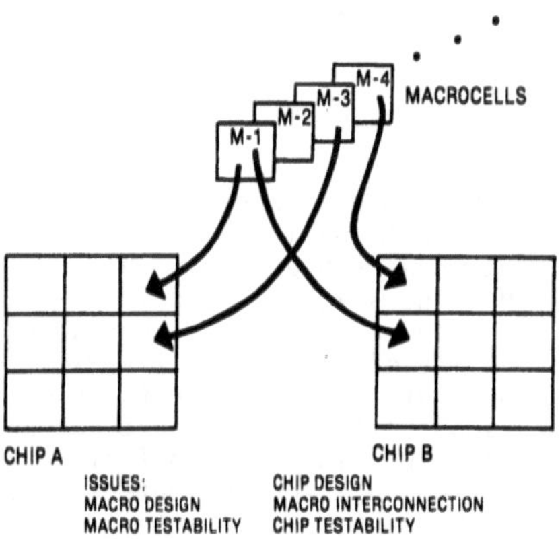

Fig.16
The macrocell VHSIC design approach

Macrocells can be designed in various ways. For example, a macrocell for widespread use, such as a memory cell, could be designed very carefully, perhaps in a manpower-intensive manner, to achieve high density. On the other hand, a specialized macrocell for a one-of-a-kind application might be designed using an automated technique which would provide a low-cost but low-functional-density macrocell.

In principle, the macrocell approach brings flexibility to VHSIC chip design. However, as listed in Fig.16, there are other issues, aside from the design strategy itself, such as testability of the macro, effective interconnection of macros on a chip, and overall chip testability.

8.8 Testing

Another challenge to the VHSIC Program is integrated circuit testing. The ability to access the individual gates of an integrated circuit and therefore the ability to fully test the device decreases with increasing integration. This is because the number of input and output pins on the device has remained essentially constant while the number of gates has increased significantly. For example, an SSI device has a pin-to-gate ratio on the order of 0.5. A VLSI/VHSIC device with 10,000 gates and perhaps 200 pins has a pin-to-gate ratio that is at least two orders of magnitude smaller than the SSI device. Greater operating speeds coupled with this increased complexity make it extremely difficult and costly to test VHSIC devices in a reasonable time frame unless new design and testing methodologies are developed. Since the ability to test complex random logic devices depends upon the device architecture, future VLSI circuits will require the use of design-for-testability concepts as well as advanced testing techniques. Built-in-test schemes for fault detection, fault isolation and ultimately for fault tolerance will be required. On-chip test circuitry will allow various portions of a chip to be tested in parallel to increase the testing speed and decrease the off-chip testing complexity. Built-in redundancy and a capability for reconfiguration will allow chips to test themselves and to recover from failures. One of the purposes of the VHSIC Phase III effort is to develop and evaluate methods to test circuits both at the end of the fabrication process for determination of yield related failures as well as methods to continuously test devices in an operational environment. Regardless of the application, testing of devices of VHSIC complexity will require this problem to be addressed early in the design phase of device development.

8.9 Substrates

Finally, we turn to a somewhat mundane but nevertheless barrier problem — that is, the quality of the silicon substrate. In a qualitative sense, as minimum feature sizes are reduced, defects which were previously not debilitating become more and more debilitating since the size of the defect increases relative to the size of the active device. As a result, more attention must be given to defect density. The quality of the material varies from lot to lot, from time to time, and from manufacturer to manufacturer; however, the extremely high-quality silicon substrate material that has been achieved by Shinetsu and Sony with defect densities close to 1 defect per square mm, as illustrated in Fig.17, should be emphasized. Low defect densities have been identified by Aracor [8] as being very sensitive to two parameters: the oxygen content in the wafer and the carbon to oxygen ratio. This work is continuing under VHSIC III sponsorship in an attempt to identify the growth parameters necessary to achieve very low defect density in starting materials and to improve our capability for providing high-quality material to VHSIC contractors.

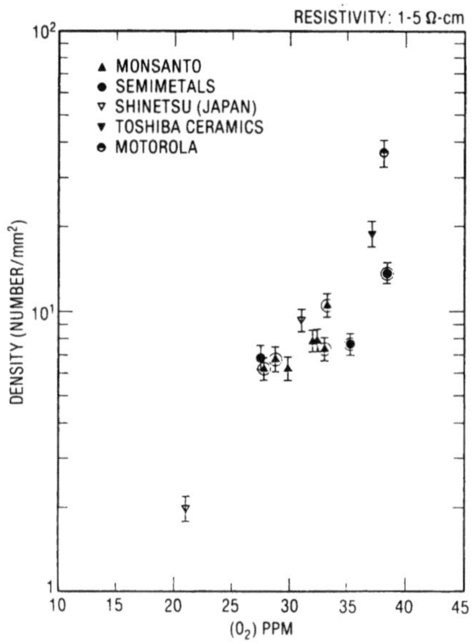

Fig.17. Defect density vs parts per million oxygen content for several domestic and foreign silicon suppliers

8.10 Summary

In summary, successful completion of the VHSIC program will result in a highly significant and broadly based payoff to the military departments, but many difficult technological and economic barriers must be overcome to reach these goals. This will require the ingenuity and joint efforts of the IC and defense systems industries, the university community, and the military departments.

References

1. G.W. Preston: "High-Speed Integration Circuits for Military Applications," IDA Paper P.1423, Nov. 1979
2. L.W. Sumney,: "VLSI With a Vengeance," *IEEE Spectrum*, Vol.17, No.4, Apr. 1980, pp.24-27
3. John Rhea: "VHSIC-Advanced Components for a New Generation of Weapon Systems," Intel. Def. Rev., Vol.6, 1980, pp.887-889
4. J.L. Prince: "VLSI Device Fundaments," in this volume
5. Robert E. Richardson, Jr.: "Modeling of Low-Level Rectification RFI in Bipolar Circuitry," *IEEE Trans. Electromagnetic Compatibility*, Vol. EMC-21, Nov. 1979, pp.307-311
6. H.L. Hughes: Private communication
7. IEEE Solid State Circuits and Technology Committee Workshop on Microprocessors and Memories, Feb. 12, 1980
8. T.J. Magee: *Interim Report on Contract N00014-80-C-0071: Silicon Quality-A Consideration for VHSI Circuit Development*

9. VLSI in Other Countries

R. I. Scace

With 2 Figures

In contrast to the other chapters of this book, here we will consider VLSI work mainly in terms of the programs and goals of other countries. Where technological issues are important, they will be discussed. For the most part, the technology for VLSI is the same everywhere; it has been dealt with in detail elsewhere.

There has been a great deal of expression in the press of a variety of opinions about VLSI work outside the United States. Here we will take a balanced approach, to give both a view of what is going on and my own opinions about how the VLSI world is expected to look in the next few years in terms of relative national capabilities.

To set the stage, there will be a short discussion of past semiconductor development programs in various countries, to see how the present structure of the semiconductor industry developed as a result. Having seen how the present has been reached, the discussion of current national development programs then leads one to see more easily how the future is likely to unfold.

The view of the semiconductor world that is presented here was developed mainly over the last five years in the course of a part-time technology assessment. That work included visits to most of the significant manufacturers in Japan in late 1977 and to a number of European ones in the previous year; the review and analysis of articles, mainly from the trade press; and an examination of the rapidly-growing array of studies of the semiconductor industry. An active participation in the industry for 25 years has given some perspective to this study.

DISCLAIMER: The content of this chapter is the sole responsibility of the author. Nothing herein is to be taken as an official statement or position of the National Bureau of Standards, the Department of Commerce, or the United States Government.

9.1 Past Major Semiconductor Programs

All of the countries in which the semiconductor industry has any size today have given government support to the development of their industry. The motivation has mainly been for defense, except in Germany and Japan where military activity is limited. In these countries, more commercial objectives have dominated.

This support has been delivered in two major ways. First, R&D programs for acquiring the basic technologies have been used, primarily in the U.S. and Japan and to a lesser degree in the U.K. Second, government purchases of state-of-the-art products for military and telecommunications applications have provided significant markets for new devices at a critical time in their life cycle. The second mechanism has been used in all the countries which we are considering.

Before we explore some specific cases, one characteristic of the industry should be noted since it has an important bearing on the relative degree of success of U.S. firms as compared with those of other countries. In the U.S., the availability of venture capital made possible the founding of a number of extremely able and aggressive firms which specialized in semiconductor device manufacturing. These companies succeeded in taking the greater part of the business from the more diversified, large firms which were the mainstays of the U.S. industry twenty years ago. Apparently the flexibility and entrepreneurial drive of the newer companies was the deciding factor. FINAN [1] has analyzed this history in detail.

Elsewhere in the world, venture capital is hard to find and this kind of development did not occur. It is not surprising that the U.S., with its cadre of firms honed to a keen edge by domestic competition, has taken and kept over half of the world market. The major producers abroad are almost all large, older and more diversified companies more like those that found it difficult to survive in the semiconductor industry in the U.S.

The concentration of these new and dynamic firms in Silicon Valley makes that area a major national resource. While competition keeps individual firms on their toes, the rapid diffusion of practical experience throughout the Bay area has a synergistic effect. There is no other place like it in the world.

Now let us briefly consider the ways in which the industry has arrived at its present state in two places, the U.S. and Japan. We will be discussing these two countries again in the VLSI context, of course.

9.1.1 Past Developments in The United States

The Semiconductor Industry Association has estimated that from 1958 through 1976 the U.S. government provided direct R&D support of $350 million to the industry and indirectly supplied another $352 million of R&D funds through purchases of products.

This is to be compared with their estimate of $1713 million of R&D money spent by the industry from its own revenues in the same period. Thus, about 30 percent of the R&D money came from the government.

This support has not come at a uniform rate by any means, as Fig. 1 shows [2]. Although this curve includes some funding for other kinds of devices than semiconductors, the trends are fairly shown. When one considers the size of the industry at a given time in comparison with the funding rate shown in the figure, it is apparent that Department of Defense (DoD) support was quite significant in the late fifties but of only minor importance in recent years.

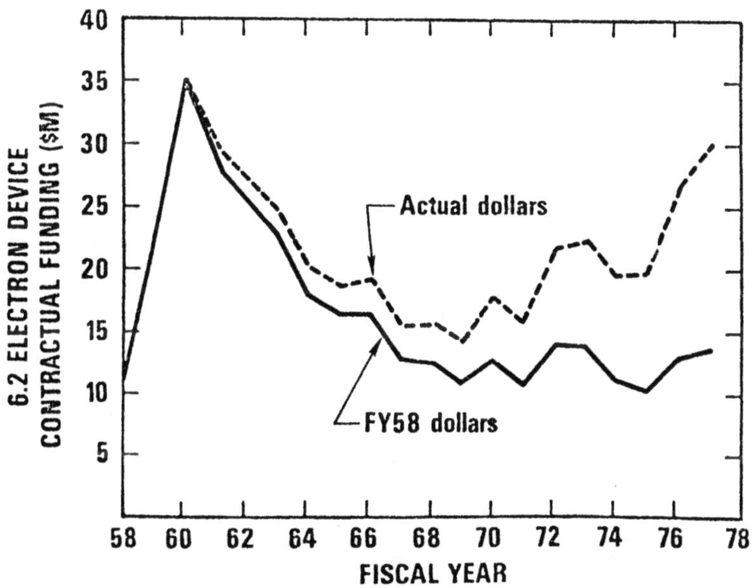

Figure 1. Department of Defense contractual funding for exploratory development of electron devices, 1958-1977

So far we have considered funding for development of every kind of device from point-contact diodes to large thyristors. Considering integrated circuits alone, DoD and the National Aeronautics and Space Administration (NASA) contributed nearly $30 million to R&D from 1959 through 1964, as Table 1 shows [3]. Moreover, during those years and continuing into the early 70's missile and space programs provided both design impetus and a market for the developing IC business [3-5]. This leverage has declined radically to the point that, today, the U.S. military market is less than 10 percent of the total U.S. semiconductor market. This is one of the factors which has led to the newly-mounted VHSIC program of DoD.

Table 1. Early U.S. government support of integrated circuit R&D, from ASHER and STROM [3]

Year	DoD	NASA	Sum
1959	$3.12 M		$3.12 M
1960	3.79 M		3.79 M
1961	6.03 M	$0.14 M	6.17 M
1962	5.11 M	0.11 M	5.22 M
1963	5.23 M	0.24 M	5.47 M
1964	4.18 M	1.36 M	5.54 M
Total	$27.46 M	$1.85 M	$29.31 M

One other major influence in the growth of the U.S. semiconductor industry must not be ignored: the computer [4]. While the British can rightly claim that Babbage invented the computer, the U.S. computer industry led by IBM made it what it is today. The positive feedback between the computer and semiconductor industries has been essential to both, and will continue to be so. This confluence has been a major factor which has so far supported U.S. leadership in both of these fields.

9.1.2 Past Developments in Japan

Japan's course of action has been quite different. Table 2 lists a number of events in the development of the semiconductor industry there, beginning with transistors under license. Nearly every U.S. firm in the business has licensed its patents or technology or both to both Japanese and European firms. The leadership and innovation were here. In parallel with these efforts, the Japanese computer industry has followed a similar course, as Table 3 shows. Since 1970, these two listings show that the Japanese government has put about $540 million into semiconductor and computer development.

Table 2. Milestones in the development of the Japanese semiconductor industry

1954	Six firms start transistor production under license.
1955	First of a series of government-aided programs.
mid-60's	Eighteen firms making semiconductors.
1968	Sony-TI joint venture for IC manufacture and patent licensing.
1971	TI buys out Sony, with MITI approval.

Table 2. Milestones in the development of the Japanese semiconductor industry (continued)

1971-72	MITI and industry rationalize IC priorities, fund joint laboratory for CAD software development.
1973	Process technologies program; $20 million to Hitachi, Toshiba, Fujitsu, and NEC.
1974	NEC-Toshiba Information Systems, Ltd., formed.
1975-77	NTT memory development program, $67 million.
1976-80	VLSI technology program, $310 million.*

*40% government, 60% industry funding.

Table 3. Milestones in the development of the Japanese computer industry

1954	Industry request to government for aid.
1955	Policy guidelines on technology imports.
1957	Electronic Industries Act: Subsidies for R&D Loans for commercialization Loans, accelerated depreciation for investment.
1957	Computer production starts.
1961	Computer output reaches $13 million (7 firms).
1961	Japan Electronic Computer Corp. established: Joint venture by the seven producers Funded by Japan Development Bank loans $78 million 1970 $94 million 1971
1966-71	Super-high performance computer project (IBM 360). $28 million from government.
1967	Sperry-Rand and Oki form joint venture.
1967	IBM permitted to manufacture in return for patent licenses (max. 35 percent of domestic output).
1970	Computer output reaches $876 million.
1971-78	Pattern information processing system project. $97 million from government.
1972-76	High-performance computers, peripherals (IBM 370). $205 million from government.

The Japanese industry has also gained strength from the remarkable growth of the consumer electronics industry in that country. The departure of these products from the U.S. is well documented [6-8], and the reasons are many: the reluctance of U.S. firms to convert from vacuum tubes to solid-state devices, the then-higher cost of U.S. labor, and the dedication of the Japanese to producing innovative, high-quality products are among them.

Yet the firms which have been the major factors in establishing Japan's present position in electronic products are, for the most part, the same kind of large, diversified manufacturers that have not generally been leaders in the semiconductor industry in this country. Table 4 gives a comparison of the major Japanese firms with three selected U.S. companies.

The reasons why these Japanese companies have done so well are not clear, and are hotly debated [9,10]. As Table 4 shows, the Japanese companies for which income data are given apparently re-invest more of their gross income than U.S. firms do; at least their net incomes are lower. The fact that the Japanese are extraordinarily hard-working people should not be ignored.

Table 4. Sales and income figures for major Japanese firms compared with three representative U.S. companies

	SALES			Net
	Total	Electronic	Semiconductor	Income
Nippon Electric	$2273	$1649	$343	1.2%
Hitachi	8022	1620	260	3.2
Toshiba	4541	817	236	NA
Matsushita	5787	3212	210	NA
Mitsubishi	2071	NA	116	NA
Fujitsu	1171	1171	77	2.8
Sony	1418	1418	NA	NA
General Electric	16240	2436	113	5.9
IBM	16580	12832	NA	14.7
Texas Instruments	1660	1378	655	5.8

1976 figures; sales in millions. NA=information not available.

Sources: Annual reports, Electronic News, Dataquest.

It is clear that there has been a concerted effort on the part of both Japanese industry and the government to develop a strong computer industry and the semiconductor capabilities which are needed to support it. Japan is a country with strictly limited natural resources, and for its economic survival it must develop an array of manufacturing industries which will allow it to make a living from the added value given to imported materials in making products which are salable to others. The electronics industry is an example of this kind, quite possibly the most important one.

The semiconductor industries of both Japan and the U.S. serve vigorous domestic markets and have good competitive positions in their foreign ones. In contrast, European companies have limited home markets, since national buying preferences still persist in European countries. No European firm with the exception of Philips has really succeeded in transcending its home boundaries; hence, only Philips has been successful in the same sense as the larger U.S. and Japanese companies [11]. Siemens, however, appears to be making an attempt to emulate Philips in this respect in that it has acquired a number of electronic component firms, including semiconductor ones, in the U.S.

9.2 Present National Semiconductor Programs

Having brought ourselves up to the present in this way, let us now look at the work going on in each of the significant countries in the business, in turn. Europe usually gets little attention in discussions of this topic, so we will begin there.

9.2.1 United Kingdom

The U.K. has, in the past, supported work in advanced areas through either the military laboratories or through the Post Office, which is responsible for telecommunications. In 1978, a $1 million program was started to improve electron-beam fabrication facilities [12]. British capabilities in equipment of this kind have long been strong.

In addition, a $200 million program* to develop their abilities in circuit design and LSI/VLSI manufacturing has been announced [13]. The first $46.5 million has been used to support the formation of a new firm, Inmos [14], which has established a development facility in the U.S. and is staffing it in part with British expatriates who evidently prefer to stay here. The main manufacturing site will be at Bristol in the U.K. Complete details of the British plan have not been published.

Plans have also been proposed [15] for the addition of material covering both semiconductor and computer technology to the curriculum in British schools, to build technological competence from an early age. About $8 million has been allocated, starting in 1979, for a five-year program. Further proposals deal with retraining programs for industrial workers to improve the supply of suitable labor.

MACKINTOSH [15] also details a number of small programs, totaling $45 million in government funds, to develop the abilities of existing semiconductor firms in the U.K.

*With the change from a Labour to a Conservative government in 1979, this program is being re-examined.

The U.K. is among the best in III-IV materials work, and it would not be surprising to see some of this effort go toward exploiting that strength. The programs described above include Plessey, which has done noteworthy work on very high-speed bipolar devices and is relatively strong in memories, and Mullards which, since it is a part of Philips, can draw on the technical strengths of the various branches of that firm.

9.2.2 France

France has had a series of relatively small aid programs in place since early in 1978, totalling $63 million, to stimulate its electronic industry [16]. In computers there have been many government actions designed to build a viable domestic industry, culminating in the formation of CII-Honeywell Bull in 1976 as a final condensation product from at least three smaller firms. The inclusion of Honeywell in the combination exemplifies a pattern of American participation which has been followed in the semiconductor program which has only recently been fully established.

That effort was announced as "Plan Composants" in 1977, with $133 million allocated over a five-year period [17]. The program is really all for integrated circuits, in spite of the name. The first $44 million was evenly split between RTC (La Radiotechnique Compelec, the Philips activity in France) and Thomson-CSF for bipolar work. RTC will develop LSI TTL and ECL devices for computers and telecommunications, while Thomson-CSF is doing medium-scale linear products for consumer and automotive applications. This brings further Philips technology into the French industry.

Second, late in 1978 a joint effort was established involving Motorola, Thomson, and EFCIS (a joint venture of Thomson and the French atomic energy agency CEA) to bring Motorola's bipolar technology to Thomson and its MOS skills to EFCIS. The latter is at present the major SOS house in Europe, and will further benefit from work being done at French government laboratories.

The agreement with Motorola also provides for a two-way technology exchange and the manufacture of equivalent products both in France and in the U.S. A four-inch processing facility will be set up at EFCIS with Motorola technical support. This combination of forces has been allocated the second $44 million slice of the total.

Finally, the remaining third of the funds has been granted to a joint venture between National and Saint-Gobain-Pont-a-Mousson for a plant to make NMOS and CMOS, the location of which has not yet been announced [18,19]. Still another U.S.-French combination, which was bidding for the last part of the project, was between Harris and Matra. They have not abandoned their efforts to find funding, however.

An additional French program combining CEA and Rhone-Poulenc, a large chemical firm, is aimed at developing a domestic supplier for silicon, lithium niobate, sapphire, and garnets. The CEA laboratory LETI already has the reported ability to make 50-mm garnet, II-VI materials, and CdTe among others [20]. Ground was broken late in 1979 for a new research center at Grenoble, near a number of French integrated circuit manufacturing facilities. The center is a part of CNET, the national center for telecommunications studies, and will concentrate on IC technology.

Some time will be required to see what the outcome of all this will be. The major IC houses in France are Sescosem (part of Thomson-CSF), RTC, Texas Instruments, and Motorola. All of these have complete facilities, including circuit design capabilities. A fair assessment of French ability in VLSI at present is that, at best, it is even with the U.S. in a few areas (primarily from the two American firms), but that overall it is a year or two behind.

9.2.3 Federal Republic of Germany

Information regarding German efforts is less detailed. A program which was to run from 1974 through 1978, funded at about $150 million, was aimed at the development of LSI IC's, optoelectronics, materials, production technology, and basic understanding. This program has been extended for another two years to cover more application-oriented work, particularly for watches and appliances [21]. More recently a $39 million program has been awarded to Wacker Chemitronic, the large silicon manufacturer, for work on solar-grade silicon. The objective of this work is the production of polycrystalline material with an oriended grain structure, by use of a directional solidification technique.

A further $21 million has gone to AEG-Telefunken for solar cell development. A Japanese source [22] has stated that the German government has established a $360 million program for computer development over the next five years, with $50 million of that sum to support VLSI work, but no further details were given.

Siemens, AEG-Telefunken, ITT Intermetall, and Valvo (Philips) all have major integrated circuit facilities and programs in Germany. In addition, IBM has a large plant at Böblingen which undoubtedly contributes to the technological health of the German industry. There is, backing all this, a number of first-class university research institutes which turn out good work and skilled graduates. One can gain some perspective in this regard from a perusal of the 374 references cited by BOSCH of the Ruhr Universität Bochum [23].

As in France, the German IC industry is not at the forefront of VLSI technology, with the presumed exception of IBM. AEG-Telefunken has been experiencing serious financial difficulties which may be expected to limit its ability to exploit its strengths in integrated circuits. Germany does not otherwise have a strong computer industry; it does have substantial competence in consumer circuits. German firms are quite strong in high-power industrial electronics, but VLSI per se is not essential for these applications.

9.2.4 Italy

The Italian government is supporting its semiconductor industry with $148 million for R&D work over the period 1978-1981 [24]. Of this sum, $90 million is in the form of grants and the balance is in financing at low interest rates. The areas in which these funds are being applied are:

LSI/VLSI technologies,
semiconductor device production automation,
linear ICs,
power and high frequency discrete devices,
optoelectronics, bubble memories, and solar cells.

The principal beneficiary of this program is SGS-ATES, which has considerable capabilities in linear and MOS IC manufacture. SGS-ATES has an agreement with Zilog of the U.S. relating to second-sourcing of Zilog microprocessors and the distribution in Europe of development systems and training materials.

9.2.5 The Netherlands

The Netherlands evidently does not find it necessary to support its semiconductor industry (Philips), since its neighbors have done so as described above. Nevertheless, Philips has done very well, particularly in circuits for consumer applications. The Philips research laboratories have made steady contributions to the fund of knowledge from the early years of the transistor, on a wide range of topics. Philips has substantial abilities in manufacturing automation, and has developed its own electron-beam equipment for direct writing on wafers. Since Philips is so thoroughly multinational, the firm enjoys relative freedom from being considered a Dutch company and therefore has entree to the European market as no other European company does.

Viewing the European picture in toto, none of the programs described above has had time to make appreciable impact, particularly in the VLSI area. One does not know yet whether they will. The motivations of the various countries for spending large sums of money on semiconductor electronics are not entirely those of national pride, though that is a factor.

The European market is between 20 and 25 percent of the world total, but only 8 percent of the world's production comes from European companies [19]. Most of this imbalance is served by U.S. companies, many of which operate complete facilities in Europe. European governments view their home ground as an opportunity which they would rather exploit for themselves, and they are making a serious effort to get the competence needed to do that.

9.2.6 Japan

By any measure, the Japanese programs in VLSI are the most significant ones outside the U.S. There are two: the well-known VLSI program itself and another, about half as large, in the Nippon Telephone and Telegraph (NTT) Electrical Communication Laboratories (ECL) for memory development.

One should not forget that the U.S. is almost unique in having a privately-owned and -operated telecommunications system. In most other countries, including Japan, this is a governmental function. When comparing the sources of R&D funding in the U.S. with those elsewhere, one properly should consider the programs of the Bell System, GT&E, Western Union, Comsat, and similar firms as part of the public effort, not the private one. This is somewhat overstated, since telecommunication equipment manufacturing is done everywhere by private industry and the R&D work that these firms do is thus in the private sector. But the point remains that a fair comparison requires a somewhat different way of grouping the public and private efforts than Americans usually tend to do.

Figure 2 illustrates the broad organization of the VLSI program in Japan. The Ministry of International Trade and Industry (MITI) contains, among others, the Electrotechnical Laboratory. This laboratory has close working ties with the NTT lab, ECL, since NTT is a public corporation rather like the one which runs the U.S. Postal Service. The VLSI Research Association is funded jointly by MITI and the participating companies. Its Cooperative Laboratories are charged with the generic R&D responsibility. The other two labs, NEC-Toshiba Information Systems (NTIS) and Computer Development Laboratories (CDL), do the more applied, product-specific work.

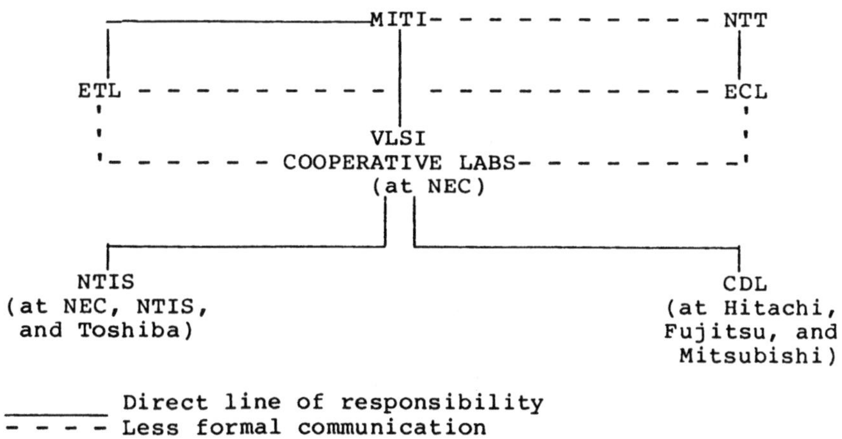

Figure 2. Organizational relationships in the VLSI program in Japan.

The physical location of the work and the staffing are not so straightforward. In Japan, professional employees do not move readily from one employer to another. How, then, does one put in place a large but temporary organization to do such a program? The people are borrowed, to return when the program is done. Even the physical facilities are borrowed. The Cooperative Lab is in a new building at the site of Nippon Electric Company's (NEC) research laboratory. It is physically fenced off from the remainder of the grounds to isolate the employees borrowed from NEC's competitors from the rest of the facility.

The Director of the program, Dr. Yasuo Tarui, is borrowed from ETL as are several other technical managers and staff members. In this lab, as well as the others, the responsibility for the management is carefully and equitably shared by employees of the participating firms.

Both of the other two labs exist in several locations. They are paper entities, since the borrowed staff works in borrowed space which conveniently happens to be right where they were all along. The management of these laboratories must be an exercise in frustration, since they are scattered all around the western and southern suburbs of Tokyo and as far away as Osaka 300 miles to the south. Many opportunities for technical cross-fertilization are thus lost.

The major work topics are listed in Table 5, with further areas of concentration shown in Table 6. One should keep clearly in mind the fact that this program is entirely devoted to digital circuitry for computers. Much of the basic work, however, applies as well to large-scale integration of any kind. Lowdefect crystals, clean facilities, low-stress processing, and sub-micrometer imaging are essential for all.

Table 5. Major responsibilities of the labs in the VLSI program

Cooperative Laboratory

- Micropatterning
- Super-clean facility
- Crystal growth
- Wafer processing
- Basic and common design, test, and evaluation technology

Computer Development Lab and NTIS

- Design technology
- Process know-how
- Test and evaluation know-how
- Logic and memory technology related to specific system architecture

Table 6. Areas of concentration of the VLSI program

e-beam and X-ray imaging
Super-clean facilities
Large-diameter perfect crystals
Improved crystal evaluation techniques
Processing of large-diameter slices
Oxide growth and removal techniques
Logic design, simulation and layout
Device design, test and evaluation

The program has been in operation since late 1976, and is to end on 31 March 1980. The visit at which some of the information presented here was learned took place in late 1977 when the program was just getting well under way. Most of the equipment was running and staffing was essentially complete. But it was too early to expect much output.

The goals of the program were then the subject of much speculation in the press; a list said to be those goals is given in Table 7. At the VLSI lab there was much distress expressed about the press coverage of the program and objections to this list as being their goals. Their position was that they were conducting a program of basic scientific and technical studies which would be communicated in due course through the literature and at meetings. What was seen and heard on our visit and what has transpired since seem entirely consistent with their view.

Table 7. Alleged goals of the VLSI program (see text)

Reduce logic delay time from 1 to 0.5 ns per gate
Raise logic complexity from 100 to 1500-2000 gates/chip
Reduce memory access time from 50 to 0.5 ns
Raise memory complexity from 4-16K to 1000K bits/chip
Reduce line widths from 5 to 0.5 μm
Reduce flaws in slices from 300 to 10/cm^2
Increase wafer diameter from 3-4 to 6-8 in.

Some of the goals in Table 7 are clearly unrealistic for achievement by the end of the program. Wafers of 6-8 inch diameter of suitable quality for VLSI work will not be available by then, even in the laboratory. A memory access time of 0.5 ns will not be reached. If one scales the definition of a flaw down by a factor of ten (paralleling the reduction in line width), then the reduction of the frequency of occurrence of "flaws" by a factor of thirty in four years is a difficult task. The other goals in this list could be achieved, or very nearly so. The list is far too short; many other equally important items must also be considered. These factors lend credence to the argument that this is not a list of the detailed goals of the VLSI program, although those goals must address at least the topics given in Table 7.

Another way to view the Japanese VLSI program is by considering what is done with the funds. In this instance, about $200 million was earmarked for equipment, $23 million for labor, $20 million for materials, and $67 million for "other". The total planned professional time is close to 500 man-years.

There are secondary effects implied. Much of the equipment which has been bought has involved advances in the state of the art. For example, when the program began Toshiba had already built an electron-beam mask-writing system. This system, while able to create 0.5 micrometer details, did not have beam-shaping capabilities and took roughly a day to generate one mask level. Toshiba's second-generation raster-scan machine is about a factor of four faster; it was installed at the time of our visit to the VLSI lab. A new vector-scan machine with variable beam-shaping capability, by Japan Electron Optics Laboratory (JEOL), was nearly done and was to be installed in the VLSI lab soon. An additional development, from the NTT program, is the world's only commercially-available X-ray align and expose equipment, by Nikon [25]. The equipment money spent with outside suppliers has drawn with it additional funds from the suppliers to develop advanced equipment for sale to others. In this way the leverage of such programs is extended, leading to a larger increase in national capability than one would expect from the basic program alone.

Taken as a whole, the program would be expected to produce the results listed in Table 8. Since it is a publicly funded program, the new knowledge which results is more generally available than if the work were done entirely with private money. Other firms in Japan have the privilege of licensing the patents which result; negotiations to this end have been undertaken to provide such a license to IBM, which manufactures in Japan. Nearly 300 patents have so far been registered [26].

Table 8. Anticipated impact of the VLSI program

- Sharing of technology among the major manufacturers
- Support state-of-the-art computer development
- Improve international competitive ability of
  - device manufacturers
  - computer manufacturers
  - electronic products generally
  - equipment manufacturers
- Develop domestic expertise in advanced manufacturing equipment, especially electron-beam systems
- Equip the five participating firms with the most advanced equipment available

In a number of cases, Japanese manufacturing equipment suppliers were already at the state of the art, and their products are in wide use in the U.S. Such examples as Canon and Nikon aligners, Kokusai and Shinkawa bonders, and several makes of dicing saws are often seen. In Japan, one also sees Auger spectrometers (NEC-Varian), laser scribers (NEC), molecular beam epi (Ulvac), plasma processing (Tokuda Seisakusho), tape automatic bonders (Dai-Nippon), and final test equipment (Takeda Riken, Ando); these are preferred by many Japanese device houses because they work as well as American equipment and the local service is better. One should also mention that there seems to be little competition for U.S. suppliers in certain areas: implanters, evaporators, epi reactors, and CAD equipments were mostly from this country in the ten manufacturing locations which were visited. Cameca, a Thomson-CSF subsidiary, has most of the SIMS business; the VLSI lab had one and ECL is installing the latest model [27].

In the device area, in late 1977 a number of examples of VLSI products were already evident, at least in small numbers. Two labs were then working at the 2-μm line-width level, using optical methods. Devices seen in plant visits included a 10 by 15 mm CCD camera retina with full TV resolution, 4K static RAMs, 64K CCD memories, a 64K dynamic RAM, 90K bubble memories, and a 16-bit isoplanar SOS microprocessor developed as part of the pattern recognition program. Many of these devices have been described at device meetings in the U.S. since that time. An 84-lead package was in production in one plant, and another firm was using a multilayer hybrid substrate to mount up to 64 ICs by TAB methods for standard computer circuit modules. These exemplify the capabilities which existed at the start of the VLSI program. Both NEC and Fujitsu have been marketing their products in the U.S. with a growing degree of success; both have facilities for device manufacture either in operation or planned for the U.S. Both firms have been quick to introduce competitive products which perform well.

More recently, NEC has announced a 135K-bit bubble memory [28], NTT is using a 64K bubble memory in its electronic telephone switching systems and has a 256K bubble memory in the lab at ECL [29]. ECL has also developed a 128K MOS ROM, e-beam programmed, on a 20-sq mm chip [30]. Already at the 1978 International Electron Devices Meeting, four out of seven papers on MOS VLSI memory were from VLSI program member companies. The same population (4 of 7) in the session on device applications of new technology was also from these companies [31]. Hitachi, at the 1979 International Solid State Circuits Conference, reported development of a 5122-gate I2L logic array and a 1K RAM with 5.5 ns access time, the fastest at the time [32]. Additional VLSI program developments have been discussed at more recent meetings, as the program of this year's International Solid State Circuits Conference abundantly shows. Minor items also show up: a closed meeting on crystal processing and evaluation was held last summer, organized by a senior research staff member of the VLSI lab [33].

The aim of these Japanese efforts is to develop their capabilities in making computers. Various reports show increased activity; the impact of the VLSI program is evident. NTIS has announced a new low-end computer that competes with IBM's System/38 and 4300 series, the ACOS 250. The latter is claimed to be 30 to 40 percent less expensive for the same performance. The CPU logic is ECL with 550 gates per chip and TTL-level logic at the pins. I/O controllers use 3300-gate NMOS devices. Deliveries were to start in July 1979 [34]. Hitachi makes an IBM-compatible computer, the ITEL Advanced System/6, which replaces the IBM 3-31 and 3032. Fujitsu's M-180-II is roughly equal to an IBM System/370 model 158. Fujitsu is also an investor in Amdahl, and builds the main frames for the Amdahl machines [35]. Large-scale ECL in special packages to permit 5-W dissipation is used in both the Fujitsu and Amdahl machines.

Of the two groups of companies in the VLSI program, the CDL troika of Fujitsu, Hitachi, and Mitsubishi is addressing the IBM-compatible machines while NTIS has non-IBM-compatible hardware. This distinction predates the VLSI program for some of their older products. NTIS was actually formed a couple of years before the start of the VLSI effort with the objective of fusing the product lines of NEC and Toshiba into one rationally-planned one, and had gone a long way toward that goal.

At this writing, the VLSI program was almost at its end. A new four-year effort had been announced, with an apparent emphasis on computer technology. The detailed goals and other information on the implementation of the program have not been discussed. It will be interesting to see how this activity will fit in with the previous ones.

9.2.7 Korea

The Korea Institute of Electronics Technology has recently received $60 million from the Korean Government for a long-range project to promote the semiconductor industry there [36]. The funds are for R&D and pilot production of microprocessors and other ICs. The first bipolar IC developments are to appear in mid-1980, with MOS memory and microprocessors following in early 1981.

Korean firms have in the past largely confined their work to device assembly operations. Two companies have wafer fabrication facilities at present, Korea Electronics (with partial ownership from Toshiba) and Samsung Electronics, which bought out the U.S.-owned firm Integrated Circuits International. Samsung now makes CMOS watch circuits and linear ICs for consumer applications.

9.3 Future Prospects

At this point, the current activities of the various countries in VLSI have been discussed. Some viewers of this scene stop here and wring their hands, with the feeling that all this activity must work to the detriment of the U.S. industry. I do not share this opinion. There are many U.S. activities of the same nature going on, as they must unless the domestic semiconductor houses have lost their will to compete. Consider the following factors.

The U.S. industry did business worldwide in 1979 at about a $6500 million annual sales rate. Expenditures for R&D typically range from 10 to 15 percent of sales; let us assume 12 percent as an average. That implies that U.S. firms are investing about $780 million annually in development work. The total annual sum spent for all the programs which we have reviewed above is in the neighborhood of $250 million. This is government money, presumably above that which the firms themselves spend. Collectively the companies from outside the U.S. will have about $4500 million in sales for 1979, so there is another $540 million in company-funded R&D, bringing the total to $790 million. The two sums are remarkably close to one another.

The figure quoted for the sales of U.S. firms does not include captive producers, the two biggest of which are IBM and Western Electric. We shall have to make an informed guess here, since these figures are not public. Let us suppose that WE is roughly the size of Fairchild, or about $500 million. IBM is larger, probably in the range of $1000 million. The contribution to R&D from these two firms then is at least $180 million, giving a U.S. total without any government funding of nearly $1050 million annually for R&D. The U.S. industry is certainly well enough funded to do its homework. Beyond that, U.S. government funding has been running at about a $30 million annual rate, with a roughly equal amount announced for the new VHSIC program for the next six years.

As a second point, let us consider if the U.S. industry is vigorous enough to hold its own. I maintain that it is. The U.S. market is the most intensely competitive one in the world. It is used as a model of a free-market industry. Elsewhere in the world such conditions do not hold. Either by national purchasing preference or by other kinds of influences, most of the other firms in the world have been sheltered to some degree. The fact that U.S. firms hold over half of the European market supports the view that they can compete effectively outside the U.S.

It is true that U.S. companies have not had the same degree of success in Japan, but other factors are at work there. For one, the Japanese distribution system is quite different from that in the U.S., and unless considerable pains are taken it is difficult for foreign firms to gain access to potential customers. Also, much of the Japanese market is captive. Both Sony and Matsushita make most of their own devices, for example, as do many of the large, vertically-integrated companies such as Hitachi and Toshiba.

Although one hesitates to mention it, there have been in the past significant barriers of other kinds to the entry of foreign semiconductors. With the change in value between the dollar and the yen, prices to Japanese customers of American products have become more favorable and sales of U.S. components in Japan have been increasing. One effect of the recently concluded Tokyo round of trade agreements has been to reduce significantly many of the barriers to entry of semiconductors into Japan. The principal market which apparently still is closed is for telecommunication applications.

Further, American firms have learned a lesson in quality control from the Japanese. Ironically, while U.S. authors wrote the book on quality control, Japanese manufacturers read and practiced the subject. Where there have been direct market confrontations in the past few years, such as in Europe and more recently in this country, Japanese quality control in many cases was superior to American. This difference has been corrected to some degree, and U.S. firms are now taking more care in this respect. It is interesting to observe the extent to which U.S. MIL standards are taken as the basis for quality control procedures both in Europe and in Japan. While MIL methods were widely used in the U.S. in former times when military sales were a significant part of the semiconductor business, and though they played an important role in improving the reliability of devices, the practice of this discipline had apparently not been pursued so diligently in the U.S. recently.

There are still lessons to be learned from the Japanese about participatory management methods, and how to get the involvement of all employees in quality programs. These techniques can be transplanted to this country successfully, as the instance of the Quasar plant bought by Matsushita from Motorola clearly shows [37].

When considering the possible impact of the various government-funded programs which have been mentioned above, one should remember that the effectiveness of such programs varies greatly. In those instances in which the objectives of the government and of the industry are in consonance, the probability of success of a program is often high. When the objectives differ, the success rate drops substantially. Thus, the Japanese have had relatively good results from their past programs, and will probably continue to enjoy similar good fortune. In the U.S., the early government-supported programs were timely and aimed in the direction in which the industry wanted to go. They had a significant positive effect. There is concern in some quarters regarding the success of the VHSIC program, since some firms see the goals of the program as not in their best interest from a commercial viewpoint.

In this light, one might expect the Japanese and Korean programs to be successful, at least in the technical sense. The German efforts may or may not have a significant effect. One has concern about the remaining European plans, since the proportion of government involvement is high in proportion to the size of the local industry and because the record of performance of the companies involved (at least the indigenous ones) has not been generally outstanding. Whether or not any of these national programs will have a significant effect on the relative performance of one country's semiconductor industry as compared with another depends on many factors other than technical ones, and the future is correspondingly harder to foresee.

Finally, if one examines the sources of innovation in the semiconductor industry, they are more often than not found to be in the U.S. In this context it is not so important where an idea was first conceived, but where it was put into practical use. The major efforts to make electron-beam mask and wafer writing a reality were put forth by Bell Labs, IBM, and Texas Instruments at large cost. Ion implantation, to which the Russians made many early contributions, was put into place here; American equipment predominates in use. The many ingenious tricks which have made MOS devices much faster and more competitive with bipolar products have mainly come from U.S. firms under the pressure of intense competition. One should recall that few major innovations have resulted in the past from government-funded programs. Thus the impact of such programs on future innovative abilities will probably be slight.

It is true that the Japanese are highly motivated and have their priorities clearly identified. Government and industry share many common goals. Their products are well-designed and of excellent quality. Where businesses have not been inclined to compete sharply, Japan has been very successful in carving out a large market, as exemplified by the U.S. steel and consumer electronics industries. But we are not dealing here with an environment of that kind.

In the absence of a major change in the operating rules (a move toward protectionism everywhere, for example), I predict that U.S. companies will hold their own. The Japanese are exceptionally able competitors. They may be able to penetrate the European market to some degree, but will probably yield in return some of their own home market. In developing countries, where industrialization is proceeding and the consumer market is growing, there will be mighty struggles for market share. U.S. firms may well look to their laurels there, since in many cases (South America, for example) Japan has a good start.

It is a healthy situation. It will be exciting. There will be many hot words spoken, and those who believe them all will be greatly confused. But one primary idea should be kept in mind. Competition is what has given the semiconductor industry its incredible vitality. Competition is not to be feared, but welcomed. In spite of the cries of alarm from some, I have confidence that the U.S. semiconductor manufacturers will meet this challenge as they have met so many others, with ingenuity and vigor.

References

1. Finan, W., The International Transfer of Semiconductor Technology Through U.S.-Based Firms, NBER Working Paper 118, National Bureau of Economic Research, New York, NY (December 1975).

2. 1977 Annual Report, Advisory Group on Electron Devices.

3. Asher, N. J., and Strom, L. D., The Role of the Department of Defense in the Development of Integrated Circuits, IDA Paper P-1271, Institute for Defense Analyses, Arlington, VA (May 1977).

4. Linville, J. G., and Hogan, C. L., Intellectual and Economic Fuel for the Electronics Revolution, Science 195, (4283), 1107-1113 (18 March 1977).

5. Bullis, W. M., Government Programs on Advanced Technology and Manufacturing Techniques: Comments on U.S.A., Japan, and Europe, National Bureau of Standards NBSIR 78-1482 (June 1978).

6. The U.S. Consumer Electronics Industry, U.S. Department of Commerce, Washington, DC (September 1975).

7. The Impact of Electronics on the U.S. Calculator Industry 1965-1974, U.S. Department of Commerce, Washington, DC (November 1975).

8. The U.S. Calculator Industry Since 1974, U.S. Department of Commerce, Washington, DC (May 1977).

9. Threats and Opportunities: How to Face Today's Challenges, conference at Washington, DC, December 7-8, 1977.

10. U.S.-Japan Semiconductor Seminar, Palo Alto, CA, November 14, 1978.

11. Mackintosh, I. M., Large-Scale Integration: Intercontinental Aspects, IEEE Spectrum 15, 51-56 (June 1978).

12. Anon., £1 Million for Electron-Beam Fabrication, Physics in Technology 9, 43 (1978).

13. Smith, K., Britain Plans to Aid IC Makers, Electronics 51 (10), 88-89 (May 11, 1978).

14. McCarthy, P., U.K. to Put $46.5M Initially in IC Plant, Electronic News, July 24, 1978.

15. Microelectronics into the 80's, Mackintosh Associates, Ltd., Luton, England (1979).

16. France Plans Aid Program for its Electronics Industry, Electronic News, January 23, 1978.

17. A Plan for the Development of Integrated Circuits in France, Ministère de l'Industrie et de la Recherche, Paris (August 1977).

18. French Aid to National Venture May Kill Harris/Matra, Electronic News, January 1, 1979.

19. New French Integrated Circuit Programs, Semiconductor International, March 1979, p. 20.

20. A New Source of Electronic Materials for the European Semiconductor Industry, Semiconductor International, March 1979, p. 20.

21. Electronic Components Programme of the German Government, EEC Committee on Scientific and Technical Research, paper XII/558/74-E, Brussels.

22. Ouchi, A., U.S.-Japan Semiconductor Issues: The Japanese Point of View, U.S.-Japan Semiconductor Seminar, Palo Alto, CA, November 14, 1978.

23. Bosch, B. G., Gigabit Electronics - A Review, Proc. IEEE 67, 340-379 (1979).

24. Maselli, G., Istituto Mobiliare Italiano, private communication (1979).

25. Lyman, J., Lithography Chases the Incredible Shrinking Line, Electronics, April 12, 1979, pp. 105-116.

26. Japanische VLSI-Lizenz für IBM, Elektronik 13, 36 (November 23, 1978).

27. Wafers Cause Boom in Ion Analyzers, Electronics, January 4, 1979, p. 64.

28. NEC Will Launch 135-kilobit Bubble, ibid.

29. Araki, K., Nippon TT Advances Bubble Memory Density, Digital Design, June 1978, p. 12.

30. MOS VLSI ROM by Electron Beam Drawing, Japan New Product and Technical Newsletter, March-April 1978, p. 8.

31. IEEE International Electron Devices Meeting Technical Digest, Washington, DC, December 4-6, 1978.

32. Capece, R. P., Design, Process Innovations Expand Logic, Memory Frontiers, Electronics, February 15, 1979, pp. 123-128.

33. Mentioned in International Report, Solid State Technology, September 1978, p. 10.

34. NEC Takes on IBM's New Mainframe Generation, Electronics, March 1, 1979, p. 68.

35. Mitsubishi Announces IBM-Compatible Computers, Electronics, August 31, 1978, p. 48.

36. Hataye, J., Korea Allots $60M for IC R&D, Output, Electronic News, October 22, 1979, p. 54.

37. Garino, D. P., Takeover by Japanese Hasn't Hurt After All, Quasar Workers Find, Wall Street Journal, October 10, 1978.

Addenda

Chapter 3

Since the appearance of the first edition, several new developments in lithography have occurred. High-speed logic chips with impressive speed and 1-micrometer design rules have been reported by Bell Laboratories: ring oscillators with 30-ps stage delay, a counter operating at 2.5 GHz, and a 750-MHz digital-to-analog converter. The Bell X-ray lithography system with a Pd source was used for all six levels to expose resist in a multilevel resist structure [94].

A new third-generation electron beam pattern generator has been disclosed by IBM [95]. The EL3 machine is used in a Quick-Turnaround Time automated line for custom logic (PLA) circuits [96]. Throughput is 10 to 20 four-inch wafers/h with 1-micrometer geometries. The system has a variable beam shape. Bipolar microprocessors and MOS memories with approximately 1-micrometer least dimensions have been reported from Texas Instruments; these circuit patterns were also electron beam written [97,98].

The unity magnification electron image projection system developed at Mullard has been enlarged to allow printing on 4-inch wafers [99]. In the area of ion lithography, a 10:1 reduction projection system has been demonstrated [100].

A third-generation optical full-field projection printer for five-inch wafers is available from Perkin Elmer. Resolution is 1 micrometer, and the system is as expensive as a wafer stepper.

Additional References

94. M.P. Lepselter: Technical Digest IEDM, p.42, 1980
95. J.Z. Chen, W.B. Chen, T.S. Jen, J. Hutt: IBM J. Res. Dev. *25*, 142 (1981)
96. R.D. Moore, G.A. Caccoma, H.C. Pfeiffer, E.V. Weber, O.C. Woodard: Sixteenth Symposium on Electron, Ion and Photon Beam Technology, Dallas, May 26, 1981
97. P. Shah, G. Pollack, R. Miller, R. Love, W. Lee, S. Wood, R. Robbins, G.L. Varnell: Sixteenth Symposium on Electron, Ion, and Photon Beam Technology, Dallas, May 26, 1981
98. S.A. Evans, S.A. Morris, L.A. Arledge, Jr., J.O. Englade, C.R. Fuller: IEEE Trans. El. Dev., *ED-27*, 1373 (1980)
99. R. Ward, A.R. Franklin, P. Gould, M.J. Plummer: Sixteenth Symposium on Electron, Ion, and Photon Beam Technology, Dallas, May 26, 1981
100. G. Stengl, R. Kaitna, H. Loschner, P. Wolf, R. Sacher: J. Vac. Sci. Technol. *16*, 1883 (1980)

Chapter 5

Since the publication of the first edition, much progress has been made toward the goal of achieving ultrahigh-speed LSI or VLSI with GaAs integrated circuits [53]. Many of the more exotic GaAs device technologies [54] have received considerable attention, such as GaAs IGFET or MOSFET approaches [55,56]. The wide bandgap heterojunction gate (HJFET) device approach for achieving larger enhancement-mode logic swings (as discussed in Sect.5.2) has been further pursued and has nearly reached the ring-oscillator stage of development [57,58]. HJFET's of a new type, variously termed modulation-doped FETs or high electron mobility transistors (HEMTs), have been fabricated with molecular beam epitaxy (MBE) [59]. They are designed for low-temperature operation, and have shown electron mobilities in their GaAs channels as high as 50,000 cm^2/Vs at 77K, which could lead to very low dynamic switching energy VLSI. A very promising device approach for achieving extremely high-speed ($\tau_d \sim$ 1-10 ps) VLSI is the GaAs heterojunction bipolar transistor [54]. This device, also a candidate for fabrication by MBE, should have the bipolar transistor virtue of extremely uniform V_{BE} thresholds so important for low logic swing VLSI, while achieving extremely high ($f_\tau \sim$ 200 GHz) performance levels [53,54]. Even more exotic MBE-fabricated device structures have been proposed for ultrahigh-speed, ultralow dynamic switching energy logic, based on the "ballistic" electron dynamic effects discussed in Sect.5.1 (just after Fig.3). GaAs planar doped-barrier transistors have been demonstrated [60] which, by achieving electron velocities approaching 5×10^7 cm/s through the base region, should offer extremely high speeds with very low power supply voltage operation (possibly under 0.5 V) for extremely low dynamic switching energies. Many of these new GaAs device structures are, of course, quite complicated and difficult to fabricate, and hence are many years behind the "standard" GaAs MESFET processes for practical LSI or VLSI implementation.

The "plain vanilla" GaAs MESFET integrated circuit technology has seen great progress as well. E-MESFET GaAs logic circuits have shown ring oscillator gate delays as low as 30 ps at room temperature or 17.5 ns at 77K using L_g = 0.6 μm MESFETs [61]. As discussed in Sect.5.3.8, we have developed in the DARPA/Rockwell GaAs LSI program, a super planar fabrication process capable of optimizing two or more types of devices as required for sophisticated LSI/VLSI circuit approaches. This process has demonstrated the capability for achieving tight parameter control on low-power D-MESFETs and high-speed switching diodes, and for achieving good yields on lower LSI complexity (hundreds of gates) circuits (up to 20% on 260-gate 5 × 5 bit parallel multipliers). The Schottky diode-FET logic circuit approach used in this work has recently demonstrated gate propagation delays as low as τ_d = 52 ps (at $P_D \sim$ 1 mW/gate), low power ($P_D \sim$ 100 μW to 2 mW/gate), high densities (NOR gate areas as low as 600 μm^2), and outstanding $P_D\tau_d$ products (as low as 16 fJ in ring oscillators, or 95 fJ even in large MSI parts). The simplicity, high density, and large device parameter margins of SDFL, along with its speed and power characteristics, make it ideal for ultrahigh-speed VLSI. We have demonstrated scores of different SDFL circuits, from SSI to LSI [53]. We have fabricated 1008-gate LSI parts, 8 bit × 8 bit latched parallel multipliers, and have obtained fully working samples of these parts both on the smaller horizontal Bridgman-grown GaAs substrates and the new 3-in. liquid-encapsulated Czochralski (LEC) GaAs substrate material. We have demonstrated that even in these LSI parts, the inherent GaAs speed is maintained, with SDFL NOR gate delays as low as 150 ps obtained in the 8 bit × 8 bit parallel multiplier chips (corresponding to a 5.25-ns multiply time).

Based on the results to date, extension of this planar GaAs IC technology to the capability for 10,000-gate VLSI circuits seems a reasonable goal, and efforts to achieve this goal are in progress.

Additional References

53. R.C. Eden, B.M. Welch: "Ultra High Speed GaAs VLSI: Approaches, Potential and Progress," in *VLSI Electronics: Microstructure Science, Volume 3*, ed. by N. Einspruch (Academic, New York 1981)
54. H. Kroemer: Heterostructure bipolar transistors and integrated circuits. IEEE Proc. *70*, No.1 (1982), and other papers in this special issue on Very Fast Solid State Technology
55. J.G. Yu, A.G. Rode, T.G. Ruttan: "A Planar Ion-Implanted GaAs MISFET With Almost Hysteresis-Free Characteristics," 39th Annual Device Research Conference, Santa Barbara, June 1981, Paper VA-6
56. T.L. Andrade, N. Braslau: "GaAs Lossy Gate Dielectric FET," 39th Annual Device Research Conference, Santa Barbara, June 1981, Paper VA-5
57. T.J. Maloney: "AlGaAs Heterojunction Gate GaAs FETs by Organometallic and Molecular Beam Epitaxy," 39th Annual Device Research Conference, Santa Barbara, June 1981, Paper VA-8
58. T. Mimura, S. Hiyamizu, T. Fujii, K. Nanbu: "A New Field-Effect Transistor with Selectively Doped GaAs/N-$Al_xGa_{1-x}As$ Heterojunctions," Jpn. J. Appl. Phys. Lett. *19*, 225-227 (1980)
59. T. Minura, S. Hiyamizu, H. Idhikawa, T. Misugi: "An Enhancement-Mode High Electron Mobility Transistor for VLSI," High Speed Digital Technologies Conf., Paper III-5, San Diego, Jan. 14, 1981
60. R.J. Malik, M.A. Hollis, L.F. Eastman, D.W. Woodard, C.E.C. Wood, T.R. AuCoin: "GaAs Planar Doped-Barrier Transistors," 39th Annual Device Research Conference, Santa Barbara, June 1981, Paper VA-7
61. T. Mizutani: "Gigabit Logic Operation with Enhancement-Mode GaAs MESFET IC's," IEEE MTT-S Workshop on Gigabit Logic for Microwave Systems, May 1980. Data presented for Lg = 0.6 μm 15-stage inverter ring oscillators with room temperature speeds up to τd = 30 ps at P_D = 1.9 mW/gate or τd = 63 ps at P_D = 100 μW/gate, and 77K speeds up to τd = 17.5 ps at P_D = 9.2 mW/gate. Earlier, Lg = 0.8-μm data were presented in: T. Mizutani, N. Kato, M. Ida, M. Ohmori: High-speed enhancement-mode GaAs MESFET logic. IEEE Trans. MTT-*28*, 479-483 (1980)

Subject Index

MIX
Papier aus verantwortungsvollen Quellen
Paper from responsible sources
FSC® C105338

If you have any concerns about our products,
you can contact us on
ProductSafety@springernature.com

In case Publisher is established outside the EU,
the EU authorized representative is:
Springer Nature Customer Service Center GmbH
Europaplatz 3, 69115 Heidelberg, Germany

Printed by Libri Plureos GmbH
in Hamburg, Germany